252

Topics in Current Chemistry

Topics in Current Chemistry

Recently Published and Forthcoming Volumes

Anion Sensing
Volume Editor: Stibor, I.
Vol. 255, 2005

Organic Solid State Reactions
Volume Editor: Toda, F.
Vol. 254, 2005

DNA Binders and Related Subjects
Volume Editors: Waring, M.J., Chaires, J.B.
Vol. 253, 2005

Contrast Agents III
Volume Editor: Krause, W.
Vol. 252, 2005

Chalcogenocarboxylic Acid Derivatives
Volume Editor: Kato, S.
Vol. 251, 2005

New Aspects in Phosphorus Chemistry V
Volume Editor: Majoral, J.-P.
Vol. 250, 2005

Templates in Chemistry II
Volume Editors: Schalley, C.A., Vögtle, F., Dötz, K.H.
Vol. 249, 2005

Templates in Chemistry I
Volume Editors: Schalley, C.A., Vögtle, F., Dötz, K.H.
Vol. 248, 2005

Collagen
Volume Editors: Brinckmann, J., Notbohm, H., Müller, P.K.
Vol. 247, 2005

New Techniques in Solid-State NMR
Volume Editor: Klinowski, J.
Vol. 246, 2005

Functional Molecular Nanostructures
Volume Editor: Schlüter, A.D.
Vol. 245, 2005

Natural Product Synthesis II
Volume Editor: Mulzer, J.
Vol. 244, 2005

Natural Product Synthesis I
Volume Editor: Mulzer, J.
Vol. 243, 2005

Immobilized Catalysts
Volume Editor: Kirschning, A.
Vol. 242, 2004

Transition Metal and Rare Earth Compounds III
Volume Editor: Yersin, H.
Vol. 241, 2004

The Chemistry of Pheromones and Other Semiochemicals II
Volume Editor: Schulz, S.
Vol. 240, 2005

The Chemistry of Pheromones and Other Semiochemicals I
Volume Editor: Schulz, S.
Vol. 239, 2004

Orotidine Monophosphate Decarboxylase
Volume Editors: Lee, J.K., Tantillo, D.J.
Vol. 238, 2004

Long-Range Charge Transfer in DNA II
Volume Editor: Schuster, G.B.
Vol. 237, 2004

Long-Range Charge Transfer in DNA I
Volume Editor: Schuster, G.B.
Vol. 236, 2004

Spin Crossover in Transition Metal Compounds III
Volume Editors: Gütlich, P., Goodwin, H.A.
Vol. 235, 2004

Spin Crossover in Transition Metal Compounds II
Volume Editors: Gütlich, P., Goodwin, H.A.
Vol. 234, 2004

Spin Crossover in Transition Metal Compounds I
Volume Editors: Gütlich, P., Goodwin, H.A.
Vol. 233, 2004

Contrast Agents III

Radiopharmaceuticals
From Diagnostics to Therapeutics

Volume Editor: Werner Krause

With contributions by
R. Alberto · C. J. Anderson · A. Boschi · J. D. G. Correia · A. Duatti
D. S. Edwards · W. P. Li · S. Liu · L. A. Meyer · A. Paulo · S. P. Robinson
I. Santos · V. Sharma · L. Uccelli · D. Piwnica-Worms

Springer

The series *Topics in Current Chemistry* presents critical reviews of the present and future trends in modern chemical research. The scope of coverage includes all areas of chemical science including the interfaces with related disciplines such as biology, medicine and materials science. The goal of each thematic volume is to give the nonspecialist reader, whether at the university or in industry, a comprehensive overview of an area where new insights are emerging that are of interest to a larger scientific audience.

As a rule, contributions are specially commissioned. The editors and publishers will, however, always be pleased to receive suggestions and supplementary information. Papers are accepted for *Topics in Current Chemistry* in English.

In references *Topics in Current Chemistry* is abbreviated Top Curr Chem and is cited as a journal.

Visit the TCC content at springerlink.com

Library of Congress Control Number: 2004112157

ISSN 0340-1022
ISBN 3-540-22577-3 **Springer Berlin Heidelberg New York**
DOI 10.1007/b94549

Springer is a part of Springer Science+Business Media
springeronline.com

Printed in The Netherlands

Cover design: KünkelLopka, Heidelberg/design & production GmbH, Heidelberg
Typesetting: Fotosatz-Service Köhler GmbH, Würzburg

Printed on acid-free paper 02/3141 xv – 5 4 3 2 1 0

Volume Editor

Prof. Dr. Werner Krause
Berlex
International Project Management Oncology
PO BOX 1000
Montville, NJ 07045-1000, USA
werner_krause@berlex.com

Editorial Board

Topics in Current Chemistry also Available Electronically

For all customers who have a standing order to Topics in Current Chemistry, we offer the electronic version via SpringerLink free of charge. Please contact your librarian who can receive a password for free access to the full articles by registration at:

springerlink.com

If you do not have a subscription, you can still view the tables of contents of the volumes and the abstract of each article by going to the SpringerLink Homepage, clicking on "Browse by Online Libraries", then "Chemical Sciences", and finally choose Topics in Current Chemistry.

You will find information about the

- Editorial Board
- Aims and Scope
- Instructions for Authors
- Sample Contribution

at springeronline.com using the search function.

Preface

Radiopharmaceutical chemistry is currently in a transition state. It has come a long way from using radioisotopes as they are, e.g. iodine isotopes, via the synthesis of organometallic complexes to molecular imaging with the administration of highly complicated radiolabelled biomolecules and to very recent applications of targeted therapeutic radiopharmaceuticals, the "magic bullets" of Ehrlich, labelled with radioisotopes.

The process of transition can also be derived from the still-valid definition of radiopharmaceuticals as "radioactive drugs that, when used for the purpose of diagnosis or therapy, typically do not elicit any physiological response from the patient". Apart from being a contradiction in itself, since any therapy should result in a physiological response, otherwise we would be using a placebo, experience with the newly introduced radiopharmaceutical therapeutics, Zevalin and Bexxar, radiolabelled monoclonal antibodies for the treatment of non-Hodgkin lymphoma, has shown that these drugs are indeed capable of eliciting significant beneficial, therapeutic effects but also result in physiological responses experienced by the patient as adverse reactions. Both, the beneficial and the undesired effects are expressions of pharmacological activity, an aspect that is new in the field of radiopharmaceuticals.

This volume on radiopharmaceutical contrast agents, which also covers new developments on radiopharmaceuticals used as therapeutic drugs is the third of a series on diagnostics following a volume that was exclusively dedicated to magnetic resonance imaging and a "mixed" volume on optical, ultrasound, X-ray and radiopharmaceutical agents.

The transition from diagnosis to therapy is also reflected in the contributions to this volume, where the first chapters are dedicated exclusively to diagnosis and the chapters at the end of the book cover both aspects, dealing with compounds for diagnostic imaging and with drugs used for therapy.

The first chapter of this volume is dedicated to bioorganometallic chemistry with special focus on diagnostic imaging using technetium complexes, particularly Tc-CO complexes. The state of the art in this field is described in great detail covering the basics of Tc complexes and their synthesis and characterization, new Tc compounds, and requirements for practical applications. The last part of this chapter deals with labelling of small molecules for the use in CNS imaging and with labelled biomacromolecules for cancer applications. In the second chapter, complexes of technetium and rhenium are reviewed with special emphasis on heterofunctional phosphines and borates (scorpionates). Synthetic procedures, results on reactivity, and structural characterisation are supplemented with data on the labelling of biomolecules.

Nitrido complexes of technetium and rhenium are covered in the third chapter. An extensive discussion of electronic configurations and of the geometry of these complexes and their spatial requirements at the molecular level is provided. Applications in molecular imaging include cardiac investigations and studies in degenerative diseases.

Chapter 4 addresses Tc labelling of small biomolecules using bifunctional coupling agents, in particular 6-hydrazinonicotinamide (HYNIC). Activities relevant for successful labeling are discussed, starting from the conjugation procedure, the protection of the hydrazine moiety, via development of a suitable formulation to examples for the diagnosis of various diseases. Additionally, new techniques for the characterization of technetium-labelled drugs at the tracer level are discussed.

In the fifth chapter, one of the major problems in cancer therapy, the development of multi-drug resistance (MDR) is addressed. In MDR, cancer patients who have responded to their therapy, develop resistance against this treatment making further administration useless. Unfortunately, this phenomenon is not restricted to individual drugs but, instead, resistance is developed across different classes of drugs without the patient ever having been treated with them. As a consequence, not only is treatment with the drug against which MDR has initially developed futile but also with a complete array of other anti-cancer drugs. Analysis whether or not MDR plays a role in the patient's reaction to treatment is an important piece of information for the physician. Technetium complexes useful for PET or SPECT analysis of this phenomenon are extensively discussed in this chapter.

Chapter 6 deals with the radiolabelling of somatostatin analogs for both diagnosis and therapy of cancer. Somatostatin is a small peptide of 14 amino acids regulating the release of certain hormones such as growth hormone, TSH or prolactin. Somatostatin receptors are expressed in several organ systems. In some tumor types, the receptor is overexpressed and may be used for diagnostic and therapeutic purposes. The review describes the production of radionuclides suitable for labelling somatostatin analogs and their radiochemistry and gives details on preclinical and clinical studies with these compounds.

The last chapter is exclusively dedicated to radiotherapy. It describes adhesion molecules out of the integrin family labelled with yttrium-90 via a chelator. Integrins are proteins that are expressed on epithelial and endothelial cells facilitating cellular adhesion on membranes. Some tumor types show high expression of integrin $\alpha_v\beta_3$ making this protein a useful target for the detection and therapy of these tumors. The chapter focusses on radiopharmaceutical design, selection of the radionuclide and the bifunctional chelator and on the pharmacokinetics of yttrium-labelled integrin $\alpha_v\beta_3$ radiopharmaceuticals.

The current transition state clearly provides chances for new directions in this exciting field of research. The future will show whether and when a new volume on this topic is needed.

Montville, October 2004 Werner Krause

Contents

Contents of Volume 221
Contrast Agents I
Magnetic Resonance Imaging

Volume Editor: Werner Krause
ISBN 3-540-42247-1

Contents of Volume 222
Contrast Agents II
Optical, Ultrasound, X-Ray and Radiopharmaceutical Imaging

Volume Editor: Werner Krause
ISBN 3-540-43451-8

Top Curr Chem (2005) 252: 1–44
DOI 10.1007/b101223

New Organometallic Technetium Complexes for Radiopharmaceutical Imaging

Roger Alberto (✉)

Institute of Inorganic Chemistry, University of Zürich, Winterthurerstrasse 190, 8057 Zürich, Switzerland
ariel@aci.unizh.ch

Abstract It is the objective of this review to introduce the reader to the current status of organometallic chemistry, oriented towards the development of novel ^{99m}Tc-based imaging agents. This type of inorganic chemistry has nowadays become part of so-called bioorganometallic chemistry.

The article describes the current state of research and development of radiopharmaceutical imaging agents based on organometallic moieties and complexes. Organometallics are, in general, not considered to play an important role in inorganic medicinal chemistry. An introductory section describes the requirements for practical application of a radiopharmaceutical to underline in the second part that currently available compounds might well be the basis for the development of completely new imaging agents. The introduction gives also a short outline of basic organometallic technetium compounds as well as organometallic aqua ions from other elements which are currently under investigation in the new field of bioorganometallic chemistry.

In the following section, major focus is put on technetium–CO complexes. The compound $[^{99m}Tc(OH_2)_3(CO)_3]^+$ is the prototype of a precursor, the special behaviour of which is based on its organometallic character. The synthesis and some fundamental properties are discussed. Within this section substitution reactions of $[^{99m}Tc(OH_2)_3(CO)_3]^+$ are presented according to the denticity of the ligands. The reactions are critically highlighted in respect of their versatility for the development of novel radiopharmaceuticals. Combining monodentate and bidentate ligands leads to the potential of a new [2+1] mixed-ligand concept.

In the last section, the labelling of biomolecules with organometallic moieties is described. Since most of the organometallic complexes are still being researched this section focuses on concepts, and some of the most important developments towards new CNS receptor ligands and labelled peptides and proteins are given. This section aims at motivating researchers from radiopharmaceutical chemistry to experience new approaches which are probably superior to the currently available ones. Throughout the text, examples from other organometallic compounds are described; however, since the carbonyl concept is the most advanced, the review is oriented towards the chemistry of $[^{99m}Tc(OH_2)_3(CO)_3]^+$.

Keywords Technetium · Rhenium · Bioorganometallics · Carbonyl · Radiopharmaceuticals · Labelling

1 Introduction

The artificial element technetium was discovered by Perrier and Segré in 1937 [1]. In those days, nobody suspected that this element would become an essential part of medical health care some 40–50 years later. At that time, X-rays, which had been discovered by Röntgen some 40 years earlier, were the most important tool for diagnosis by imaging. The application of radionuclides had already been introduced by deHevesey but this was more a toy than a serious instrument. In fact patients usually suffered severely from the radiation of administered radionuclides such as radium or thorium and, not surprisingly, a number of them even died as a consequence of the treatment. Nowadays, this situation has changed completely. With growing knowledge about the interaction of ionizing radiation with tissues the properties of a radionuclide to be useful for and not dangerous for a patient became more defined. In parallel, the development of sensitive cameras for the detection of single photons increased. A selection of radionuclides whose decomposition characteristics would allow high-resolution images in single-photon-emission computer tomography without the burden of a high dose to patients became possible. Among these few versatile radionuclides, ^{99m}Tc is probably the most important one and a large portion of nuclear medicine diagnostics is carried out with it. The main reasons for the versatility of ^{99m}Tc are not only its decay characteristics but also its price and availability. Costs of public health care are steadily increasing and play a major role in the selection of a particular diagnostic procedure. The cost for a diagnostic amount of ^{99m}Tc is very low compared with that for other radionuclides, which might even have better decay characteristics. Another important feature of ^{99m}Tc is its availability. In the 1960s researchers at Brookhaven National Laboratories developed a generator in which ^{99}Mo, the mother of ^{99m}Tc, is loaded on an alumina column [2–4]. The radionuclide ^{99}Mo decays with a 67-h half life time to ^{99m}Tc, which can be continuously eluted with saline from the generator. Mother and daughter are in a so-called secular equilibrium. Brookhaven National Laboratories did not patent the generator, anticipating that the system would not have any application in the future. Considering the number of generators sold so far, it has become obvious that a large amount of money did not reach the inventors owing to this wrong decision.

Price and availability are the favourable features that help to keep the clinical importance of ^{99m}Tc alive. The disadvantages arise from the chemical side owing to the artificial nature of technetium and the difficulty of combining a transition metal with targeting biomolecules. From a chemical point of view, other radionuclides are more versatile, in particular the positron emitters ^{11}C, ^{18}F and ^{123}I, which are most important in positron emission tomography and are strongly competing with single-photon-emission computer tomography despite other disadvantages not discussed here. Whereas a number of so-called technetium-essential radiopharmaceuticals have been successfully introduced onto the market, labelled vectors such as peptides or brain receptor imaging agents are still subjects of research and, with a very few exceptions, in phase I clinical trials only. The radiopharmaceutical chemistry of technetium-based coordination compounds has been reviewed without considering organometallic compounds systematically [5–9]. A selection of perfusion agents introduced onto the market and some ^{99m}Tc-labelled targeting vectors in phase I clinical trials are given in Scheme 1.

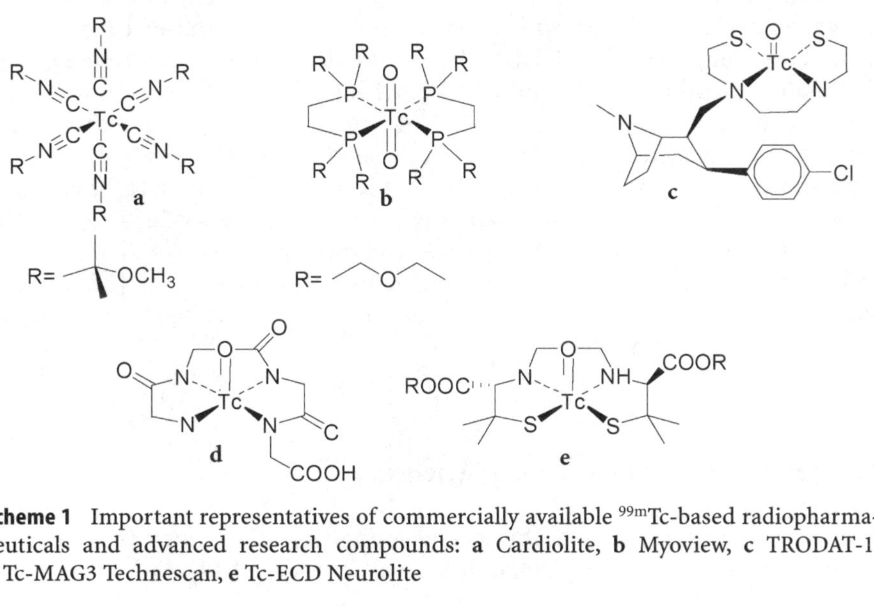

Scheme 1 Important representatives of commercially available ^{99m}Tc-based radiopharmaceuticals and advanced research compounds: **a** Cardiolite, **b** Myoview, **c** TRODAT-1, **d** Tc-MAG3 Technescan, **e** Tc-ECD Neurolite

Labelling of targeting vectors is the main object of modern radiopharmaceutical research. Inorganic chemistry is challenged to find convenient ways to introduce physiologically stable ^{99m}Tc cores into biomolecules without affecting their bioactivity. The labelling of biomolecules must inherently be possible with the facilities of hospitals and the skills of their personnel. Beyond these very fundamental requirements, targeting radiolabelled biomolecules accumulate, in general, in nontargeted organs such as kidney or liver; hence, getting clear images from whole body scans is spoiled. Since the biodistribution of an unlabelled biomolecule is not sufficiently well known, quantification of organ uptake becomes obvious only after labelling the biomolecule.

Besides the essential requirements pointed out before, it now becomes the task of the radiopharmaceutical chemist to change the nature of the tagged complex in a way that accumulation in nontargeted organs is reduced and in targeted organs enhanced. Hence, the attached ^{99m}Tc complex has to influence the biological behaviour of the native biomolecule in a positive direction. From a pharmaceutical point of view, such a procedure is called "compound finding" and is comparable to what is performed with nonradioactive organic compounds to enhance, for example, the therapeutic index whilst reducing side effects, and allow the administration of higher amounts of the drug. Consequently, compound finding in radiopharmacy means altering the structure of a metal complex to achieve a better in vivo distribution. This is a new situation and demands a flexible metal core with respect to the chelators forming stable complexes under physiological conditions.

Radiopharmaceutical chemistry is a topic in life science and the related coordination chemistry must be aqueous chemistry. Accordingly, the main focus related to the labelling of biomolecules with ^{99m}Tc is classical "Werner-type" chemistry. Coordination chemistry of technetium has recently been the focus of several reviews [10–16]. As is obvious from Scheme 1 the vast majority of complexes comprises the "$[Tc{=}O]^{3+}$" core in which technetium is in the oxidation state +V. Although a hypothetical aqua ion of $[Tc{=}O]^{3+}$ does not exist, it can be stabilized with weak mulitdentate ligands present in large excess. They can then be replaced via ligand exchange to give stronger multidentate chelators. Another core that merits attention is "$[Tc{\equiv}N]^{2+}$", also with technetium in oxidation state +V and similar chemical behaviour in respect of aquo ions as in $[Tc{=}O]^{3+}$ [17–20]. Besides these two central moieties, the mixed-ligand [3+1] concept based on the $[Tc{=}O]^{3+}$ core is widely applied in research [21, 22]. The approach is problematic for certain biomolecules since in vivo instability of complexes has been observed.

2 Organometallic Complexes in Life Science

At a first glance it seems that the special properties of *M*–C bonds are not involved in biological systems. The well-known coenzyme B12, in which a cobalt-carbon (to adenosyl or methyl) is present, seems rather to be the exception that confirms the rule. In the recent past however, more and more metalloenzymes that contain *M*–CO fragments have been structurally characterized. These enzymes represent a classical sort of organometallic compound [23–27]. It is probably not just by chance that CO is applied as a coligand in biology and it can be anticipated that more organometallic enzymes will be discovered in the near future.

Owing to the pure aqueous nature of radiopharmaceutical chemistry, organometallic compounds are rare in this field and in life science, in general. It is even a fact that nonradioactive drugs in inorganic medicinal chemistry rarely contain a *M*–C bond. Recent research on cytotoxic agents with $[(C_6H_6)Ru(OH_2)_3]^{2+}$ represents a valuable research extension in the organometallic direction [28–32]. Actually, to our knowledge there is only one compound containing *M*–C bonds which has successfully been introduced onto the (radio)pharmaceutical market. This concerns Cardiolite, a compound which will be discussed later and is shown in Scheme 1. On the other hand, the rapidly growing research field of "bioorganometallic chemistry" is focused on the application of organometallic compounds in life science [33]. Bioorganometallic chemistry will not be discussed here but it merits emphasis that organometallic compounds such as ferrocene, $[Cp^*Rh(OH_2)_3]^{2+}$, $[(C_6H_6)Ru(OH_2)_3]^{2+}$, $[(Cp^*)_2MoX_2]$, where Cp* is pentamethylcyclopentadienyl, and others proved to be superior to coordination compounds (Scheme 2). Many water- and air-stable organometallic compounds have been described in detail and in respect of their biological versatility. Still, the use of organometallics in life science remains a niche although, for example, ferrocifene and some cytotoxic ruthenium-based agents have progressed to phase I clinical trials.

In radiopharmaceutical chemistry, the only exception is the Tc(I) isonitrile complex $[^{99m}Tc(CN\text{–}R)_6]^+$ (*R* is methoxy isobutyl isonitrile) which has found its way to routine clinical application for myocardial imaging [34, 35]. This organometallic complex is probably the most successful radiopharmaceutical

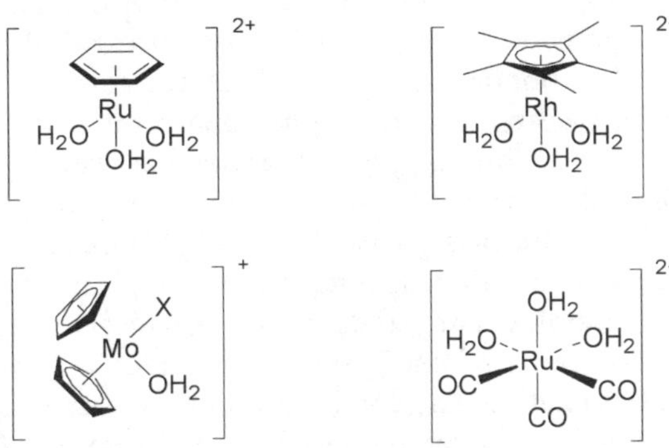

Scheme 2 Examples of organometallic aqua ions that are currently under investigation for application in life science

ever produced. The complex merits special attention not only owing to its radiopharmaceutical application but, probably more importantly, owing to its surprising preparation and unexpected properties. Originally, the complex was formed in quantitative yield within a short time after heating and in the presence of the isonitrile ligand and dithionite as a reducing agent. It is remarkable that a low-valent organometallic compound is thermodynamically favoured under such conditions and that the complex is completely stable over the pH range and at high temperature. It is a drawback of the stability that the monodentate ligands can hardly be substituted by other ligands and, thus, the complex can be taken as a starting material for further complexes. The synthesis matches with the conditions as required from routine application and in vitro and in vivo stabilities are excellent. The stability must be of kinetic origin and this allows application of a compound that contains exclusively single bonds. This structural feature would not be expected with classical coordination compounds of higher valences.

The preparation of low-valent complexes such as $[^{99m}Tc(CN{-}R)_6]^+$ from water might still initiate research in the field of organometallic compounds for life science. The paradigm that organometallic compounds are not compatible with conditions imposed by biological systems could clearly be repudiated with this compound.

2.1 Particular Properties of Organometallic Compounds

Organometallic complexes applied in life science must be stable in water and air. Additionally, in radiopharmaceutical chemistry they have to be prepared directly from $[^{99m}TcO_4]^-$ in a one-pot synthesis from aqueous saline. Low-valent organometallic complexes are, in general, closed shell and obey the 18-electron rule as, for example, in $[^{99m}Tc(CN{-}R)_6]^+$ as mentioned before. These two important features essentially govern all complexes under investigation in life science, they have 18 electrons and have d^3 or d^6 electronic configuration. Often, the organometallic ligands are not involved in pH equilibria. The electronic configuration is responsible for kinetic stability towards full or partial ligand replacement. The M–C bonds are reasonably stable and prevent dissociative substitution and the lowest unoccupied molecular orbital is too high in energy to enable an interchange or an associative reaction pathway. This makes, for example, d^6 organometallic complexes robust if the ligand bond energies are sufficiently high, which is the case for CO or CN–R, for example. In contrast to Werner-type ligands, organometallic ligands are not prone to protonation and the corresponding complexes are stable over a broad pH-range. Werner-type ligands, as applied for the Tc(V) cores, are σ and/or π donors which can be protonated and are pH-sensitive. The combination of donors is limited since, for example, weak donors such as thioethers have to be combined with a set of electronically "correct" ligands in order to achieve thermodynamic stability. As a drawback of the robust nature of organometallic complexes, they are more flexible owing to the electronic flexibility of the CO ligand. Strong donation can be compensated by better π backbonding and an incoming π acceptor by stronger donation from the CO or other ligands, respectively. Complexes obeying the 18-electron rule make complexes with a mixed classical/nonclassical ligand sphere very robust as well. Consequently, monodentate ligands are feasible for stable binding of a technetium fragment to a targeting vector. This contrasts with complexes in oxidation state +V, where stability must be achieved by multidentate chelators. Complexes with tridentate or lower-dentate ligands are, in general, not sufficiently stable. Aiming at electronically robust complexes of technetium must focus on oxidation states +IV (d^3) and +I (d^6). Tc(IV) is problematic since, at least on the macroscopic level, TcO_2 or other hydrolysed species represent a thermodynamic trap. Tc(I), on the other hand, is an electron-rich centre and π acceptors such as CO or CN–*R* must be present in order to compensate the high electron density.

An aqua-ion of technetium would be most convenient but its existence is unlikely since aqua-ions in the second transition-metal row are rare. On the other hand, acceptor ligands can stabilize low-valent metal centres and form a kind of aqua ion (e.g. $[Tc(OH_2)_3(CO)_3]^+$) and donors can stabilize medium-valency centres for the same purpose and the existence of $[Cp^*Rh(OH_2)_3]^{2+}$ is a nice example for the latter case [36].

The electronic flexibility of organometallic moieties is the important feature for their application in life science. The difficulties of introducing a new organometallic-based concept are manifold. Organometallic compounds are considered to be air- and water sensitive. Still, they represent a great opportunity to explore new possibilities in biological fields as outlined in the following, especially with the $[^{99m}Tc(CO)_3]^+$ moiety but also with others.

2.2 Organometallic Technetium Complexes

This section gives a short overview of the most relevant organometallic technetium complexes. More details about the compounds applied in radiopharmaceutical imaging will be given in later sections. The organometallic chemistry of technetium has been reviewed on several occasions [13, 14, 37, 38]. The main motivation for exploring technetium chemistry is the application in radiopharmacy; hence, organometallic chemistry is rather underdeveloped. In the early days of technetium chemistry, fundamental types of compounds were synthesized and characterized but no studies about their behaviour in water appeared. Since the chemistry of technetium in its low oxidation states strongly resembles that of its higher homologue rhenium, not much effort was undertaken to really explore more details. It has to be emphasized at this point that technetium does not only differ gradually from rhenium but principally in many of its compounds and reactions.

Homoleptic $[Tc_2(CO)_{10}]$ can be synthesized in autoclaves, a procedure that is difficult to reproduce owing to safety regulations [39–41]. $[Tc(CO)_6]^+$ is available as well and can even be synthesized from water as recently published [42]. Most of the organometallic chemistry starts with $[Tc_2(CO)_{10})]$ and, for example, the important Tc(I) starting material $[TcBr(CO)_5]$ can be produced directly by bromination [43]. All kinds of cyclopentadienyl (Cp) complexes or complexes that comprise the *fac*-$[Tc(CO)_3]^+$ moiety can be prepared starting from $[Tc_2(CO)_{10})]$ [44]. Oxidation of $[CpTc(CO)_3]$ gives *cis*- and *trans*-$[TcBr_4(CO)_2]^-$, in which technetium is in oxidation state +III and this is the highest oxidation state in which CO remains coordinated [45, 46]. A limited number of other complexes that contain classical organometallic ligands such as alkyls and alkenes are known. Important starting compounds are shown in Scheme 3.

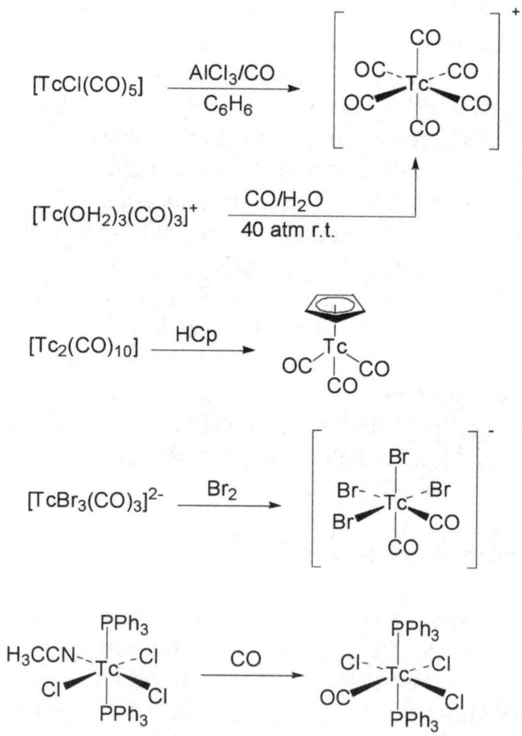

Scheme 3 Basic organometallic compounds of technetium useful as starting materials for further syntheses

An interesting reaction in which a metal carbonyl was formed from an in situ CO source and not from free CO was published in 1982 [47]. Baldas et al. aimed at the preparation of a dithiocarbamate Tc(III) complex using formamidine–sulfinic acid as a reducing agent. Instead, they isolated a seven-coordinate Tc(III) complex $[Tc(CO)(S_2CNR_2)_3]$ containing a CO ligand that formed upon degradation of formamidine–sulfinic acid (Scheme 4). Although not studied in more detail and not extended to other studies, the procedure is

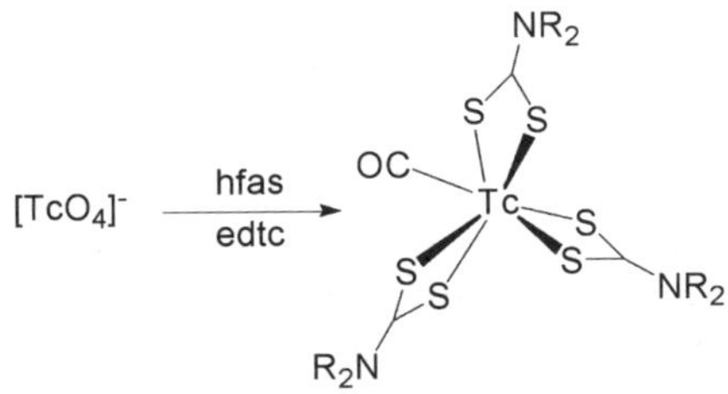

Scheme 4 First carbonyl complex of Tc(III) prepared with an in situ CO source

remarkable. The authors propose that the formation of CO and its binding to technetium is metal-mediated and that the binding does not occur by the coordination of the free CO generated. This mechanistic picture, if true, is probably also valid for the formation of the *fac*-$[Tc(CO)_3]^+$ moiety as discussed in a later section. Other approaches for in situ CO formation are known (e.g. HCl and formic acid), however, the respective conditions are not compatible with a one-step procedure as required for routine use in life science. As a model, the procedure described by Baldas et al. deserves more attention since carbonyl complexes are versatile starting materials. They need to be prepared without pressure and, for safety reasons, without free CO but with an in situ CO source. Formamidine–sulfinic acid might be a reasonable alternative.

3
Technetium Carbonyl Complexes in Water

Precursor complexes that are versatile for the labelling of targeting molecules must comprise ligands which can be exchanged for incoming, biomolecule-bound chelators. These ligands are preferentially water molecules since water is the solvent of biology and medicine. Organometallic fragments with water ligands are rare, probably owing to the lack of attempts to prepare them. A limited number of reactions have been described in the literature comprising some of the cations described earlier but also $[Ru(OH_2)_3(CO)_3]^{2+}$. A review of this interesting class of compounds has been given by Koelle [48]. $[Ru(OH_2)_3(CO)_3]^{2+}$ has some typical features as a precursor for aqueous organometallic coordination chemistry: three more or less tightly bound CO ligands and three labile water ligands. Extrapolating to group 7 transition metals, we can expect that the analogous complexes exist and that they are even stabler. This led to the development of the chemistry of *fac*-$[Tc(OH_2)_3(CO)_3]^+$, whose basic chemistry and potential applications in imaging are described.

3.1
The $[Tc(OH_2)_3(CO)_3]^+$ Complex, Basic Properties

The original synthesis of $[^{99}TcCl_3(CO)_3]^{2-}$ was performed in tetrahydrofuran (THF) at 1 atm CO and directly from $[NBu_4][TcO_4]$ with $BH_3{\cdot}THF$ as a reducing agent [49]. The exact reaction mechanism remains unclear. It consist of a 6-e^- reduction and concomitant coordination of three CO ligands. No intermediates were found. The dianion can be dissolved in water and the three chlorides readily exchange for solvent molecules as evident from IR spectroscopy. Only at high Cl^- concentration ($[Cl^-]$>2 M] is some evidence from NMR spectroscopy obtained that one chloride re-coordinates. Aqueous solution of $[Tc(OH_2)_3(CO)_3]^+$ are completely stable over the full pH range although at high pH reversible hydrolytic di- tri- and tetramerization occurs [50, 51]. Thermal or hydrolytic cleavage of the CO ligands through the formation of a metallacarboxylic acid as in case of the ruthenium analogue (water gas shift reaction) could not be observed, which implies that the CO ligands cannot be substituted by an incoming ligand. This is true for incoming Werner-type ligands; however, strong trans-effect ligands such as CO can replace technetium-bound CO as shown in Scheme 5 and this has been proven experimentally with ^{13}CO. Mechanistically, an incoming CO ligand replaces first one of the coordinated H_2O, which results in the formation of an intermediate complex $[Tc(OH_2)_2(CO)_4]^+$. Owing to the strong trans-effect, one of the two trans COs is then cleaved and either the original compound, $[Tc(OH_2)_3(CO)_3]^+$, or the isotope-labelled analogue, $[Tc(OH_2)_3(^{13}CO)(CO)_2]^+$, is regenerated. Complexes $[Tc(OH_2)_2(CO)_4]^+$ and $[Tc(OH_2)(CO)_5]^+$ can be observed with ^{99}Tc NMR but both are not stable under these conditions and are true intermediates present in a steady-state concentration after an extended period of time. If CO pressure is released, rapid back reaction to $[Tc(OH_2)_3(CO)_3]^+$ occurs, behaviour known for the structurally related ruthenium complex. Experiments with ^{13}CO showed a very nice and stepwise exchange of ^{12}CO bound to rhenium or technetium by ^{13}CO. The reaction can be followed by ^{13}C and ^{99}Tc NMR and a picture of the reaction is given in Scheme 5.

Surprisingly, after about 2 weeks at moderate pressure, significant amounts of $[^{99}Tc(CO)_6]^+$ could be detected in the ^{99}Tc NMR. The signal is weak owing to the fact that in the solution most of the hexacarbonyl complex precipitated as $[^{99}Tc(CO)_6][ClO_4]$ from the 1 M $HClO_4$ solution. This method presents a convenient strategy to prepare the binary Tc(I) carbonyl complex and is much easier than the original procedure [39, 42].

The water exchange rate is very important for the ligand exchange reactions described later in this section and has been determined for the corresponding rhenium complex $[Re(OH_2)_3(CO)_3]^+$. If the water self-exchange rate is too slow, a labelling process is not expected to occur at a reasonable rate; hence, the precursor is not convenient for radiopharmaceutical purposes. The self-exchange rate is strongly pH dependent. This can be expected since the exchange follows a classical base-catalysed mechanism. The pK_a value of $[Re(OH_2)_3(CO)_3]^+$ has

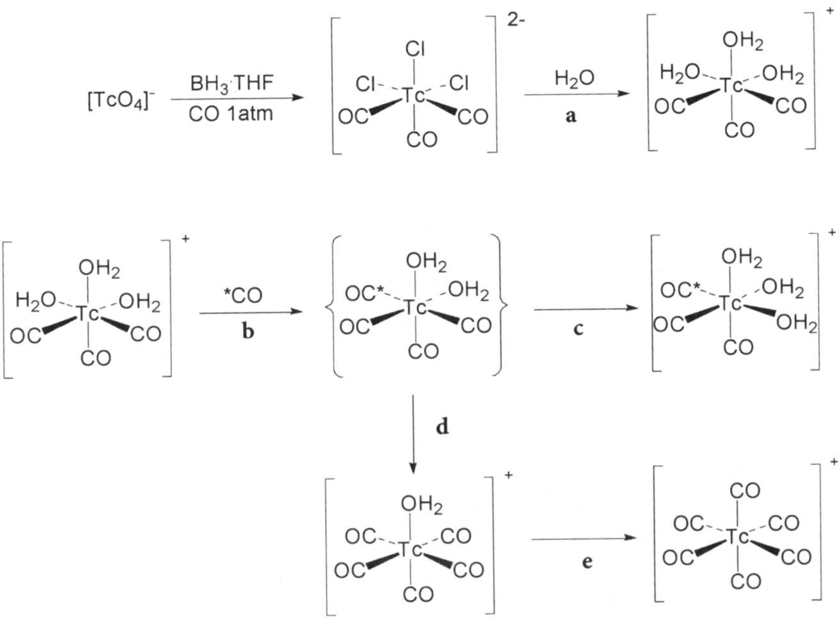

Scheme 5 **a** Synthesis of $[^{99}Tc(OH_2)_3(CO)_3]^+$, CO exchange with ^{13}CO, **b** formation of $[^{99}Tc(OH_2)_2(^{13}CO)(CO)_3]^+$ and **c** re-formation of $[^{99}Tc(OH_2)_3(^{13}CO)(CO)_2]^+$ or **d** formation of $[^{99}Tc(OH_2)(CO)_5]^+$ and **e** $[^{99}Tc(CO)_6]^+$, respectively

been estimated to be about 7.8. Deprotonation of one water yields a hydroxo complex which stabilizes the transition state in a D or I_d ligand exchange mechanism. From the data it is clear that the substitution will be slow, but reasonably fast for labelling processes at physiological pH and at elevated temperature [52]. The corresponding experiments for technetium are currently being performed but it can be estimated that the reactions are about 10–100 times faster according to the general trends when going from the second to the third transition-metal series.

Besides the stability of the CO ligands towards ligand exchange, oxidation stability in aqueous solution is, in particular, remarkable. Cyclovoltametric experiments showed that the oxidation potential must be higher than 1.2 V in water. As described later, the ^{99m}Tc compound is moderately air sensitive at high dilution. $[Re(OH_2)_3(CO)_3]^+$ and $[^{99}Tc(OH_2)_3(CO)_3]^+$ are stable when exposed to air.

These basic properties are favourable for application in imaging. The $[Tc(OH_2)_3(CO)_3]^+$ complex can also be described as a semi aqua ion. One half of the coordination sphere is shielded, whereas the other half is prone to substitution with classical or nonclassical ligands. Three ligands of small size and low molecular weight are tightly bound, while three water ligands can be exchange by incoming chelators. To yield physiologically stable complexes, the resulting compounds must be robust.

Before outlining the coordination properties of $[^{99}Tc(OH_2)_3(CO)_3]^+$ in more detail, the synthesis of the corresponding precursor $[^{99m}Tc(OH_2)_3(CO)_3]^+$ merits attention. A precursor is only useful if quantitative aqueous preparation in a hospital or a clinical research institute can be done in one step under reasonable conditions. Such requirements are not routine for organometallic compounds. CO is merely soluble in water and other classical ligands such as C_5H_6 (HCp) rapidly decompose. Isonitriles are reasonably stable but will, as demonstrated earlier, occupy all available positions on the metal centre. This does not leave a coordination site available for subsequent ligand substitution and labelling. The preparation of $[^{99m}Tc(OH_2)_3(CO)_3]^+$ starts from $[^{99m}TcO_4]^-$ comprising a six-electron reduction and concomitant coordination of three COs. Differently from organic solvents, many side reaction pathways are now possible, since stable hydroxo and oxo species are known in intermediate oxidation states. Still, the switch from BH_3·THF to reasonably water stable $[BH_4]^-$ enables the straightforward synthesis of $[^{99m}Tc(OH_2)_3(CO)_3]^+$ in one step [53]. The presence of 1 atm CO and heating to 95 °C for 30 min is sufficient. The mechanism has not been elucidated but it represents a challenge not only in respect of this particular reaction, but also to transfer the principle to other transition elements. The reaction conditions are depicted in Scheme 6.

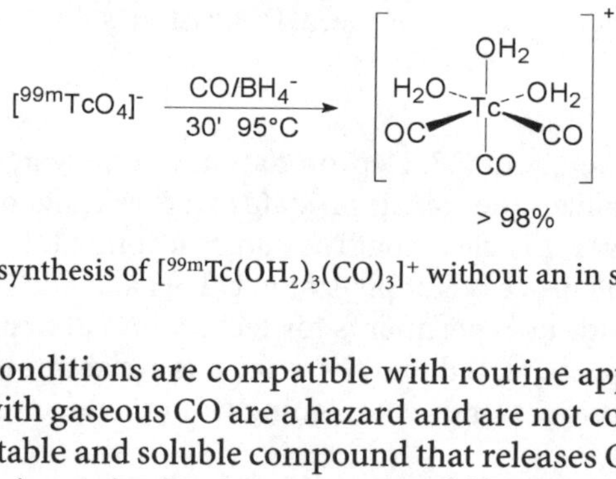

Scheme 6 Standard synthesis of $[^{99m}Tc(OH_2)_3(CO)_3]^+$ without an in situ CO source

This reaction conditions are compatible with routine application. However small vials filled with gaseous CO are a hazard and are not convenient for safety reasons. A water-stable and soluble compound that releases CO under well-controlled conditions, i.e. an in situ CO source, would be the solution to the problem. Molecules or slats with such properties are rare. Malone and coworkers [54, 55] and Carter and Parry [56] described the synthesis and properties of so-called boranocarbonate $[H_3BCO_2]^{2-}$ (BC), which does release CO under acidic conditions. BC is formed from H_3BCO under alkaline conditions. The CO adduct of BH_3 is difficult to prepare and to handle and a synthesis was developed starting directly from H_3B·THF. Bubbling CO through a THF solution carries H_3BCO continuously off and this can then be hydrolysed as in the original procedure [57]. This allowed the preparation of bulk amounts of $Na_2[H_3BCO_2]$ or $Na[H_3BCO_2H]$ as shown in Scheme 7.

BC parallels the behaviour of metallacarboxylic acids which could potentially also serve as in situ CO sources along the same pathway. BC comprises two functions in one anion: the reducing function of $[BH_4]^-$ and CO as a ligand.

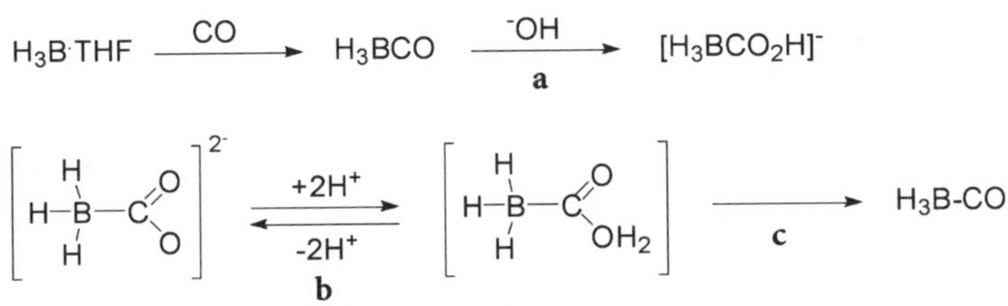

Scheme 7 **a** Schematized preparation of boranocarbonate, **b** protonation equilibrium of boranocarbonate and **c** dehydration to boranocarbonyl, which decomposes under release of CO

BC is reasonably stable at neutral-to-alkaline pH but decomposes rapidly in acidic media. The reaction of $[^{99m}TcO_4]^-$ with BC at pH 11–12 yields quantitatively the organometallic aqua ion $[^{99m}Tc(OH_2)_3(CO)_3]^+$. Mechanistically, it is possible that BC acts as a ligand which binds to the technetium centre. Hydride transfer followed or paralleled by reduction occurs concomitantly with CO coordination. The X-ray structure of a model with K^+ as the metal is shown in Fig. 1.

BC is stable in aqueous solution and can be lyophilized. This is the basis for a kit formulation which is nowadays commercially available under the trade name Isolink (Mallinckrodt Med.). The precursor $[^{99m}Tc(OH_2)_3(CO)_3]^+$ can be synthesized in quantitative yield by adding generator eluate to the vial and subsequent heating to 98 °C for 20 min. The reaction yield does not depend on the total amount of activity. The complex $[^{99m}Tc(OH_2)_3(CO)_3]^+$ is stable for hours if kept under N_2 but is slowly reoxidized when exposed to air. In contrast, all complexes in which the three water ligands are substituted as discussed later are thermally very stable even when exposed to air. It merits attention at this point that the rhenium homologue $[^{188}Re(OH_2)_3(CO)_3]^+$ cannot be prepared along the same route. Since rhenium is more difficult to reduce and reacts, in

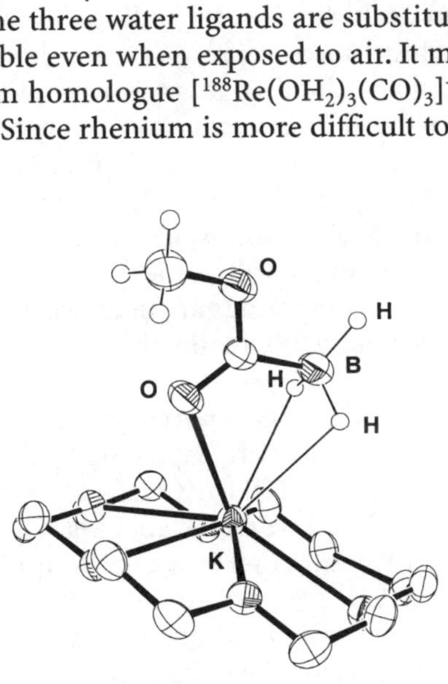

Fig. 1 X-ray structure of $[K(18crown\text{-}6)][H_3BCOOCH_3]$

general, much slower, the technetium procedure gives only minor yields of $[^{188}Re(OH_2)_3(CO)_3]^+$. In this case, $H_3B{\cdot}NH_3$ is the reducing agent and the reaction must be carried out in the presence of H_3PO_4 at acidic pH. So far, this formulation excludes an instant kit formulation but this is not considered to be a limiting issue for therapeutic applications [58].

The reducing agent BC is likely to be a general compound and it deserves more attention for the preparation of mixed water–CO complexes on the macroscopic level of other transition-metal cations. We observed in preliminary experiments the formation of $[^{99}Tc(OH_2)_3(CO)_3]^+$ on the macroscopic level directly from water, although in moderate yields of less than 30%. The same was found for $[MoO_4]^{2-}$, which gives mixed H_2O–CO complexes of so far unknown composition. The general behaviour of BC appears to point towards a new field in aqueous organometallic chemistry.

3.2 Reaction with Werner-Type Ligands

3.2.1 Monodendate Ligands

The incentive of the chemistry of $[^{99(m)}Tc(OH_2)_3(CO)_3]^+$ is the fact that it can be considered as a normal aqua-ion with only three available coordination sites. Since the complexes resulting from water substitution are kinetically stable, almost any type of chelator forms coordination compounds with the *fac*-$[^{99(m)}Tc(CO)_3]^+$ moiety. This is equally true for mono- bi- and tridentate chelators regardless of whether the hard–soft condition for ligand and metal centre matches. This behaviour also represents the major impact of using $[^{99m}Tc(OH_2)_3(CO)_3]^+$ for the labelling of targeting molecules for imaging purposes. As outlined later it is of importance to conjugate a metal complex with similar physicochemical properties such as hydrophilicity, size, and charge to a biomolecule. For the traditional "$[Tc(V)=O]^{3+}$" approaches, the choice of the ligand is limited, whereas for Tc(I) almost any donating molecules can be used as ligands. Some minor limitations will be mentioned at the appropriate places. In this section coordination compounds are in focus. It will be shown later how the observed general coordination chemistry can be extrapolated to biomolecules for receptor targeting.

Some selected reactions with different type of ligands that are of relevance for radiopharmaceutical chemistry are depicted in Scheme 8.

It has been observed that $[^{99m}Tc(OH_2)_3(CO)_3]^+$ interacts directly with serum proteins, resulting in partially irreversible labelling. The major potential coordination sites in proteins are thioether side chains in methionine and the imidazole group in histidine. The primary amine group in lysine or other groups such as the amides of the protein backbone might be weakly coordinating as well but are unstable and therefore less significant [59]. Accordingly, the reaction with monodentate ligands such as imidazole and thioethers results in a

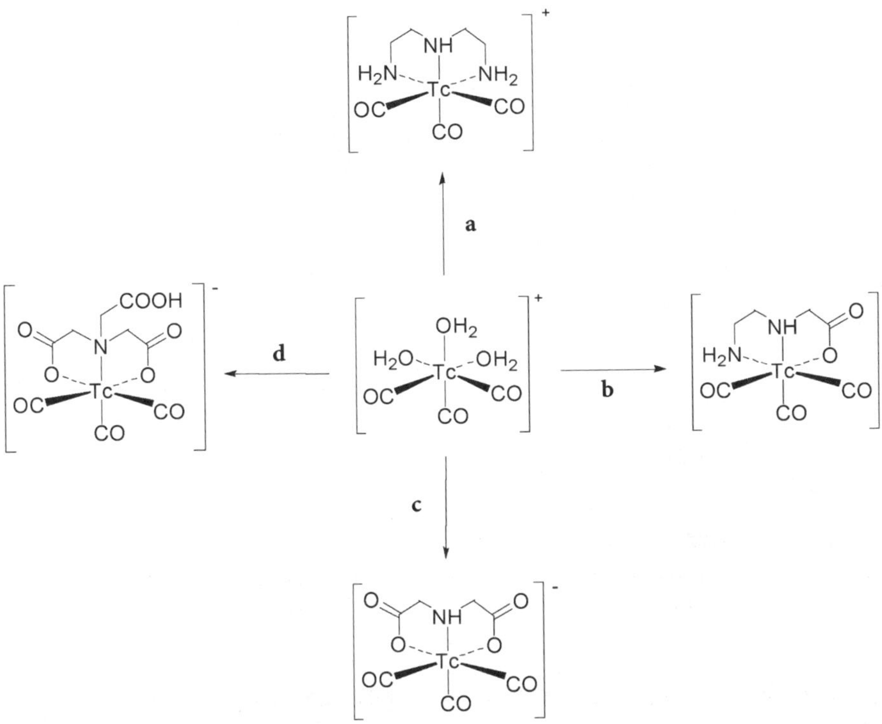

Scheme 8 Reaction of $[^{99m}Tc(OH_2)_3(CO)_3]^+$ with simple tridentate N,O ligands in aqueous solution: **a** with diethylenetriamine, **b** with ethylenediamine monoacetic acid, **c** with imino–diacetic acid and **d** nitrilotriacetic acid

stepwise substitution of one, two and eventually three monodentate ligands to the *fac*-$[^{99m}Tc(CO)_3]^+$ moiety. If such a reaction with monodentate ligands is performed in saline with $[^{99m}Tc(OH_2)_3(CO)_3]^+$ it is often encountered that the reaction is fast for the first two incoming ligands but then becomes very slow. The reason for this observation is the occupation of the third coordination site with a chloride from the solution. After substitution of two ligands, the resulting cation readily attracts an anion, preferred for charge-neutralizing reasons. It has been shown that this Cl^- is in equilibrium with water and prevents the fast substitution of a third incoming ligand. The equilibria are shown in Scheme 9 [60, 61].

The complexes formed are stable and the monodentate ligands do not exchange for H_2O or halides. Ligand-exchange studies of $[Re(im)_3(CO)_3]^+$ with a large excess of fully deuterated imidazole did not show any replacement of the coordinated imidazole even at elevated temperature. Imidazole is a potent chelator. It was shown that recombinant scFv which often bear a histidine tag can conveniently be labelled just by mixing the protein with $[^{99m}Tc(OH_2)_3(CO)_3]^+$ [62, 63]. The exact coordination geometry around the technetium-centre is unknown but coordination must occur at the histidine

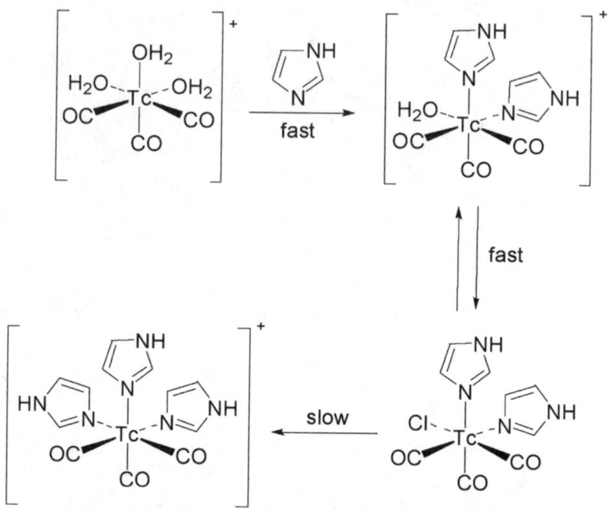

Scheme 9 Equilibria of $[^{99(m)}Tc(OH_2)_3(CO)_3]^+$ with monodentate ligands (imidazole) in aqueous solution. The end product is the tris-substituted complex $[^{99m}Tc(im)_3(CO)_3]^+$, whereas the bis-substituted intermediate $[^{99m}TcCl(im)_2(CO)_3]$ is the kinetic product

tag since labelling of the same protein without his-tag gave very low yields only.

Monodentate thioether molecules, isonitriles, nitriles and phosphines react in the same way and yield stable bis- and tris-substituted complexes [49, 64]. The drawback of such complexes with monodentate ligands is the halide exchange, which is slow but still occurs and results in physiological media in unspecific serum protein binding. The exchange rate is likely to depend on the type of monodentate ligand which influences the electronic properties of the metal and, thus, the strength of Cl^- binding to the technetium-centre. As will be outlined later, monodentate ligands are still versatile when combined with a bidentate ligand in the so-called mixed-ligand [2+1] approach.

One recent example in the context of the monodentate ligands is the reaction with purin bases which shall be discussed here. The N7 nitrogen of guanine is a potential coordinating site, electronically comparable to the one found in imidazole. In guanine the carbonyl oxygen O6 might sterically interfere with the coordinated COs. Reaction of $[^{99}Tc(OH_2)_3(CO)_3]^+$ with 9Me-guanine and 7Me-guanine showed that a total of two purin bases can still coordinate despite some sterical interference (Scheme 10). The X-ray structures of the corresponding complexes could be elucidated and are shown in Fig. 2. Clearly, N7 and N9 coordinate to the rhenium-centre in a head-to-tail conformation. The coordination of the second guanine is an equilibrium, whereas the first guanine is strongly coordinated to the metal [65]. The implication of this observation is obvious: $[^{99m}Tc(OH_2)_3(CO)_3]^+$ can be used to directly label oligodeoxynucleotides or more simply, guanine. Interestingly, no coordination occurs to

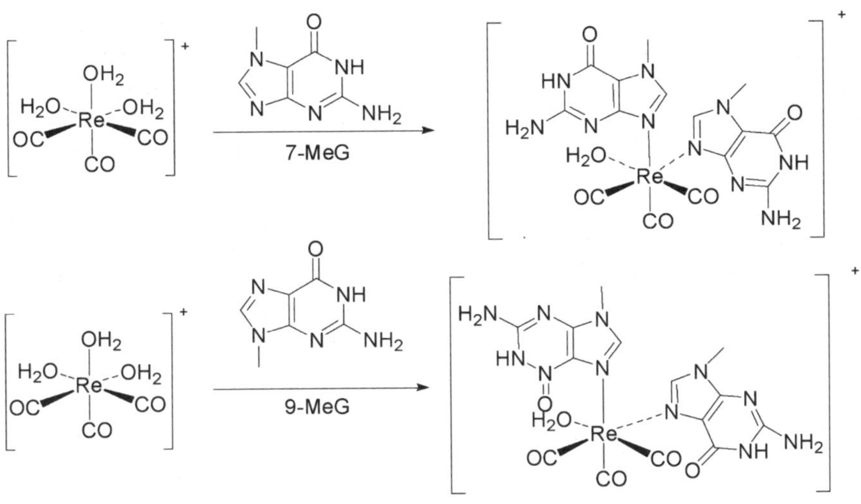

Scheme 10 Reaction of $[Re(OH_2)_3(CO)_3]^+$ with 7-MeG and 9-MeG in water

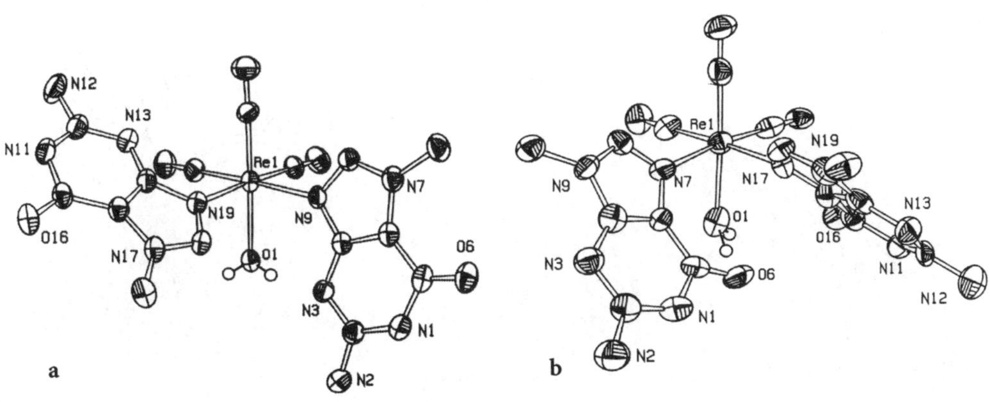

Fig. 2 X-ray structures of **a** $[Re(OH_2)(7\text{-MeG})_2(CO)_3]^+$ and **b** $[Re(OH_2)(9\text{-MeG})_2(CO)_3]^+$

adenine or the pyrimidin bases, which follows observations made with, for example, Pt^{2+}, which also shows a strong preference for guanine.

3.2.2
Bidentate Ligands

Two monodentate ligands can be combined to one bidentate ligand. According to the chelate effect, the complexes are expected to be even stabler. This is not so crucial since it is almost impossible to replace even monodentate ligands but more importantly, the rate of coordination is expected to be faster. Hence, the ligand concentration to achieve quantitative labelling or complex formation will be lower. In terms of preference, bidentate ligands are similar to monodentate ones. Many possible combinations of donor atoms have been described. The bidentate ligand can be neutral or anionic, resulting in neutral complexes of general composition [^{99m}TcCl(L^2)(CO)$_3$] and [^{99m}Tc(OH$_2$)(L^2)(CO)$_3$], respectively. In the latter example, the solvent molecule (H_2O) remains coordinated. A number of potential bidentate ligands are shown in Scheme 11 and a representative X-ray structure is shown in Fig. 3. The bidentate ligands are very strongly coordinating and cannot be replaced. As in the case of the complex type [^{99m}TcCl(L)$_2$(CO)$_3$], the Cl^- or the water ligand can be replaced by other incoming ligands. This might lead in certain cases to cross-reactivity to other potential coordinating sites in physiological media [66]. Bidentate ligands are fast ligands, in particular if they are of anionic nature such as, for example, picolinic acid.

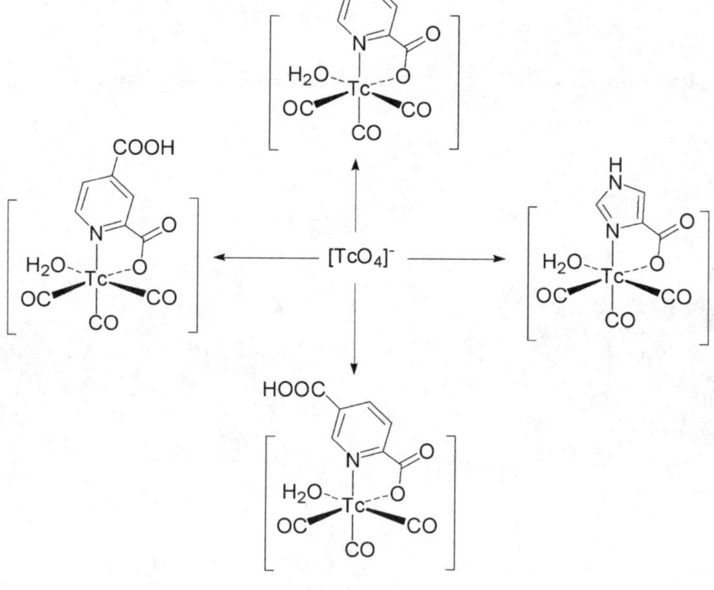

Scheme 11 Direct synthesis of [^{99m}Tc(OH$_2$)(L^2)(CO)$_3$] from [^{99m}TcO$_4$]$^-$ in water in the presence of [H$_3$BCO$_2$H]$^-$, 30 min, 95 °C, where L^2 represents bidenate anionic ligands such as imidazole–carboxylic acid, picolinic acid or 2,4(5)-dipicolinic acid

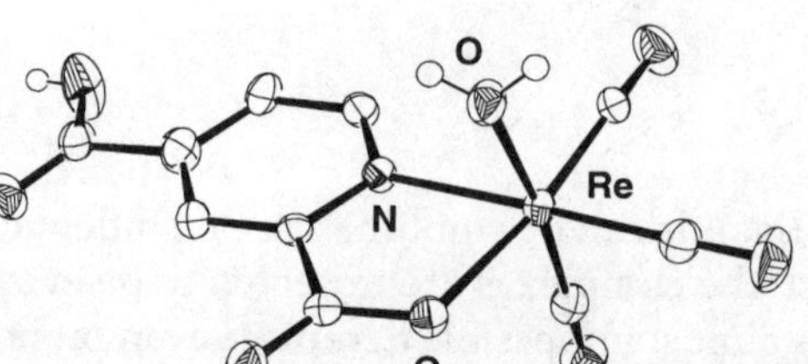

Fig. 3 X-ray structure of [Tc(OH$_2$)(2,4-dipic)(CO)$_3$]

Since very low concentrations of biomolecules are crucial for the labelling of receptor binding molecules the rate of complex formation is important. Ideally, the ratio biomolecule-to-technetium ratio is close to 1. This is rarely the case for the routinely available methods. For bidentate ligands, the lower limit is about 10 μM to achieve quantitative labelling within a reasonable time period (typically 30 min at 100 °C). In comparison with other methods a concentration of 10 μM can still be considered to be very good. Remarkably, a full one-pot preparation of complexes of the formula $[^{99m}TcCl(L^2)(CO)_3]$ or $[^{99m}Tc(OH_2)(L^2)(CO)_3]$ is feasible, if the bidentate ligands are not significantly reduced under the conditions applied in the labelling kit, which is the case for most of the simple bidentate ligands discussed in this section. These complexes can be prepared by adding generator eluate and a ligand solution of appropriate concentration to the vial, yielding quantitative complex formation upon heating. One could expect that hard ligands efficiently stabilize intermediate oxidation states and prevent further reduction to Tc(I) but this has never been observed. It underlines the hypothesis that reduction takes place in one step from coordinated BC.

3.2.3
The $[2+1_B]$ and $[2_B+1]$ Mixed-Ligand Concept

The water ligand in $[^{99m}Tc(OH_2)(L^2)(CO)_3]$ is only loosely bound and can be exchanged for another monodentate ligand. This observation led to a novel mixed-ligand concept besides the well-known [3+1] approach developed by Pietzsch et al. [21]. In the $[2+1_B]$ approach, an anionic, bidentate ligand is coordinated as described previously. Then a monodentate ligand replaces the last H_2O molecule on the technetium-centre, yielding a complex of the general formula $[^{99m}Tc(L^1)(L^2)(CO)_3]$. The monodentate ligand L^1 can be attached to a biomolecule which allows convenient labelling. The bidentate ligand L^2 is variable and, consequently, the physicochemical properties of the labelled biomolecule are also variable. As in normal compound finding, the structure and the properties of the (radio)pharmaceutical are flexible and allow systematic screening in order to optimize, for example, the target-to-nontarget ratio. Versatile bidentate ligands are commercially available. Two of them, imidazole–carboxylic acid and monopicolinic or dipicolinic acid, are particularly interesting since a second ligand coordination resulting in compounds of composition $[^{99m}Tc(\eta^1\text{-}L^2)(\eta^2\text{-}L^2)(CO)_3]$ should not be possible for steric reasons. This was indeed not the case for imidazole–carboxylic acid and monopicolinic acid and was not even observed for structurally more flexible bidentate ligands such as the amino acids serine or alanine. Exclusive formation of $[^{99m}Tc(OH_2)(L^2)(CO)_3]$ was observed, which implies an electronic preference of the remaining coordination site. This hypothesis is confirmed by comparing the rate of formation for $[^{99m}Tc(L^1)(L^2)(CO)_3]$ for different types of monodentate ligands L^1. Aromatic amines such as imidazole and pyridine react fast but still require a relatively high concentration of 0.1 mM to go to completion

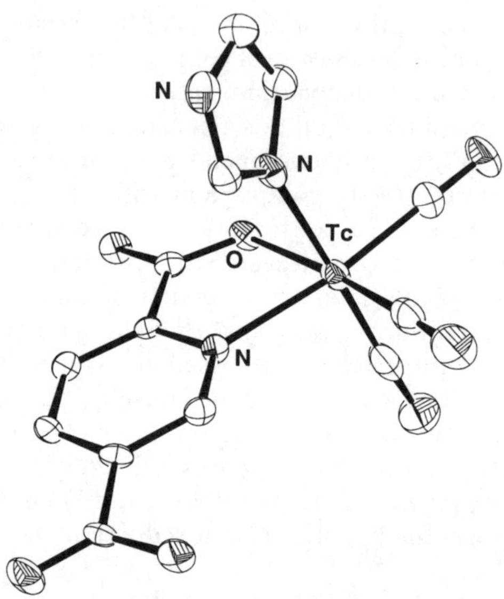

Fig. 4 Example of a [2+1] mixed-ligand complex: X-ray structure of [Re(im)(2,5-dipic)(CO)$_3$]

within 30 min at 100 °C. The most efficient ligand group is isonitriles, for which 10 μM is sufficient for quantitative complex formation. The X-ray structure for a typical complex is given in Fig. 4. Imidazole or pyridine can conveniently be introduced in biomolecules without extensive protecting group chemistry. Depending on the biomolecule, higher temperatures can be employed and allow the fast preparation of labelled biomolecules. These procedures apply, in particular, for peptides, which, in general, are thermally stable. Systematic screening becomes possible and allows fast and efficient development of radiopharmaceuticals.

The mixed-ligand concept can be altered to [2_B+1], in which the biomolecule is attached to the bidentate ligand. In this case, the variable portion is represented by the monodentate ligand, which determines the physicochemical properties of the radiopharmaceutical. Most of the peptides are synthesized with solid-phase techniques and the corresponding ligand can be introduced as the last step in the synthesis. The bidentate ligand is, for example, a normal histidine at the N-terminus and the α- and γ-amine groups provide a strong ligand. The bidentate ligand L^2 is preferentially an anionic ligand since the Cl^- substituent in the case of neutral L^2 is more difficult to replace. We have synthesized a dipicolinic acid based bidentate ligand. This ligand can be coupled through alkylation of the N-terminus to a peptide. Subsequent labelling of this molecule occurred at 10 μM concentrations and monodentate ligands could be coordinated to the sixth site. The principles of the [2+1_B] and the [2_B+1] concept are shown in Scheme 12.

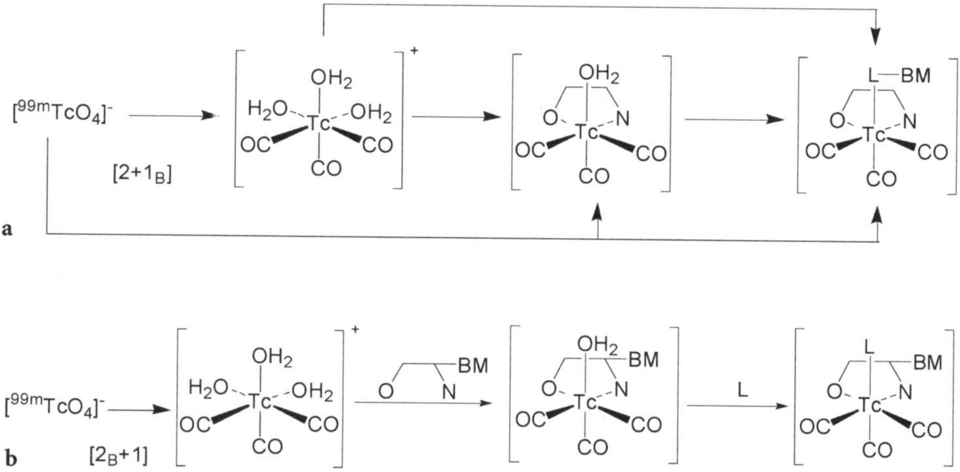

Scheme 12 Two different mixed-ligand approaches based on $[^{99m}Tc(OH_2)_3(CO)_3]^+$: **a** the $[2+1_B]$ approach, in which the bidentate ligand is coordinated first, followed by the labelling of the targeting biomolecule (*BM*) with a pendant monodentate ligand (the synthesis of the intermediate can be performed in one step), and **b** the $[2_B+1]$ approach, in which the bidentate ligand is attached to the BM and the monodentate ligand represents the variable portion

One of the features of the mixed-ligand concept is the exclusive formation of one single complex. This result is general for reactions with $[^{99m}Tc(OH_2)_3(CO)_3]^+$ and is a consequence of the high kinetic stability of all the complexes formed. Lability of the bi- or the monodentate ligand would clearly result in statistical mixtures of all possible complexes but even a very high excess of monodentate ligand did not replace the bidentate chelator. In a further step, it is even possible to prepare the mixed-ligand complexes in one single step directly from the kit and the two ligands L^1 and L^2. Adding a solution of L^1 and L^2 to the lyophilized kit gave only the mixed-ligand complex $[^{99m}Tc(L'^1)(L^2)(CO)_3]$. Again, if the ligands and the biomolecules resist the reductive conditions, there is the possibility of preparing the compounds in one single step.

The [2+1] approach with the *fac*-$[^{99m}Tc(CO)_3]^+$ moiety can be used for mimicking the structure of certain small biomoelcules (Scheme 13). Coordination of serine as the bidentate ligand and guanine as the monodentate ligand gives a structure that roughly resembles guanosine. The X-ray structure of the corresponding rhenium complex could be elucidated and is shown in Fig. 5. The ribose moiety is replaced by the five-membered ring of the coordinated serine, whereas the guanine coordinates directly through N9 to rhenium. Whether enzymes in which guanosine is the substrate still recognize this copy is currently under investigation.

Bidentate ligands have been applied in combination with various targeting molecules. Recently, fatty acids have been labelled by introducing a bidentate

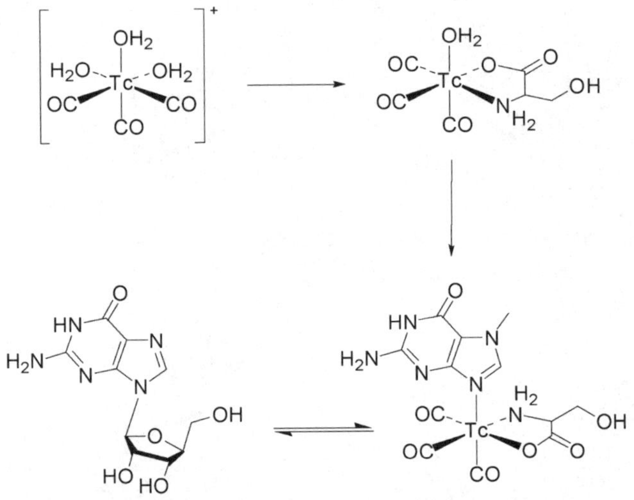

Scheme 13 Mimicking organic structures: coordination of one serine and 7-MeG to $[Re(OH_2)_3(CO)_3]^+$ resembles in its overall structure that of guanosine

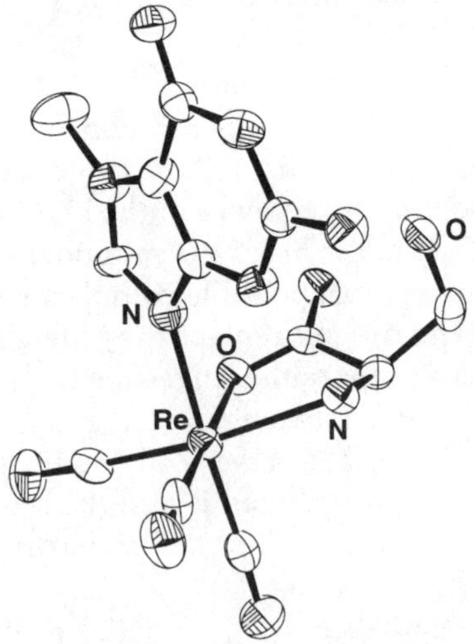

Fig. 5 X-ray structure of $[Re(9\text{-}MeG)(serin)(CO)_3]$, a potential mimic for guanosine

Schiff base type chelator at one of the terminal groups of the C18 chain. Schiff bases have also been introduced in WAY, a piperazine-based serotonergic receptor binding molecule (Scheme 14). For both molecules labelling could be performed at low concentrations and, depending on the spacer length between the receptor binding part and the chelator, affinity was fully retained [67]. Unfortunately, brain uptake was too small to make this compound relevant for CNS receptor density imaging.

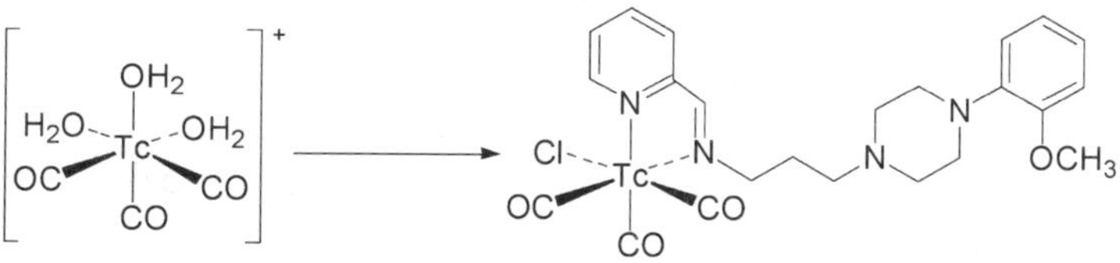

Scheme 14 Labelling of a serotonergic receptor ligand from the aryl–piperazine type, derivatized with a Schiff-base type bidentate chelator

3.2.4 Tridentate Ligands

As previously mentioned, the lowest possible level of vector concentration is essential for future application of imaging agents. If the concentration for quantitative labelling of the targeting molecule is high, then purification from cold, unlabelled material is required. Depending on the type of receptor binder, the upper concentration is between 1 and 10 μM. At such low concentrations it is sometimes difficult to get one single product. For mechanistic reasons, heating for all the common approaches is necessary to achieve complete labelling within a short period of time. Even the very efficient 6-hydrazinonicotinamide (HYNIC) approach needs heating below a certain concentration level to afford quantitative labelling [6, 68, 69]. The labelling efficiency is not only a question of stability but also of the reaction pathway. Classical coordination chemistry with the *fac*-$[^{99m}Tc(CO)_3]^+$ moiety follows a D or an I_d mechansim, using Langford–Gray nomenclature. The formation of an outer sphere complex is therefore an important step in the labelling procedure. Anionic bidentate ligands are more efficient than neutral ones probably owing to an increased outer sphere stability constant. Correspondingly, tridentate anionic ligands should be even more efficient, a behaviour that could indeed be observed. Although tridentate ligands do not provide significantly higher thermodynamic stability, their reaction rate is much faster than that of bidentate ligands. Since all the complexes are kinetically stable, the rate of reaction becomes the decisive step. In that respect tridentate ligands are favoured. Tridentate ligands provide an additional advantage in that they shield the technetium-centre perfectly from cross-reactivity as is often the case for $[^{99m}Tc(OH_2)(L^2)(CO)_3]$ or $[^{99m}TcCl(L^2)(CO)_3]$. The drawback of tridentate ligands is the request for protecting group chemistry since several cross-reacting functional groups may be present.

Research with tridentate ligands has been carried out for several fundamental coordinating groups, pyrazolyl-borates, aliphatic triamines and histidine-type ligands. Other ligands have also been investigated and some of the structures are shown in Scheme 15.

These tridentate ligands are coupled to biomolecules although the organic chemistry is not routine. All the ligands shown are very efficient for labelling reactions and concentrations of 1 μM or even lower are sufficient to afford

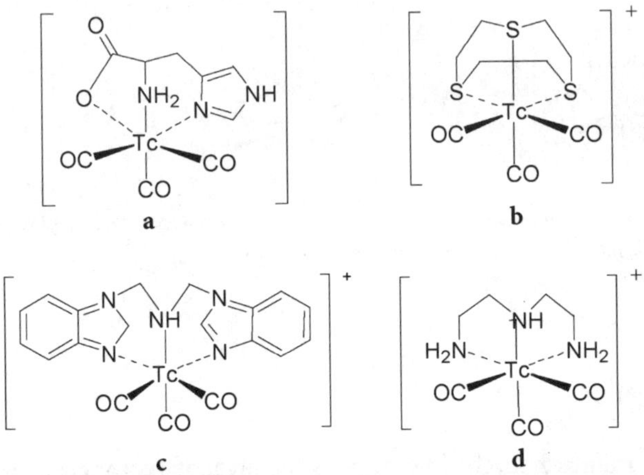

Scheme 15 The *fac*-[Tc(CO)$_3$]$^+$ core with different types of tridentate chelators: **a** histidine, **b** 1,4,7-trithiacyclononane, **c** bis-benzimidazole amine and **d** diethylenetriamine

quantitative complex formation at 98 °C after 30 min. The imidazole–thione borohydrides are particularly fast ligands [70]. Besides the efficiency of labelling, they have the very interesting feature of agostic hydrogen coordination. An X-ray structure is shown in Fig. 6. The reaction in aqueous solution yields the complex in which the sulphurs from two imidazoles and one hydride are coordinated. Originally it was expected that a halide would occupy the sixth site; however, reaction in water or organic solvents afforded only a complex with the shown structure. The replacement of the agostic hydrogen with other incoming monodentate ligands proved to be impossible in water and could only be

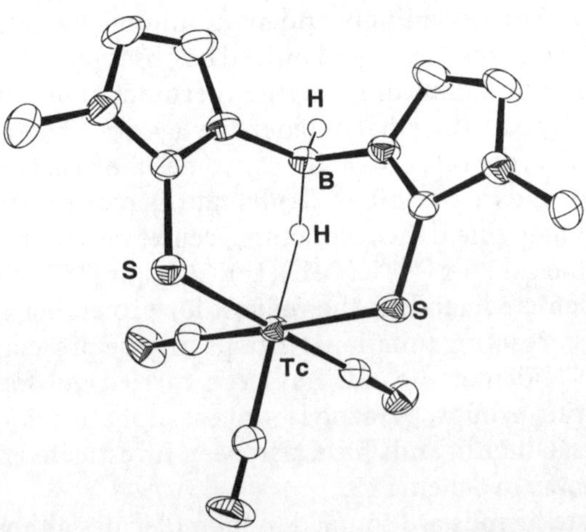

Fig. 6 Unusual tridentate coordination with an agostic hydride; X-ray structure of [Tc{$H_2B(Sim)_2$}(CO)$_3$]

achieved in organic solvents under drastic conditions [71]. It is obvious that the derivatization of biomolecules with this type of chelator will result in radiopharmaceutical precursors that are very efficiently labelled.

Since of natural origin, histidine is a versatile tridentate chelator if not coupled to a biomolecule by the amine or the carboxylic function through peptide formation. As discussed later, histidine has been introduced in the peptide neurotensin through alkylation at the α-NH_2 group and amide formation with the N-terminus of neurotensin. This gives a highly efficient tridentate, histidine-based ligand. Keeping the histidine part intact for coordination requires derivatization at the imidazole ring. The ε-amino group is particularly versatile and additional alkyl chains bearing amino, hydroxo, halide or carboxylic functions could be introduced in fully protected histidine. Subsequent deprotection results in coupled histidine as a tridentate chelator. The reaction pathway is outlined in Scheme 16.

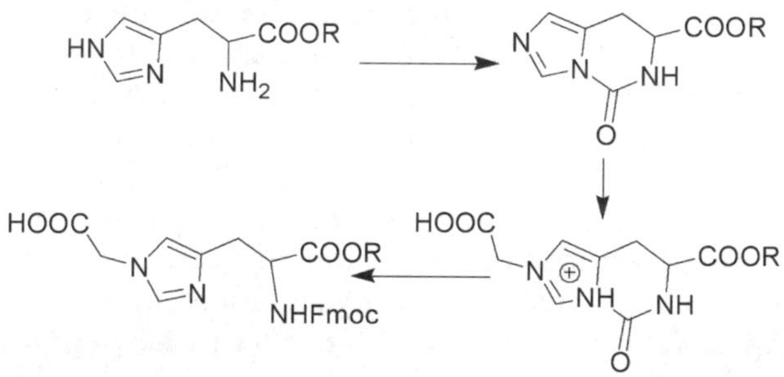

Scheme 16 Introduction of an acetic acid group at N^{ε} in histidine to yield an N^{α} and carboxylate-protected histidine ready to be coupled to a biomolecule

These histidine derivatives are easily introduced in biomolecules. The complex $[^{99m}Tc(his)(CO)_3]$ is hydrophilic and can be combined with hydrophilic peptides. The high-performance liquid chromatography retention times of the native and the labelled peptide do not differ significantly. Histidine is a very efficient ligand and quantitative labelling occurs, in general, at the micromolar level. The complexes are physiologically stable and well characterized. Large quantities of the ligand precursor can be synthesized from cheap starting materials and can then be introduced in biomolecules by applying standard peptide coupling techniques. Since histidine is not sensitive for reduction, the complex $[^{99m}Tc(his)(CO)_3]$ can also be prepared in one single step directly from $[^{99m}TcO_4]^-$. If the biomolecule is stable under these conditions, the same can be applied for a one-step labelling procedure. The X-ray structure analysis of a model complex is shown in Fig. 7 [72].

Surprisingly, aliphatic tridentate amines also showed a high tendency for coordination to the *fac*-$[Tc(CO)_3]^+$ core. It is expected that the very basic aliphatic amines are protonated and fully or partially cleaved at physiological pH. This did not occur, demonstrating the high kinetic stability of the Tc–N bond. Like

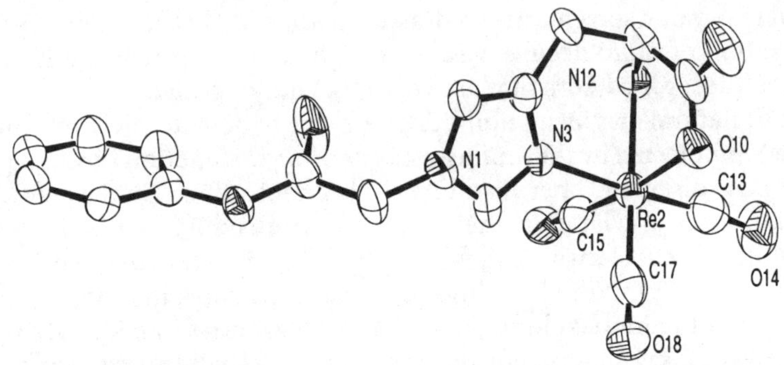

Fig. 7 X-ray structure of a complex with tripodal histidine and derivatization at N^{ε}

histidine, aliphatic tridentate amines afford coordination at concentrations as low as 1 μM in the ligand or the biomolecule. The resulting complexes of general composition $[^{99m}Tc(N3)(CO)_3]^+$ are cationic and highly hydrophilic. One or two of the terminal primary amines can be replaced by pyridines or imidazoles without changing the charge of the complex but making it more lipophilic if required. Tridentate aliphatic amines can conveniently be introduced into biomolecules. Commercially available tetramines such as "trien" can be protected and coupled through peptide bond formation to a carboxylic acid. The terminal amines can also be reacted with aldehydes and reduced to get directly the tridentate ligand as shown in Scheme 17. Of course, alkylation at any of the amines is possible as well.

As shown in Scheme 17, a tridentate amine was coupled to an intercalating aromatic molecule such as pyrene or anthrachinon and was labelled with ^{99m}Tc.

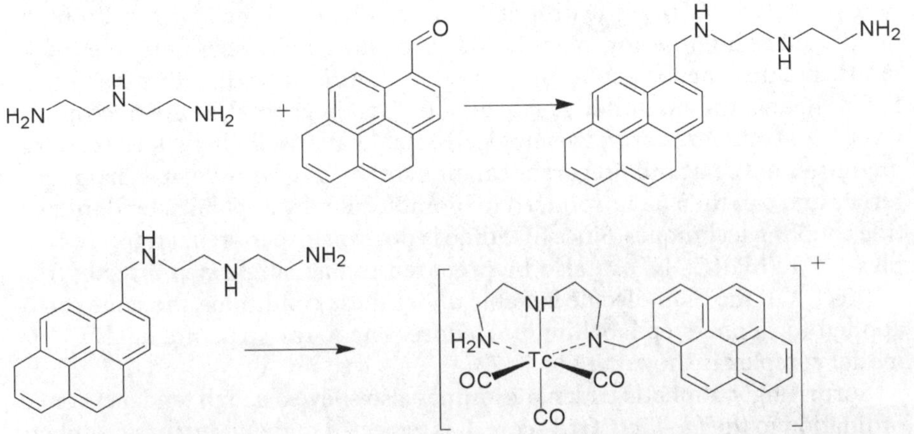

Scheme 17 Triamine ligands in combination with an intercalating moiety yields a highly stable cationic complex

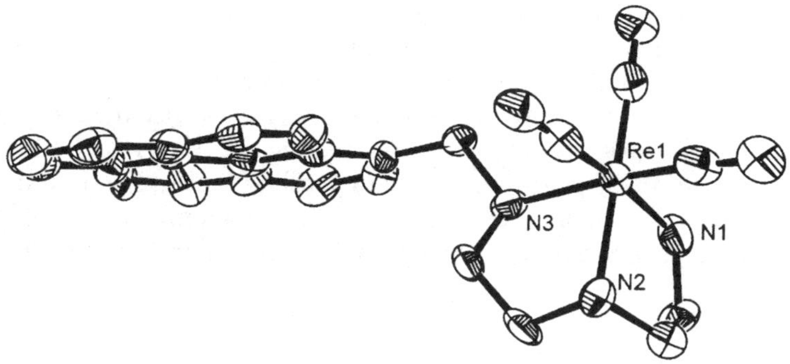

Fig. 8 X-ray structure of a cationic complex comprising a tridentate aliphatic amine ligand and an intercalating pyrene moiety

The X-ray structure of the rhenium homologue is given in Fig. 8. These labelled intercalators exhibit strong interaction with plasmide DNA and it was shown that the Auger electrons emitted from the decay of ^{99m}Tc induced double strand breaks in DNA [73]. Since Auger emitters are frequently discussed as potential therapeutic nuclides, ^{99m}Tc could probably be applied for the same purpose [74–77]. Coupling a nucleus-localizing peptide to the other terminus of the triamine ligand resulted in a trifunctional molecule,: one function to coordinate, one to intercalate and one to transport the radionuclide into the nucleus. Enhanced radiocytotoxicity was found when the intercalating portion was combined with such an appropriate peptide.

The stability constant of intercalation depends, of course, on the charge of the molecule. Since tridentate amines yield cationic complexes, for simple electrostatic reasons interaction with the negatively charged DNA backbone is enhanced.

Derivatives of aliphatic amines have an interesting drawback. It was observed that tertiary amines involved as ligand atoms in coordination to the *fac*-[^{99m}Tc(CO)$_3$]$^+$ moiety cleaved during the labelling reaction, affording a coordinating secondary amine. If the tertiary amines link to a biomolecule, cleavage of the biomolecule from the ^{99m}Tc-complex occurs; hence, tertiary amines should not be part of a chelator. Cleavage could be observed regardless of the nature of the two other donators and only the extent of cleavage varied as a function of these two additional groups. Selective cleavage was transferred to solid-phase bound chelators. Metal-mediated cleavage then allows the preparation of essentially no carrier added complexes or biomolecules. Solid-phase-based preparation of radiopharmaceuticals of very high specific activity has also been described in the context of Tc(V) chemistry in which a Au–S bond cleaved upon coordination of the [Tc=O]$^{3+}$ core [78, 79]. Metal-mediated cleavage of solid-phase ligand bonds provides a very elegant and simple way to high specific-acitivity radiopharmaceuticals. Furthermore, application on a routine basis is convenient since filtration is sufficient to get a radiopharmaceutical of high radiochemical and chemical purity. In the case of *fac*-[$^{99m}Tc(CO)_3]^+$ it could be shown that cleavage occurs exclusively with technetium but not with rhenium and that the active state must be an intermediate in which the tridentate ligand is not yet fully coordinated. Complexes with the completed ligand sphere were perfectly stable towards cleavage of the tertiary amine bond. Typical yields of these processes varied depending on the ligand and reaction conditions between 10 and 50% relative to the total activity of the ^{99m}Tc precursor. Examples of the molecules involved are shown in Scheme 18 [80].

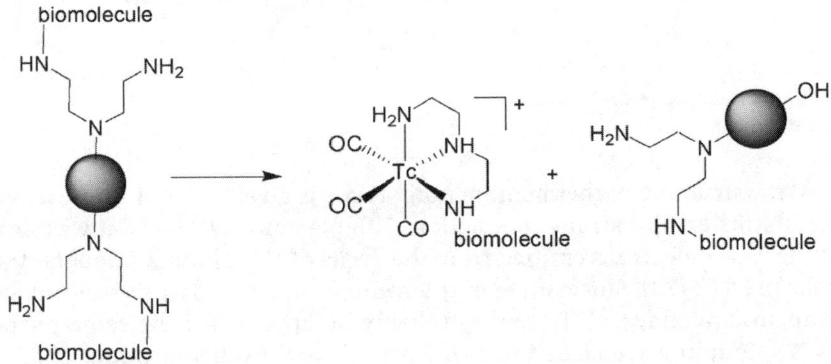

Scheme 18 Metal-mediated cleavage of ^{99m}Tc-labelled biomolecules from a solid-phase support (*grey spheres*). Unlabelled biomolecules remain bound, whereas labelled ones only are partially cleaved

It is likely that the Lewis acidity of the *fac*-[$^{99m}Tc(CO)_3]^+$ moiety is the reason for cleavage. It is assumed that one remaining water molecule is acidified and attacks intramolecularly the tertiary N–C bond, resulting in a hydroxo group on the cleaved carbon and a secondary amine coordinated to technetium.

3.2.5 Reaction with Nonclassical Ligands

Among the nonclassical organometallic ligands, $[C_5H_5]^-$ (Cp^-) plays an essential role for several reasons. It was shown in the early days of bioorganometallic chemistry that complexes of the cymantrene type [Cp*M*(CO)$_3$] (*M* represents group 7 elements) are very stable in physiological media [81, 82]. Cp^- is one of the smallest ligands with a low molecular weight that stably occupies three coordination sites. Furthermore, Cp^- can be derivatized to introduce targeting biomolecules; the corresponding chemistry is well known from classical organometallic chemistry. Apart from these advantages, a major disadvantage is the stability of free Cp^- in particular in water. Since radiopharmaceuticals have ultimately to be prepared from water, Cp^- is probably not very versatile in this solvent. Attempts to prepare cytectrene (the technetium analogue of cymantrene) go back to the early 1990s. Wenzel [83] reported a so-called double ligand transfer reaction and succeeded for the first time in the preparation of [$Cp^{99m}Tc(CO)_3$] if we do not consider the preparation of the same complex by decaying the molybdenum analogue reported by Fischer and Schmidt [44]. The pioneering reaction of Wenzel is shown in Scheme 19.

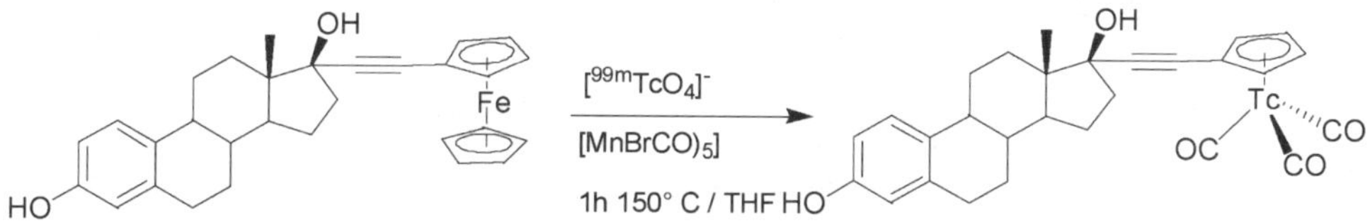

Scheme 19 First preparation of a [$Cp^{99m}Tc(CO)_3$] labelled biomolecule by the so-called double ligand transfer approach

Apart from the reaction conditions, cymantrene was formed concomitantly in this reaction and could not easily be separated from cytectrene. Progress in the Wenzel approach was reported by the group of Katzenellenbogen [84]. They developed the double ligand transfer reaction by reaction of [$^{99m}TcO_4$]$^-$ with $CrCl_2$ as a reducing agent, [$Cr(CO)_6$] as a CO source and ferrocene as a Cp source. Under these conditions, cytectrene could be synthesized in reasonable yields without getting cymantrene at the same time. The difficulty of this approach is the presence of chromium and the still relatively harsh condition of 150 °C in MeOH. The same group reported the labelling of octreotide with cytectrene and performed biological studies [85]. Since cytectrene is rather lipophilic, accumulation in the liver was observed and no real advantages over other approaches were stated. Clearly, cytectrene is probably most interesting for lipophilic biomolecules such as CNS receptor ligands and to a lesser extent only for peptides or antibodies.

The main problem with Cp^- is the insolubility in water (since it is readily protonated to HCp), its high pK_a value of about 14 which prevents its deprotonation on the aqueous pH-scale and its instability, resulting in fast dimerization and polymerization. The pK_a value could be decreased by the introduction of an electron withdrawing group. Acetyl–Cp (ACp) represents such an acidified cyclopentadiene. ACp was described in the early literature but the pK_a was not determined. Later, it was found that the pK_a value of Acp was around 8.7 [86]. Acp is water-soluble and reasonably stable. The reaction with [$^{99}Tc(OH_2)_3(CO)_3$]$^+$ in phosphate buffer resulted in the formation of [$^{99}Tc(ACp)(CO)_3$] in 10–40% yield. The yield is not quantitative owing to the hydrolytic formation of [$Tc_4(\mu^3\text{-}OH)_4(CO)_{12}$], which is the competing side reaction to the formation of cytectrene. Since tetramerization does not occur in highly dilute solutions, the same reaction was carried out with ^{99m}Tc and re-

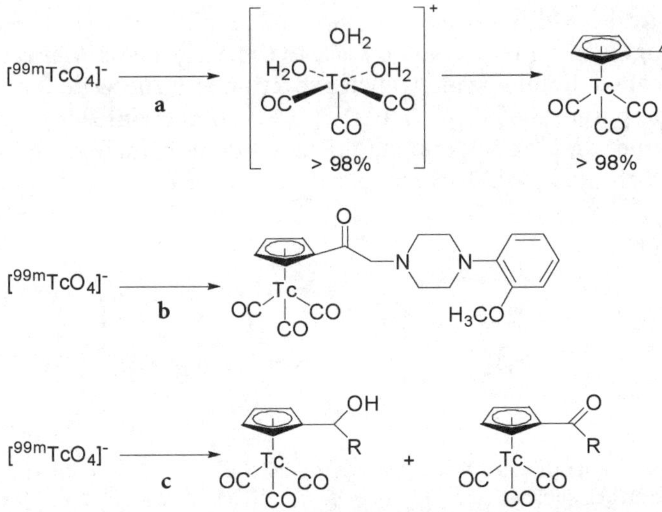

Scheme 20 Fully aqueous preparation of a [Cp^{99m}Tc(CO)$_3$] labelled biomolecule, enabled by the carbonyl group directly attached to the cyclopentadienyl-ring: **a** two-step approach, **b** direct one-step procedure and **c** observed by-product from carbonyl reduction by boranocarbonate

sulted in quantitative formation of [^{99m}Tc(ACp)(CO)$_3$] after 30 min at 95 °C. The reaction requires a relatively high concentration of ligand but further optimization of the conditions will probably reduce this factor. The procedures are shown in Scheme 20 [87].

The acetyl group can act as an anchoring group for biomolecules. The general strategy of reacting NaCp with an ester results in the formation of a Cp–CO–biomolecule conjugate. This approach has been described with the serotonergic receptor ligands WAY, one of the most thoroughly investigated CNS receptor ligand. Some of the compounds are shown in Scheme 20. Reaction with [^{99}Tc(OH$_2$)$_3$(CO)$_3$]$^+$ in phosphate buffer gave the corresponding ^{99}Tc complexes. The X-ray structure of a model complex is shown in Fig. 9. Labelling with ^{99m}Tc gave the corresponding complexes in quantitative yield, but again at relatively high concentrations of 10^{-4}–10^{-3} M. The structure–activity relationship made clear that a two-carbon spacer is too short and the in vitro receptor affinity decreased significantly. On the other hand, the complex with a four-carbon spacer showed full retention of the affinity. Whether the main obstacle for ^{99m}Tc carrying CNS receptor ligands, the blood brain barrier, can be overcome needs to be proven [88].

The approach of using acetylated Cps has one limitation. If the C=O group and an additional donator in, for example, the biomolecule can form a six-membered chelate, this mode of coordination is preferred and slippage from this η^2 coordination to η^5 coordination on the Cp takes place, if ever, very slowly.

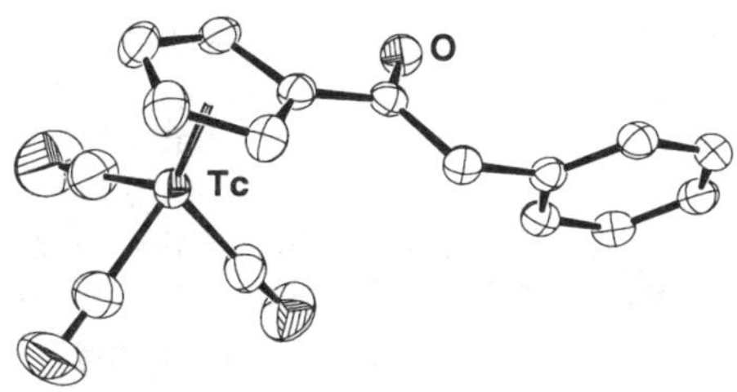

Fig. 9 Preparation of a cyclopentadienyl (*Cp*) complex from water: X-ray structure of $[(PhCO–Cp)Tc(CO)_3]$

Recently, it was shown that ferrocene derivatives can be used directly as a Cp^- source. Ferrocene is introduced in a biomolecule through a carbonyl group and the bioconjugate is then reacted with $[^{99m}Tc(OH_2)_3(CO)_3]^+$ in a dimethyl sulphoxide (DMSO)/water mixture at 95 °C. The corresponding Cp compound is produced in good yield. The advantage of this procedure is the easier handling of the ferrocene precursor compared with that of the biomolecule with an attached free and deprotonated Cp. On the other hand, the reaction conditions are still quite rough and the demand for DMSO might cause problems in the practical application. It seems that DMSO is solely required to achieve sufficient solubility of the highly lipophilic ferrocene precursor [89].

Other ligands compatible with aqueous conditions and providing additional *M*–C bonds are CO and isonitriles. As discussed earlier, at ambient pressure of CO, no more than three COs will coordinate to Tc(I) as a consequence of the trans effect. At elevated pressure, however, it is possible to synthesize the binary carbonyl complex $[^{99m}Tc(CO)_6]^+$ [42]. This could be shown by ^{99}Tc NMR spectroscopy and by isolation of the complex.

More relevant to radiopharmaceutical imaging are isonitriles since they can be derivatized with a biological vector. Complexes of general composition $[^{99m}Tc(CN–R)_3(CO)_3]^+$ can be prepared by mixing the precursor with the corresponding isonitrile. The coordination of three biomolecules is not desirable. The potential behind compounds of formulation $[^{99m}Tc(CN–R)_3(CO)_3]^+$ is rather the synthesis of novel perfusion such as myocardial imaging agents. It has been shown by two groups that $[^{99m}Tc(CN–R)_3(CO)_3]^+$ with *R* representing ^{t}Bu has comparable biological properties as Sestamibi [90]. The drawback is the two-step synthesis, which is not convenient for routine use. Only much better biological behaviour would justify a replacement of the current market leader $[^{99m}Tc(CN–R)_6]^+$.

Pioneering work has recently been performed at the interface between boron neutron capture therapy and diagnostic imaging agents. If the carborane 3-isocyano-1,2-dicarba-*closo*-dodecaborane is reacted with $[ReBr_3(CO)_3]^{2-}$ in an organic solvent, the three bromides are replaced by three carboranes to yield the highly lipophilic complex $[Re(CN–R)_3(CO)_3]^+$, in which *R* represents the *closo*-carborane. The reaction was also performed with ^{99m}Tc and the analogous compound was formed in quantitative yield. This opens a new avenue to carry a high boron density molecule to a target and to quantify and to visualize the distribution of boron, an analytic task that is not easily verified with nonradioactive boron compounds [91]. The replacement of one CO ligand by one isonitrile ligand was observed during the reaction resulting in the formation of $[Re(CN\text{-}R)_4(CO)_2]^+$. The same group reported the very elegant synthesis of a carborane analogue of cytectrene. The nido-carborane $[C_2B_9H_{12}]^-$ reacted with $[Re(OH_2)_3(CO)_3]^+$ in slightly alkaline solution to produce $[Re(\eta^5\text{-}C_2B_9H_{11})(CO)_3]^-$ in very good yield [92]. Although not described explicitly, it can be expected that ^{99m}Tc behaves the same way. The complex can be considered as an analogue of cytectrene. In fact, the carboranes have also been derivatized with a propionic acid residue which allows the introduction

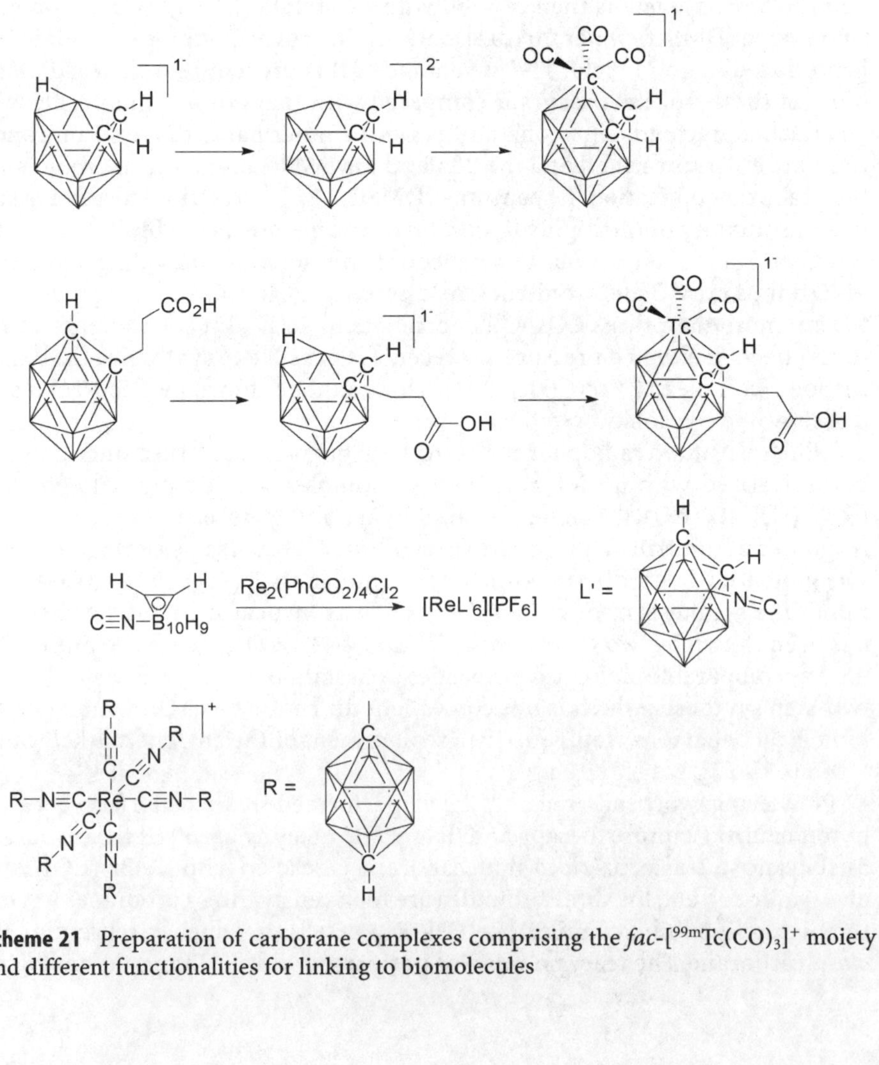

Scheme 21 Preparation of carborane complexes comprising the *fac*-$[^{99m}Tc(CO)_3]^+$ moiety and different functionalities for linking to biomolecules

of a targeting biomolecule. The advantage of the carborane ligands is their stability and their low molecular weight. Although bulky at first glance, the actual size is not much bigger than that of Cp and the coordination chemistry is straightforward. Still, the complexes are highly lipophilic, which would make them the choice for brain receptor imaging, but the complexes are still anionic, which might prevent them from passing the blood-brain barrier.

For completeness, it was reported in the context of binary isonitrile complexes that the Re(III) precursor $[Re_2(OAc)_4Cl_2]$ reacted with 3-isocyano-1,2-dicarba-*closo*-dodecaboran to produce $[Re(CN\text{–}R)_6]^+$ in good yield. The X-ray structure of this complex was elucidated. Again, it remains to be shown that the same complex can be prepared with ^{99m}Tc to be useful in imaging [93]. Relevant complexes of this section are depicted in Scheme 21.

4 Labelling of Biological Vectors with Organometallic Moieties

As outlined in the Introduction, an organometallic precursor useful for the labelling of targeting molecules must be stable in water or be synthesized in situ and subsequently be stabilized with additional ligands. In most cases, the precursor is synthesized in a first step and reacted in a second step with the derivatized biomolecule. Most of the papers describe labelling chemistry with the *fac*-$[^{99m}Tc(CO)_3]^+$ moiety and a lower number of papers with Tc(III) complexes comprising one Tc–C bond to an isonitrile. Since the application of organometallic compounds in life science is recent, systematic studies are not available yet and the most important research results will be discussed in the following sections.

4.1 CNS Receptor Ligands

A good overview of CNS receptor binding ^{99m}Tc-labelled complexes became available recently [88]. To achieve a useful labelled CNS receptor ligand, the attached complex must be neutral and possess an octanol/water partition oefficient between 2 and 3. The hydrogen-bonding properties of the complex affect the ability to pass the blood brain barrier. Efforts were essentially applied to the dopamin transporter ligand DAT and the 5-HT_{1A} serotonergic receptor ligand WAY100635.

For the latter CNS receptor ligands, different types of chelators have been introduced to the basic receptor binding framework. Originally a bidentate Schiff base was introduced by the reaction of an aliphatic amine with pyridine–aldehyde. Labelling occurred at low concentration and the radioconjugate was physiologically stable. Although receptor affinity was fully retained, brain uptake was low [67]. ACp was introduced at the same position and, as described earlier, and receptor affinity was perfectly retained. No brain uptake data are available so far [87]. The HYNIC group presents a favourable coordination site for the *fac*-$[^{99m}Tc(CO)_3]^+$ moiety. Derivatization of WAY and subsequent labelling gave a well-defined complex with good binding properties but brain uptake again was too low. As is obvious from these few experiments, more systematic studies are required to improve brain uptake in particular. Other labelling methods suffer from the same difficulties but, as more compounds are described, a better structure–activity relationship is deductible.

Recently, ligands with PNN or PNO donor sets have been introduced at the alkyl chain of WAY as described before. This donor set represents a powerful alternative to other ligand types but is probably too large in size owing to the two phenyl groups attached to the phosphine. Labelling with the *fac*-$[^{99m}Tc(CO)_3]^+$ moiety occurs at medium to low concentration. The X-ray structures of the corresponding rhenium-complexes have been elucidated showing that the PNN ligand is tridentate and the PNO ligand bidentate. Interestingly, in the latter case the amide group coordinates through the carbonyl oxygen and not through the deprotonated NH group, representing one of the rare cases in which O is preferred over N [94]. The same group published rare examples of pyrazol-based ligands. It could be shown that bis-pyrazolyl-borohydrides are extremely powerful and highly lipophilic ligands. Labelling with the *fac*-$[^{99m}Tc(CO)_3]^+$ moiety occurs at low concentration [70]. Pyrazole-based-ligand frameworks have been extended to tridentate amines in which two pyrazolyl ligands are connected through a secondary amine yielding a tridentate NNN ligand. The corresponding complexes have been characterized structurally [95]. Similarly, ligands based on (mercaptoimidazolyl)borohydrides and hydro-tris-(mercaptoimidazolyl)borate have been shown to be powerful tridentate ligands as well. So far, none of these ligands have been coupled to biomolecules to prove their versatility. The coordination chemistry encountered with these ligand types is very promising for the preparation of high specific-activity radiopharmaceuticals [71].

An organometallic Tc(III) complex has been described in the context of the HT_{1A} receptor. An isonitrile group was introduced into the pendant alkyl chain of WAY. The isonitrile group then reacted with a Tc(III) centre coordinated to an umbrella-type NS3 chelator to produce a five-coordinated complex containing one Tc–C bond. Depending on the chain length, the affinity was in the nanomolar range but brain uptake was still relatively low (about 0.3%) [96].

Based on the very positive results with "TRODAT", a DAT receptor ligand with an N2S2 ligand for labelling with Tc(V), a few studies have been performed in the context of carbonyls. Cp^- was introduced at various positions in the tropane framework. It could be shown that dervatization at the bridging nitrogen retained the biological activity but that biological activity was lost at all other positions [84, 97]. In another attempt, a bidentate thioether ligand was introduced at the carboxylic acid group in the tropane framework following the structure–activity relationship of the TRODAT molecule. This molecule could be labelled with the *fac*-$[^{99m}Tc(CO)_3]^+$ moiety and the molecule was called TROTEC-1. The affinity to the receptor was fully retained (IC_{50}<1 nM) but the brain uptake was again not sufficient to allow structure-resolved brain imaging [98, 99].

The labelling of estradiol is of ongoing interest. Some lead structures are depicted in Scheme 22. Originally, ferrocene was introduced through an ethinyl spacer at position 17. Following the procedure described in the so-called double ligand transfer protocol, the *fac*-$[Tc(CO)_3]^+$ moiety was introduced using $[MnBr(CO)_5]$ as the CO source. The cytectrene-labelled estradiol was produced in 30–90% yield. Although a suboptimal synthesis, the authors could show receptor-specific binding to the uterus and tumour uptake [100]. In a more recent approach, the ferrocene approach was combined with the higher reactivity of ACp coupled to estradiol. The reaction of $[^{99m}Tc(OH_2)_3(CO)_3]^+$ with acetyl–ferrocene replaced the iron at 95 °C in DMSO/H_2O and gave the corresponding cytectrene-labelled estradiol. The presence of DMSO is not an option as DMSO is required to solubilize estradiol to a reasonable amount. Further development is expected to solve this practical inconvenience [89] Various alterations of these methods have been described [101, 102]. In general, the conditions worked out so far in the context suffer from rather rough conditions and are, in general, not compatible with demands from routine application. The only exception is probably the labelling of ACp derivatives of biomolecules which can be performed directly in water although the concentrations of biomolecules are still relatively high [86]. ^{188}Re–Cp compounds have been synthesized by the

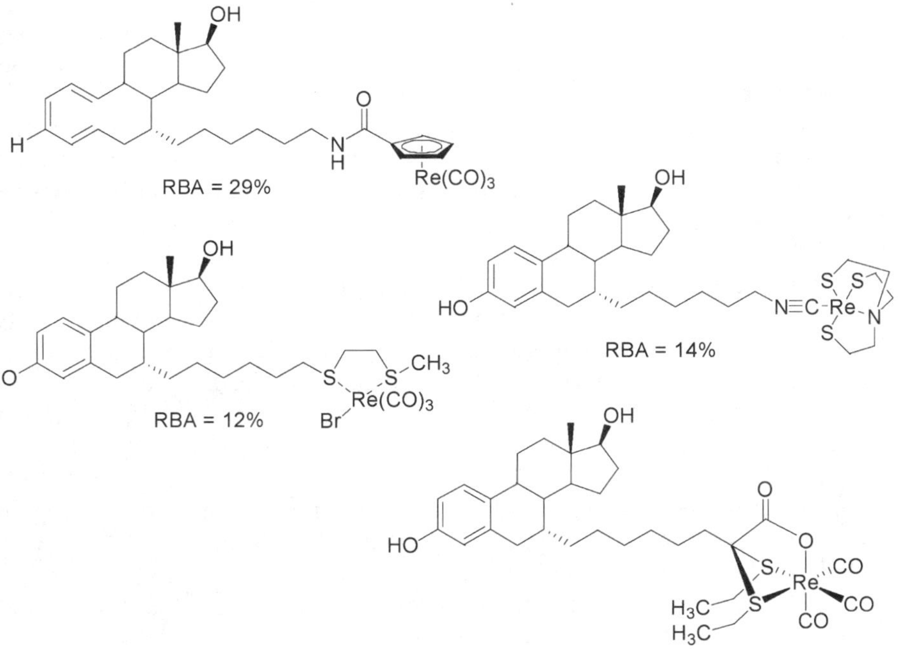

Scheme 22 Differently labelled estradiols with attached organometallic rhenium moieties. *RBA* is the relative binding affinity to the estradiol receptor

double ligand transfer approach. The in vivo metabolism was studied to prove the applicability of [(*R*–Cp)Re(CO)$_3$] for the labelling of biomolecules such as peptides or proteins. No metabolical decomposition could be observed [103].

In an earlier work, a bidentate thioether ligand was attached through an alkinyl spacer to C17 in estradiol. The corresponding Re(I) and Tc(I) complexes were characterized but no biological ^{99m}Tc data were determined [104]. More recently, bidentate SS and tridentate SSO ligands have been introduced at positions 7 and 17 in estradiol. The relative binding affinity to Erα and Erβ have been determined. The compounds were labelled with ^{94m}Tc (a positron emitter) and showed favourable estrogen receptor binding. Despite the favourable binding properties, the compounds did not show a target-tissue-directed biodistribution in vivo [105].

An extension to this work appeared recently when pyridin-2-yl hydrazine based classical chelators were introduced through an ethinyl or ethenyl moiety at position 17. Although still model complexes, this type of derivatization and complexation can easily be transferred to the corresponding ^{99m}Tc chemistry. It could be shown that the relative binding affinity was essentially retained and compared favourably with other reported estradiol-derived carbonyl complexes [106].

4.2 Peptides and Proteins

Radiolabelled peptides and proteins are currently under intense investigation since they are still considered as "magic bullets" for cancer diagnostics and therapy. Numerous publications describing labelling techniques and biological properties of Tc(V) complexes with tetradentate NS chelators have appeared. Despite these enormous efforts, it is surprising that almost none of the techniques have found their way onto the market and into routine clinical application. This fact might be related to the instability of such peptides, the unfavourable accumulation in nontargeted organs or the difficulty in preparing in one step high specific-activity radiolabelled peptides of sufficient radiochemical purity. One of the most successful concepts is probably the HYNIC approach. The only problem is the noncharacterization of the complex. Peptides are biomolecules in which the advantage of kinetically stable complexes can be shown. Introducing various types of different chelators will influence the biodistribution of the radiopharmaceutical. Since many chelators coordinate strongly to the *fac*-[^{99m}Tc(CO)$_3$]$^+$ moiety, systematic structure–activity relationships are accessible. So far, only a limited number of papers describing the labelling of peptides with organometallic complexes are available. A few selected examples will be described in the following.

In an early work, neurotensin was derivatized with histidine either through an amide bond to the carboxylic acid to produce a bidentate NN chelator or through alkylation at the N$^\alpha$-amino group under retention of the tripodal coordinating feature [107]. Labelling of neurotensin comprising the tripodal histidine occurred at very low concentrations (below 10^{-5} M) without affecting the affinity to the receptor. Biodistribution studies showed that tridentate ligands are superior to bidentate ones. A lower limiting concentration can be applied and accumulation in nontarget organs such as liver and kidney is much lower for tridentate ligands. Halide exchange for competing coordinating groups in serum might be the reason for this observation and the formation of larger aggregates. Complete shielding of the Tc(I) centre with a tridentate ligand prevents these cross-linking reactions [59, 66, 108]. Owing to the too short half life time of native neurotensin, the same strategy was applied to stabilized neurotensin. Much better biodistribution patterns could be realized and some of the derivatives have been involved in phase-I clinical trials and are currently being tested in humans [109–112].

Somatostatin and its stabilized derivatives, for example, the octreotides have been combined with organometallic ^{99m}Tc moieties. Since somatostatin derivatives are most attractive from a practical point of view, relatively many papers have appeared in this field. Of special interest was an early attempt to use cytectrene and octreotide. Octreotide was labelled at the N-terminus with $CpTc(CO)_3$ by the double ligand transfer approach in an overall 8% yield and the biodistribution was tested in rats. The labelled octreotide was found to be stable in the serum for at least 1 h and the uptake in targeted organs was specific, i.e. could be blocked by cold unlabelled octreotide. The compound possessed the important features of a radiopharmaceutical, in vivo stability, specific uptake and retention in target tissue and low accumulation in nontarget tissue such as fat and muscles. Still, the synthesis is not convenient and needs improvement [85].

In another approach, somatostatin 14-dextran was oxidized and coupled to histidine through imination. Subsequent reduction gave a glycosylated somatostatin that was labelled with *fac*-$[^{99m}Tc(CO)_3]^+$ by simple mixing. Challenging the labelled conjugate with a large excess of cysteine resulted in loss of about 25% of the label. This was probably due to unspecific binding of *fac*-$[^{99m}Tc(CO)_3]^+$ since other work showed complete stability of $[^{99m}Tc(his)(CO)_3]$ against cysteine challenge [113, 114]. Carbohydrated octreotide was coupled to picoline-aminoacetic acid. The carbohydrate makes the conjugate much more hydrophilic and an excellent biodistribution could be observed [115].

A number of other peptides have been studied in detail. Bombesin was derivatized at the C-terminus with histidine to give a bidentate NO chelator and a neutral complex after labelling. The labelled peptide fully retained the biological activity and was stable in vivo and in vitro, although the peptide as such slowly metabolized. Clearly, bidentate coordination is not optimal in respect of stability; still, labelling yield and specific activity were both very high [114]. Similarily, ethylenediamine was introduced at the N-terminus of bombesin. Specific labelling with *fac*-$[^{99m}Tc(CO)_3]^+$ could be achieved. The labelled complex exhibits subnanomolar affinity to PC-3 cells, which was demonstrated by competitive displacement assay. Accumulation in bombesin receptor containing organs was demonstrated in tumour-bearing mice models. Since the chelator is only bidentate, the third position was subsequently coordinated to the highly hydrophilic phosphine $P(CH_2–OH)_3$. This further coordination can be described as a mixed-ligand approach and resulted in significantly higher hydrophilicity of the radioconjugate and an improved biodistribution [116]. A similar investigation has also been performed with the *fac*-$[^{188}Re(CO)_3]^+$ moiety [117].

Beside these well known peptides, neuropeptide Y analogues have been labelled with the *fac*-$[^{99m}Tc(CO)_3]^+$ moiety. 2-Picolylamine-*N*,*N*-diacetic acid was coupled to the N-terminus of neuropeptide Y. The biological stability was excellent and neither cysteine nor histidine challenge replaced the label. However, since neuropeptide Y comprises a histidine in the chain, some unspecific labelling at the imidazole side chain occurred. The specific affinity of bombesin does not strongly depend on this histidine. Consequently, it was replaced by alanine, resulting in specific labelling at the ligand site only [118]. The potential of approaches with organometallic moieties was recently shown in the context of permeation tat peptides. Tat peptides are in focus since rapid translocation properties, seemingly independent of receptor-mediated endocytosis, have been observed. Tat peptides were derivatized with histidine and diethylenetriaminepentaacetic acid (DTPA). In parallel, fluorescein was introduced to enable double localization and to quantitatively measure cell uptake and tissue distribution as well as to qualitatively confirm subcellular localization by fluorescence microscopy. These dual-labelled tat peptides have been tested in vivo and in vitro. The localizations of the fluorescing agent and the radionuclides are in good agreement. No loss of label was found, confirming the high physiological stability of the organometallic moiety [119].

Proteins seem to be an attractive target for direct labelling with low-valent organometallics. On one hand, a protein offers numerous sites for anchoring an organometallic moiety but, on the other hand, monodentate coordination only is probably not sufficient to prevent cross-linking to other proteins, leading to the formation of larger aggregates. The histidine-tagged recombinant scFv proteins are an exception to this expected general behaviour. They possess a chain of 3–6 histidines at the N-terminus which are necessary for purification but not for receptor binding. This histidine tag can be considered as a multidentate ligand since two or more imidazoles from histidine can coordinate at the same time. In contrast to the direct unspecific labelling, improved stability against cross-linking is expected. This concept could be verified for a number of examples. Histidine-tagged scFv has been labelled at 37 °C to a high degree and labelling occurred specifically at the his-tag. However, endogenous histidine interfered in coordination to the histidine tag to a low degree (less than 5%). Replacement of theses histidines by mutation allowed specific labelling at the histidine tag only. Still, the procedure is convenient for the quick ^{99m}Tc labelling of this important class of targeting vectors and this will allow preliminary evaluation of some biodistribution data [62, 63].

4.3 Miscellaneous

Clearly, the labelling of targeting vectors with organometallic moieties is still at its infancy. Some other single studies have been carried out with small molecules. It is, of course, an incentive to find a way of labelling not only peptides and proteins but also smaller molecules such as amino acids. CNS receptor ligands as described at the beginning are small but a number of biologically active molecules, like glucose, are even smaller and, thus, more challenging. However, no promising results are available so far, neither for the Tc(V) approach nor for other concepts, including organometallic moieties. It is likely that positive results are not available since the challenge of labelling these molecules with ^{99m}Tc is just too big (and possibly impossible). Efforts have been undertaken in the context of folic acid which was derivatized with DTPA. DTPA is hardly expected to form stable complexes with the *fac*-$[Tc(CO)_3]^+$ moiety but reaction gave one single product in high yield and radiochemical purity. The stability of the radiopharmaceutical is excellent, exhibiting the importance of kinetic stability. The radiotracer undergoes folate receptor mediated uptake in a human carcinoma cell line in vitro and in vivo. However, as a folate-receptor-targeted radiopharmaceutical, the radiotracer does not appear to offer advantages over other folate–chelate conjugates [120].

Probably the biggest challenge is the labelling of glucose with ^{99m}Tc. Whereas the synthesis of compounds which are antagonists for hexokinase could probably be achieved (the 3D structure of hexokinase is known), it is much more difficult to transport such radiopharmaceuticals through the glucose transporter of the cell membrane. In a recent study, an imino-diacetate based chelator was introduced at C1. Labelling with *fac*-$[^{99m}Tc(CO)_3]^+$ gave one product but the glucose conjugate did not show any uptake in cells [121]. Coordination of imino-diacetate gave an anionic and hydrophilic complex which was stable in serum at 37 °C for more than 24 h.

The surfactant protein B was unspecifically labelled with *fac*-$[^{99m}Tc(CO)_3]^+$. The highly lipophilic protein spread over a hydrophilic surface and could have potential in the diagnosis of acute respiratory disease syndrome. The spreading properties of the labelled and native surfactant protein B were in coincidence, making labelled surfactant protein B a potential diagnostic radiopharmaceutical [122].

4.4 Perspectives and Conclusions

The application of organometallic complexes and moieties in life science, in general, or in radiopharmacy, in particular, is very recent. Still, bioorganometallic chemistry is progressing very fast since it offers an attractive and challenging topic besides the pathway of using classical coordination compounds. Clearly, novelty is not a sufficient argument to work in a field; however, organometallic complexes offer many advantages over coordination compounds as shown in the previous sections. Essentially, this statement was underlined with radiopharmaceuticals comprising the *fac*-$[^{99m}Tc(CO)_3]^+$ moiety since no other comparable fragments have been described or are available along a reasonable synthesis. If no organometallic imaging agent has found its way onto the market, the same holds true for classical ^{99m}Tc-based imaging agents. Comparing the amount of literature available for this part of radiopharmaceutical chemistry, we find the progress organometallics have made in this highly competitive environment is considerable. For the future it is an incentive to find other fragments with even more favourable properties than *fac*-$[^{99m}Tc(CO)_3]^+$. Corresponding model complexes can easily be drawn on paper but to synthesize them and to find a versatile synthetic approach to their production is a challenge for the inorganic chemist. If not successful in application, the reactions to be explored along this way will reveal many exciting results that are not only attractive for the radiopharmaceutical chemist but for inorganic chemistry in general. To keep the importance of ^{99m}Tc alive, such intellectual inputs and developments are urgently required.

References

1. Perrier C, Segrè E (1937) Nature 140:193
2. Richards P (1960) VII Rassegna internationale elettronica e nucleare, Roma. Atti Ufficiali 223
3. Richards P (1969) Proc ANS Topical Mtg PRNC 135:252
4. Tucker WD, Greene MW, Weiss AJ, Murenhoff AP (1958) Trans Am Nucl Soc 1:160
5. Jurisson SS, Lxdon JD (1999) Chem Rev 99:2205
6. Liu S, Edwards DS (1999) Chem Rev 99:2235
7. Dilworth JR, Parrott SJ (1998) Chem Soc Rev 27:43
8. Jain D (1999) Sem Nucl Med 29:221
9. Nosco DL, Beaty-Nosco JA (1999) Coord Chem Rev 184:91
10. Housecroft CE (1995) Coord Chem Rev 142:21
11. Housecroft CE (1997) Coord Chem Rev 162:305
12. Housecroft CE (1998) Coord Chem Rev 169:187
13. Baldas J (1994) Adv Inorg Chem 41:1
14. Sattelberger AP, Bryan JC (1995) In: Abel EW, Gordon F, Stone A, Wilkinson G (eds) Comprehensive organometallic chemistry, vol 6. Elsevier Pergamon, Oxford, p 151–166
15. Alberto R, Schibli R, Waibel R, Abram U, Schubiger AP (1999) Coord Chem Rev 192:901
16. Alberto R (2003) In: McCleverty J, MeyerTJ (eds) Comprehensive coordination chemistry II, vol 5. Elsevier Pergamon, Amsterdam, pp 127–270
17. Baldas J, Bonnyman J, Pojer PM, Williams GA, Mackay MF (1981) J Chem Soc Dalton Trans 1798
18. Abram U, Lorenz B, Kaden L, Scheller D (1988) Polyhedron 7:285
19. Bolzati C, Boschi A, Uecelli L, Malago E, Bandoli G, Tisato F, Refosco F, Pasqualini R, Duatti A (1999) Inorg Chem 38:4473
20. Bolzati C, Uccelli L, Boschi A, Malago E, Duatti A, Tisato F, Refosco F, Pasqualini R, Piffanelli A (2000) Nucl Med Biol 27:369

21. Pietzsch HJ, Spies H, Hoffmann S (1989) Inorg Chim Acta 165:163
22. Spies H, Fietz T, Pietzsch HJ, Johannsen B, Leibnitz P, Reck G, Scheller D, Klostermann K (1995) J Chem Soc Dalton Trans 2277
23. Volbeda A, Garcia E, Piras C, deLacey AL, Fernandez VM, Hatchikian EC, Frey M, FontecillaCamps JC (1996) J Am Chem Soc 118:12989
24. Happe RP, Roseboom W, Pierik AJ, Albracht SPJ, Bagley KA (1997) Nature 385:126
25. Nicolet Y, Lemon BJ, Fontecilla-Camps JC, Peters JW (2000) Trends Biochem Sci 25: 138
26. Darensbourg MY, Lyon EJ, Smee JJ (2000) Coord Chem Rev 206:533
27. Le Cloirec A, Best SP, Borg S, Davies SC, Evans DJ, Hughes DL, Pickett CJ (1999) Chem Commun 2285
28. Sadler PJ (2003) J Inorg Biochem 96:39
29. Chen HM, Parkinson JA, Parsons S, Coxall RA, Gould RO, Sadler PJ (2002) J Am Chem Soc 124:3064
30. Chen HM, Parkinson JA, Morris RE, Sadler PJ (2003) J Am Chem Soc 125:173
31. Chen HM, Parkinson JA, Bella J, Dawson A, Gould RO, Parsons S, Sadler PJ (2003) J Inorg Biochem 96:113
32. Novakova O, Chen HM, Vrana O, Rodger A, Sadler PJ, Brabec V (2003) Biochemistry-Us 42:11544
33. Jaouen G, Top S, Vessieres A, Alberto R (2000) J Organomet Chem 600:23
34. Abrams MJ, Davison A, Jones AG, Costello CE, Pang H (1983) Inorg Chem 22:2798
35. Jones AG, Abrams MJ, Davison A, Brodack JW, Toothaker AK, Adelstein SJ, Kassis AI (1984) Nucl Med Biol 11:225
36. Chen H, Ogo S, Fish RH (1996) J Am Chem Soc 118:4993
37. Schwochau K (2000) Technetium, chemistry and radiopharmaceutical applications. Wiley-VCH, Weinheim
38. Schibli R, Schubiger PA (2002) Eur J Nucl Med Mol Imaging 29:1529
39. Hieber W, Lux F, Herget C (1965) Z Naturforsch 20b:1159
40. Hileman JC, Huggins DK, Kaesz HD (1961) J Am Chem Soc 83:2953
41. Castro HHK, Hissink CE, Teuben JH, Vaalburg W, Panek K (1992) Recl Trav Chim Pay-Bas 111:105
42. Aebischer N, Schibli R, Alberto R, Merbach AE (2000) Angew Chem 39:254
43. Hileman JC, Huggins DK, Kaesz JD (1962) Inorg Chem 1:933
44. Fischer EO, Schmidt MW (1969) Chem Ber 102:1954
45. Alberto R, Schibli R, Egli A, Abram U, Abram S, Kaden TA, Schubiger PA (1998) Polyhedron 17:1133
46. Abram U, Hubener R, Alberto R, Schibli R (1996) Z Anorg Allg Chem 622:813
47. Baldas J, Bonnyman J, Pojer PM, Williams GA, Mackay MF (1982) J Chem Soc Dalton Trans 451
48. Koelle U (1994) Coord Chem Rev 135:623
49. Alberto R, Schibli R, Schubiger PA (1996) Polyhedron 15:1079
50. Alberto R, Egli A, Abram U, Hegetschweiler K, Schubiger PA (1994) J Chem Soc Dalton Trans 2815
51. Egli A, Hegetschweiler K, Alberto R, Abram U, Schibli R, Hedinger R, Gramlich V, Kissner R, Schubiger PA (1997) Organometallics 16:1833
52. Salignac B, Grundler PV, Cayemittes S, Frey U, Scopelliti R, Merbach AE, Hedinger R, Hegetschweiler K, Alberto R, Prinz U, Raabe G, Kolle U, Hall S (2003) Inorg Chem 42:3516
53. Alberto R, Schibli R, Egli A, Abram U, Kaden TA, Schubiger PA (1998) J Am Chem Soc 120:7987
54. Malone LJ, Parry RW (1967) Inorg Chem 6:817

55. Malone LJ, Manley MR (1967) Inorg Chem 6:2260
56. Carter JC, Parry RW (1965) J Am Chem Soc 87:2354
57. Alberto R, Ortner K, Wheatley N, Schibli R, Schubiger AP (2001) J Am Chem Soc 123:3135
58. Schibli R, Schwarzbach R, Alberto R, Ortner K, Schmalle H, Dumas C, Egli A, Schubiger PA (2002) Bioconjugate Chem 13:750
59. Egli A, Alberto R, Tannahill L, Schibli R, Abram U, Schaffland A, Waibel R, Tourwe D, Jeannin L, Iterbeke K, Schubiger PA (1999) J Nucl Med 40:1913
60. Seifert S, Künstler J-U, Gupta A, Funke H, Reich T, Hennig C, Rossberg A, Pietzsch H-J, Alberto R (2000) Radiochim Acta 3–4:239
61. Seifert S, Künstler J-U, Gupta A, Funke H, Reich T, Pietzsch H-J, Alberto R, Johannsen B (2001) Inorg Chim Acta 322:79
62. Waibel R, Alberto R, Willuda J, Finnern R, Schibli R, Stichelberger A, Egli A, Abram U, Mach J-P, Plueckthun A, Schubiger PA (1999) Nature Biotechnol 17:897
63. Waibel R, Stichelberger R, Alberto R, Schubiger PA, Chester KA, Begent RHJ (2000) Eur Jf Nucl Med 27:S86
64. Schibli R, Katti KV, Volkert WA, Barnes CL (1998) Inorg Chem 37:5306
65. Zobi F, Spingler B, Fox T, Alberto R (2003) Inorg Chem 42:2818
66. Schibli R, La Bella R, Alberto R, Garcia-Garayoa E, Ortner K, Abram U, Schubiger PA (2000) Bioconjugate Chem 11:345
67. Alberto R, Schibli R, Abram U, Johannsen B, Pietzsch H-J, Schubiger PA (1999) J Am Chem Soc 25:6076
68. Abrams MJ, Juweid M, Tenkate CI, Schwartz DA, Hauser MM, Gaul FE, Fuccello AJ, Rubin RH, Strauss HW, Fischman AJ (1990) J Nucl Med 31:2022
69. Liu S, Edwards DS, Looby RJ, Harris AR, Poirier MJ, Barrett JA, Heminway SJ, Carroll TR (1996) Bioconjugate Chem 7:63
70. Garcia R, Paulo A, Domingos A, Santos I, Ortner K, Alberto R (2000) J Am Chem Soc 122:11240
71. Garcia R, Paulo A, Domingos K, Santos I (2001) J Organomet Chem 632:41
72. Pak JK, Benny P, Spingler B, Ortner K, Alberto R (2003) Chem-Eur J 9:2053
73. Häfliger P, Agorastos N, Spingler B, Georgiev O, Viola G, Alberto R (2004) Chem Biochem (accepted)
74. Behr TM, Sgouros G, Vougioukas V, Memtsoudis S, Gratz S, Schmidberger H, Blumenthal RD, Goldenberg DM, Becker W (1998) Int J Cancer 76:738
75. Behr TM, Behe M, Lohr M, Sgouros G, Angerstein C, Wehrmann E, Nebendahl K, Becker W (2000) Eur J Nucl Med 27:753
76. O'Donoghue JA, Wheldon TE (1996) Phys Med Biol 41:1973
77. Howell RW (1992) Med Phys 19:1371
78. Pollak A, Roe DG, Pollock CM, Lu LFL, Thornback JR (1999) J Am Chem Soc 121:11593
79. Dunn-Dufault R, Pollak A, Fitzgerald J, Thornback JR, Ballinger JR (2000) Nucl Med Biol 27:803
80. Mundwiler S, Candreia L, Häfliger P, Ortner K, Roger A (2004) Bioconjugate Chem 15:195
81. Salmain M, Gunn M, Gorfti A, Top S, Jaouen G (1993) Bioconjugate Chem 4:425
82. Salmain M, Gorfti A, Jaouen G (1998) Eur J Biochem 258:192
83. Wenzel M (1992) J Labelled Compd Radiopharm 31:641
84. Cesati RR, Tamagnan G, Baldwin RM, Zoghbi SS, Innis RB, Katzenellenbogen JA (1999) J Labelled Compd Radiopharm 42:S150
85. Spradau TW, Edwards WB, Anderson CJ, Welch MJ, Katzenellenbogen JA (1999) Nucl Med Biol 26:1

86. Wald J, Alberto R, Ortner K, Candreia L (2001) Angew Chem Int Ed Engl 40:3062
87. Bernard J, Ortner K, Spingler B, Pietzsch HJ, Alberto R (2003) Inorg Chem 42:1014
88. Johannsen B, Pietzsch HJ (2002) Eur J Nucl Med Mol Imaging 29:263
89. Masi S, Top S, Boubekeur L, Jaouen G, Mundwiler S, Spingler B, Alberto R (2004) Eur J Inorg Chem 10:2013
90. Zhang XZ, Wang XB, Wen HT (2003) Chem J Chin U 24:21
91. Morel P, Schaffer P, Valliant JF (2003) J Organomet Chem 668:25
92. Valliant JF, Morel P, Schaffer P, Kaldis JH (2002) Inorg Chem 41:628
93. Schaffer P, Britten JF, Davison A, Jones AG, Valliant JF (2003) J Organomet Chem 680: 323
94. Correia JDG, Domingos Â, Santos IRA, Ortner K (2001) Inorg Chem 40:5147
95. Alves S, Paulo A, Correia JDG, Domingos A, Santos I (2002) J Chem Soc Dalton Trans 4714
96. Drews A, Pietzsch HJ, Syhre R, Seifert S, Varnas K, Hall H, Halldin C, Kraus W, Karlsson P, Johnsson C, Spies H, Johannsen B (2002) Nucl Med Biol 29:389
97. Cesati RR, Tamagnan G, Baldwin RM, Zoghbi SS, Innis RB, Kula NS, Baldessarini RJ, Katzenellenbogen JA (2002) Bioconjugate Chem 13:29
98. Hoepping A, Reisgys M, Brust P, Seifert S, Spies H, Alberto R, Johannsen B (1998) J Med Chem 41:4429
99. Hoepping A, Babich J, Zubieta JA, Johnson KM, Machill S, Kozikowski AP (1999) Bioorg Med Chem Lett 9:3211
100. Wenzel M, Klinge C (1994) J Labelled Compd Radiopharm 34:981
101. Saidi M, Kothari K, Pillai MRA, Hassan A, Sarma HD, Chaudhari PR, Unnikrishnan TP, Korde A, Azzouz Z (2001) J Labelled Compd Radiopharm 44:603
102. Mull ES, Sattigeri VJ, Rodriguez AL, Katzenellenbogen JA (2002) Bioorg Med Chem 10:1381
103. Uehara T, Koike M, Nakata H, Miyamoto S, Motoishi S, Hashimoto K, Oku N, Nakayama M, Arano Y (2003) Nucl Med Biol 30:327
104. Reisgys M, Wust F, Alberto R, Schibli R, Schubiger PA, Pietzsch HJ, Spies H, Johannsen B (1997) Bioorg Med Chem Lett 7:2243
105. Luyt LG, Bigott HM, Welch MJ, Katzenellenbogen JA (2003) Bioorg Med Chem 11: 4977
106. Arterburn JB, Corona C, Rao KV, Carlson KE, Katzenellenbogen JA (2003) J Org Chem 68:7063
107. Blauenstein PA, Garayoa EG, Ruegg D, Blanc A, Lutz M, Schubiger PA (2003) Cancer Biotherapy Radiopharm 18:281
108. Schubiger PA, Allemann-Tannahill L, Egli A, Schibli R, Alberto R, Carrel-Remy N, Willimann M, Bläuenstein P, Tourwe D (1999) Q J Nucl Med 43:155
109. Garcia-Garayoa E, Allemann-Tannahill L, Blauenstein P, Willmann M, Carrel-Remy N, Tourwe D, Iterbeke K, Conrath P, Schubiger PA (2001) Nucl Med Biol 28:75
110. Garcia-Garayoa E, Blauenstein P, Bruehlmeier M, Blanc A, Iterbeke K, Conrath P, Tourwe D, Schubiger PA (2002) J Nucl Med 43:374
111. Buchegger F, Bonvin F, Kosinski M, Schaffland AO, Prior J, Reubi JC, Blauenstein P, Garayoa EG, Gillet M, Schubiger PA, Delaloye AB (2003) J Nucl Med 44:23
112. Buchegger F, Bonvin F, Kosinski M, Schaffland AO, Prior J, Reubi JC, Blauenstein P, Tourwe D, Garayoa EG, Delaloye AB (2003) J Nucl Med 44:1649
113. Du J, Hiltunen J, Marquez M, Nilsson S, Holmberg AR (2001) Appl Radiat Isotopes 55:181
114. La Bella R, Garcia-Garayoa E, Bahler M, Blauenstein P, Schibli R, Conrath P, Tourwe D, Schubiger PA (2002) Bioconjugate Chem 13:599
115. Wester HJ, Schottelius M, Schwaiger M (2001) J Nucl Med 42:115p

116. Smith CJ, Sieckman GL, Owen NK, Hayes DL, Mazuru DG, Kannan R, Volkert WA, Hoffman TJ (2003) Cancer Res 63:4082
117. Smith CJ, Sieckman GL, Owen NK, Hayes DL, Mazuru D, Volkert WA, Hoffman TJ (2003) Anticancer Res 23:63
118. Langer M, La Bella R, Garcia-Garayoa E, Beck-Sickinger AG (2001) Bioconjugate Chem 12:1028
119. Bullok KE, Dyszlewski M, Prior JL, Pica CM, Sharma V, Piwnica-Worms D (2002) Bioconjugate Chem 13:1226
120. Trump DP, Mathias CJ, Yang ZF, Low PSW, Marmion M, Green MA (2002) Nucl Med Biol 29:569
121. Petrig J, Schibli R, Dumas C, Alberto R, Schubiger PA (2001) Chem-Eur J 7:1868
122. Amann A, Decristoforo C, Ott I, Wenger M, Bader D, Alberto R, Putz G (2001) Nucl Med Biol 28:243

Top Curr Chem (2005) 252: 45–84
DOI 10.1007/b101224

Rhenium and Technetium Complexes Anchored by Phosphines and Scorpionates for Radiopharmaceutical Applications

Isabel Santos (✉) · António Paulo · João D. G. Correia

Departamento de Química, Instituto Tecnológico e Nuclear, Estrada Nacional 10, 2686-953 Sacavém, Portugal
isantos@itn.mces.pt

Abstract This review presents a comprehensive overview of the most significant developments in the chemistry of technetium and rhenium complexes anchored by heterofunctionalized phosphines and by hard and soft scorpionates. The main goal is to provide the reader with chemical, radiochemical and biological knowledge which is expected to enhance the application of these types of compounds in the development of target-specific radiopharmaceuticals.

In the introductory section, a short historical outlook is presented to introduce the few ^{99m}Tc radiopharmaceuticals based on phosphorus- and boron-containing ligands already in clinical use. Generic considerations of the characteristics of these ligands are included in this section highlighting their potential usefulness in the design of biocomplexes for clinical applications.

The following section deals with the chemistry of intermediate- and low-valent rhenium and technetium complexes anchored by polydentate phosphines, which combine soft phosphorus atoms with other hard or soft donors (N, O, S). Synthetic procedures involving phosphines with denticities adjustable to a given metal fragment are described. In the last section the coordination chemistry of rhenium and technetium with poly(pyrazolyl)borates and poly(mercaptoimidazolyl)borates (hard and soft scorpionates) is described, giving special attention to high-valent rhenium and technetium oxides and to low-valent carbonyl complexes. In all these sections, synthetic procedures, reactivity studies and structural characterization are discussed, with special emphasis on those inorganic and organometallic systems that allowed, or are expected to allow, the labeling of biomolecules (e.g., small peptides and central nervous system receptor ligands). In this context, the biological properties of the already achieved biocomplexes, anchored by phosphorus- or boron-containing ligands, are also briefly discussed.

Keywords Technetium · Rhenium · Heterofunctionalized phosphines · Hard and soft scorpionates · Biomolecules · Radiopharmaceuticals

Abbreviations

BBN	Bombesin
Bp	Dihydrobis(pyrazolyl)borate
Bp*	Dihydrobis(3,5-dimethylpyrazolyl)borate
CCK	Cholecystokinin
CDO	Cyclohexanedionedioxime
CNS	Central nervous system
DPPBA	2-(Diphenylphosphanyl)benzoic acid
Dpr	2,3-Diaminopropionic acid
ESI-MS	Electrospray ionization mass spectrometry
FT-IR	Fourier transform IR
GRP	Gastrin-releasing peptide
H_2CH_2PNO	2-(Diphenylphosphanyl)-*N*-(2-hydroxyethyl)benzylamine
$HCyPN_2$	[(1*R*,2*R*)-*N*-(2-Aminocyclohexyl)]-2-(diphenylphosphanyl)benzamide
$H_2(Me_2PO_2)$	Bis(2-hydroxy-5-methylphenyl)phenylphosphine
HMPB	1,2-Bis(bis(hydroxymethyl)phosphino)benzene
HMPE	1,2-Bis(bis(hydroxymethyl)phosphino)ethane
HPLC	High-performance liquid chromatography
HPN_2	*N*-(2-Aminoethyl)-2-(diphenylphosphanyl)benzamide
$H_2P_2N_2$	*N,N'*-Bis[2-(diphenylphosphino)phenyl]propane-1,3-diamine
HPNBr	*N*-(2-Bromo)-2-(diphenylphosphanyl)benzamide
H_2PNO	2-(Diphenylphosphanyl)-*N*-(2-hydroxyethyl)benzamide
H_2PNS	2-(Diphenylphosphanyl)-*N*-(2-thioethyl)benzamide
H_3PN_2S	*N*-[*N*-(3-Diphenylphosphinopropionyl)glycyl]-*S*-cysteine
H_2PN_2S-Bz	*N*-{*N*-[3-(Diphenylphosphino)propionyl]glycyl}-L-*S*-benzylcysteine
H_2PO_2	Bis(*o*-hydroxyphenyl)phenylphosphine
HSSSH	3-Thiapentane-1,5-dithiol
HYNIC	6-Hydrazinonicotinamide
HYNIC-Ko-DPPBN-ε	(2-(Diphenylphosphino)benzoyl)-*N*-α-(6-(2-(2-sulfonatobenzaldehyde)hydrazono)nicotinyl)lysine methyl ester

HYNIC-Kp-DPPBN-ε	(4-(Diphenylphosphino)benzoyl)-*N*-α-(6-(2-(2-sulfonatobenzaldehyde)hydrazono)-nicotinyl)lysine methyl ester
ID	Injected dose
LHRH	Luteinizing hormone-releasing hormone
PBS	Phosphate-buffered saline
$Ph_2P(O)py$	Diphenyl-2-pyridyl phosphine monoxide
Ph_2Ppy	2-Pyridyl-diphenylphosphine
PhTp	Phenyltris(pyrazolyl)borate
PN	*o*-[Diphenylphosphino)benzylidene]aniline
P_2N_2-COOH	4,4-Bis[bis-hydroxymethyl-phosphanylpropyl-carbamoyl)butyric acid
PNH_2	(*o*-Aminophenyl)diphenylphosphine
$PNMe_2$	*o*-(Diphenylphosphino)-*N,N'*-dimethylaniline
PO	*o*-(Diphenylphosphino)benzaldehyde
P_1OH	(*o*-Hydroxyphenyl)diphenylphosphine
P_2OH	2-(Diphenylphosphinomethyl)-4-methylphenol
PPh_3oxaz	2-(2-Diphenylphosphinophenyl)oxazoline
P_2S_2-COOH	6,8-Bis-[3-(bis(hydroxymethyl)phosphanyl)propyl-sulfanyl]octanoic acid
PTA	1,3,5-Triaza-7-phosphaadamantane
py	Pyridine
pzH	Pyrazole
pz*	3,5-Me_2pyrazole
pzTp	Tetrakis(pyrazolyl)borate
THP	Tris(hydroxymethyl)phosphine
tim^{Me}	1-Methyl-2-mercaptoimidazole
Tp	Hydrotris(pyrazolyl)borate
Tp′	Substituted tris(pyrazolyl)borate
Tp*	Hydrotris(3,5-dimethylpyrazolyl)borate
TPPTS	Triphenylphosphine-3,3′,3″-trisulfonate

1 Introduction

Phosphines as well as hard scorpionates have played an important role in inorganic and organometallic chemistry [1, 2]. Most of the studies with these chelating ligands have been directed toward the search for complexes potentially useful in stoichiometric or catalytic transformations of organic substrates [1–7]. However, for technetium and rhenium, another major driving force for the development of their coordination chemistry is the potential applications of ^{99m}Tc and $^{186/188}$Re in nuclear medicine imaging and therapy [8–16].

Since the 1960s, phosphines, owing to their σ-donor and π-acceptor properties, have been considered as good candidates for medical applications, being used as reducing agents toward the permetalate salts $[MO_4]^-$ (*M* is Tc, Re) and also as chelating agents in the stabilization of the metal in intermediate and low oxidation states (I–V) [16–18]. A rich chemistry has been developed and, after a number of attempts to prepare cationic complexes anchored by phosphines, two myocardial perfusion imaging agents have been introduced in clinical practice: $[^{99m}TcO_2(P53)_2]^+$ [P53 is bis(2-ethoxyethyl)phosphinoethane] (Myoview) and $[^{99m}Tc\text{-}Q12]^+$ $\{[^{99m}Tc(P(CH_2CH_2OCH_3)_3)_2L]$; *L* is 1,2-bis[dihydro-2,2,5,5-tetramethyl-3(2*H*)-furanone-4-methyleneamino]ethane} (TechneScan Q12) [19, 20].

Concerning boron-containing ligands, the chemistry of rhenium and technetium, for medical applications, has been focused exclusively on technetium–boronic adducts of dioximes, which are neutral and seven-coordinate complexes formed by an elegant technetium-templated synthesis, involving reduction of $[^{99m}TcO_4]^-$ with stannous chloride in the presence of a boronic acid and a dioxime derivative. From this family of complexes, $[TcCl(CDO)(CDOH)_2BMe]$ (Cardiotec) has also been successfully applied in myocardial perfusion imaging [21] (Structure 1, myocardial perfusion imaging agents).

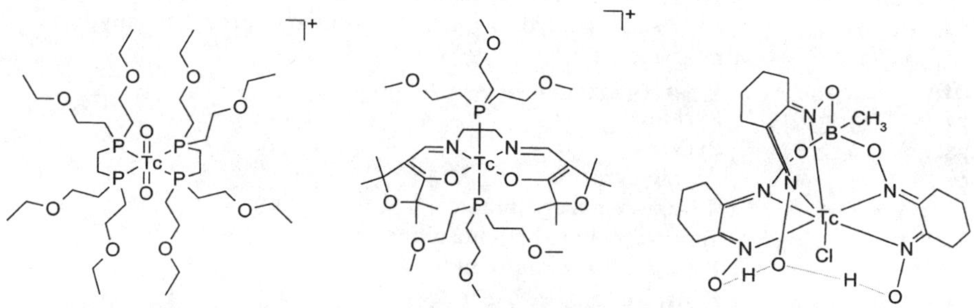

MyoviewTM TechneScan Q12TM CardiotecTM

Nowadays, there is an increasing demand for more sophisticated radioactive probes which would allow the imaging or the therapy of specific sites in the body [9–16]. To achieve this goal is a nontrivial task, one possibility being the design of bifunctional chelating agents which bind the metal and also have a site to covalently attach the targeting molecule. This biometal complex has to be stable in vitro and resistant to metabolical degradation, and at the same time the metal chelate fragment must not interfere with the receptor binding ability of the targeting molecule. Concerning heterofunctionalized phosphines, efforts to use these ligands in the development of target-specific radiopharmaceuticals have been pursued, either as bifunctional chelators or just as anchor ligands. However, the application of scorpionates for development of target specific radiopharmaceuticals is rather unexplored, despite the roughly similar chemical behavior exhibited by rhenium and technetium, and the sophisticated coordination chemistry developed for rhenium with these chelators [4, 5]. Nevertheless, scorpionates are quite attractive chelators and form quite stable metal complexes. This class of ligands can be hard or soft donors, depending on the azole rings coordinated to the boron atom, and their electronic, steric and coordinating properties can be easily tuned. Owing to this versatility the physicochemical properties of the complexes, which are determinant for pharmacokinetics and biological behavior, are easily controlled by using different coordinating azole rings and/or by introducing different substituents on the ligand framework [2, 4–7]. An interesting and challenging task for inorganic and organometallic chemists is to find new synthetic processes or methodologies adjustable to biomedical applications. Such goals can be achieved by extending or adjusting the rich inorganic and organometallic chemistry of rhenium and/or technetium to the experimental conditions needed for radiopharmaceutical applications. In most cases, this will need the disclosure of novel strategies and synthetic methodologies. An elegant example of such work was the introduction by Alberto and coworkers [13, 14, 16] of the novel and versatile organometallic aquo ion *fac*-$[M(OH_2)_3(CO)_3]^+$ (M is Re, $^{99/99m}$Tc), relevant for biomedical applications. More recently, Duatti and coworkers [9] have also introduced the $[M(N)PXP]^{2+}$ metal fragment, which certainly will also contribute to expanding the field.

In this work, the chemistry of rhenium and technetium complexes anchored by heterofunctionalized phosphines and by scorpionates will be reviewed. Special attention will be given to those chemical aspects that, according to the authors, could enhance the ability of using these ligands in the development of target-specific radiopharmaceuticals. In this context, the chemistry of the novel *fac*-$[M(CO)_3]^+$ and $[M(N)PXP]^{2+}$ metal fragments will also be mentioned, but to such an extent that minimizes any overlap with other reviews published in this special issue of *Topics in Current Chemistry*.

This contribution updates previous reviews especially the ones, recently published, covering the coordination chemistry of rhenium and/or technetium [4, 5, 14–16]. Special attention will be given to more recent achievements, but to provide some context to the current work, some overlap between this review and its predecessors is unavoidable.

2 Complexes Anchored by Phosphines

With the aim of designing prototype molecules useful for the development of radiopharmaceuticals based on rhenium and technetium, investigations have been pursued on the synthesis and chemistry of metal complexes in intermediate and low oxidation states with heterofunctionalized phosphine ligands. Polydentate phosphines, combining a soft phosphorus donor with other hard or soft donors, have been synthesized and their chemistry studied with rhenium and technetium. The most thoroughly studied heterofunctionalized phosphines combine the phosphorus donor atom with hydroxy, carboxylic, amino, amide, thioether or thiol coordinating functions. The incorporation of these functions into the phosphine framework has been achieved using aromatic or aliphatic side-arms bearing the additional heteroatom(s). These chelators have been explored either as anchors or as bifunctional chelating ligands. The results achieved in this field will be presented taking into account the oxidation state of the metal (intermediate or low) and in each case the denticity and the donor atom set of the phosphines.

2.1 Intermediate-Valent Complexes

2.1.1 Bidentate Phosphines

2.1.1.1 Mixed P, O Ligands

The potentially bidentate (*o*-hydroxyphenyl)diphenylphosphine ligand (P_1OH) reacts with $[MOCl_4]^-$ yielding the mono oxocomplex $[ReOCl_3(P_1O)]^-$ (**1**) and the bisubstituted $[MOCl(P_1O)_2]$ [*M* is Tc (**2**), Re (**3**)] (Scheme 1) [22, 23]. Complex **2** could also be obtained by a reduction/substitution reaction of pertechnetate with P_1OH in the presence of HCl, while complex **3** was also synthesized by reacting the anionic *trans*-dioxo $[ReO_2(P_1O)_2]^-$ with hydrochloric acid (Scheme 1) [23, 24].

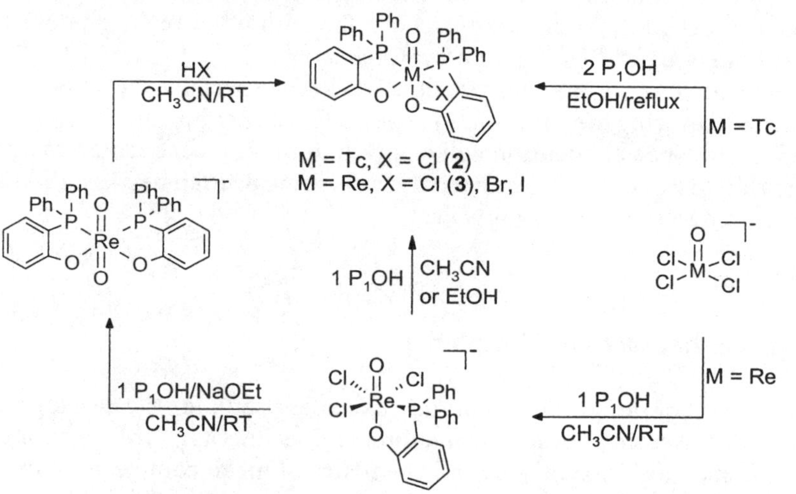

Scheme 1 Oxo complexes anchored by (*o*-hydroxyphenyl)diphenylphosphine

Reaction of $[ReOCl_3(PPh_3)_2]$ with the related 2-(diphenylphosphinomethyl)-4-methylphenol ligand (P_2OH) led to the formation of the bisubstituted species $[ReOCl(P_2O)_2]$ (**4**) [25]. The structural analysis of **2**, **3** and **4** demonstrated that these compounds display a *trans*-O–*M*=O–*cis*-P,P "twisted" octahedral geometry [22, 24, 25]. In contrast, the corresponding phenylimido complex $[Re(NPh)Cl(P_2O)_2]$ (**5**), prepared by reacting $[Re(NPh)Cl_3(PPh_3)_2]$ with P_2OH, adopts the *trans*-O–Re–NPh–*trans*-P,P arrangement, as indicated by ^{31}P NMR spectroscopy [25].

The ligand-exchange reaction of $[ReNCl_2(PPh_3)_2]$ with P_1OH was also studied. Although the formation of a five-coordinate complex might have been expected, because of the strongly trans-labilizing nitrido group and the sterically demanding ligand phenyl groups, the reaction proceeded with the formation of the six-coordinate $[ReN(P_1O)_2(PPh_3)]$ (**6**) [22].

The chemistry of the P_1OH ligand was also extended to ^{99m}Tc. In HCl media, the reduction/substitution reaction of $[^{99m}TcO_4]^-$ with P_1OH yields $[^{99m}TcOCl(P_1O)_2]$ (**2a**), which is formed only through careful control of the ligand concentration. When the same reaction was carried out in the presence of a weakly coordinating acid, such as CF_3SO_3H, $[^{99m}Tc(P_1O)_3]$ (**7a**) was obtained. In contrast to complex **7a**, which is stable in solution and in serum, **2a** is unstable, regenerating $[^{99m}TcO_4]^-$ [26]. Compound **7a** was characterized by high-performance liquid chromatography (HPLC) comparison with $[^{99}Tc(P_1O)_3]$ (**7**), which was obtained by reacting $[^{99}TcO_4]^-$ with an excess of P_1OH [27]. Similar tris-substituted Tc(III) complexes were easily obtained by reduction with other bidentate *PX*-heterofunctionalised phosphines (*X* is COOH or SH) [27,28]. The homoleptic $[Re(P_1O)_3]$ (**8**) and $[Re(P_2O)_3]$ (**9**) complexes were also reported. Their preparation involved the reaction of $[ReCl_3(MeCN)(PPh_3)]$ with the corresponding phosphine, in a 1:3 molar ratio [29]. The X-ray diffraction analysis of several of these *M*(III) complexes confirmed the coordination of three bidentate heterofunctionalised phosphines in the *mer* configuration [27–29].

Considering that replacement of the oxo by the nitrido core enhances the stability of Tc(V) toward reduction, Duatti et al. explored the chemistry of P_1OH with the isoelectronic core $[Tc{\equiv}N]^{2+}$. At the macroscopic and at the non-carrier added level, the only complexes identified were $[^{99/99m}TcN(P_1O)_2]$ (**10**, **10a**) [26]. Biodistribution studies of **10a** in rats have shown some transient heart uptake, but a significant amount of activity was also retained into the lungs [26].

The substitution lability of the halide groups in $[ReOCl_3(P_1O)]^-$ (**1**) allowed entrance to a novel family of mixed-ligand complexes of the type "3+2". As depicted in Scheme 2, the "$[ReO(P_1O)]^{2+}$" moiety can accommodate aminodithiolate ligands (**11**), other tridentate ligands containing a central amine group (**12**), model peptide fragments (**13**) and Schiff bases (**14**) [30–34]. These 3+2 compounds, **11**–**14**, are closed-shell 18-electron oxorhenium species, adopting a distorted octahedral geometry in the solid state or in solution, as demonstrated by classical spectroscopical methods, including multinuclear NMR. In all the complexes the axial positions are occupied by the oxo group and by the oxygen atom of the κ^2-P_1O ligand. In general, these 3+2 complexes are stabler against glutathione substitution than related "3+1" mixed-ligand complexes, anchored by the same type of tridentate ligands. The formation of isomers, observed for some 3+1 analogous complexes, was also prevented in the case of the 3+2 approach [30–32].

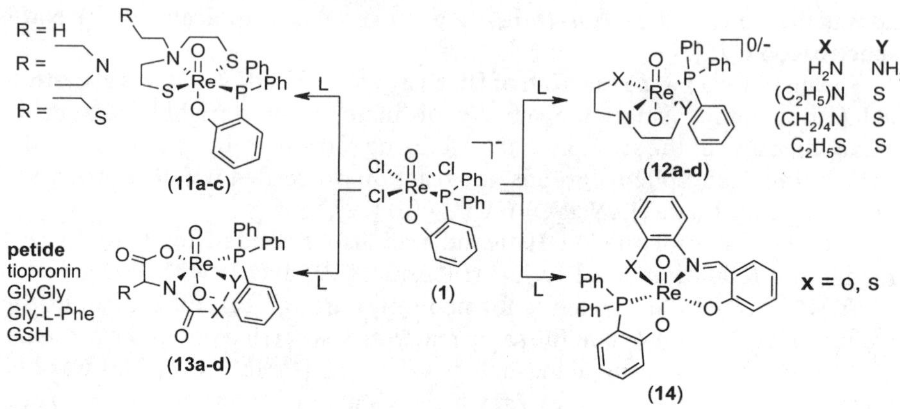

Scheme 2 Mixed-ligand complexes of the "3+2" type

Reactions of different Re(V) starting materials with 2-(diphenylphosphanyl)benzoic acid (DPPBA) gave only monosubstituted species, containing the $[ReO(PCOO)]^{2+}$ unit, even when an excess of ligand was used. This unit also allowed the synthesis of mixed 3+2 complexes with tridentate Schiff bases (Scheme 3) [33,35]. X-ray structural analysis of some of the complexes **(15–19)** confirmed the bidentate coordination mode of the monoanionic phosphino-carboxylate ligand, and a distorted octahedral coordination geometry around the metal [35].

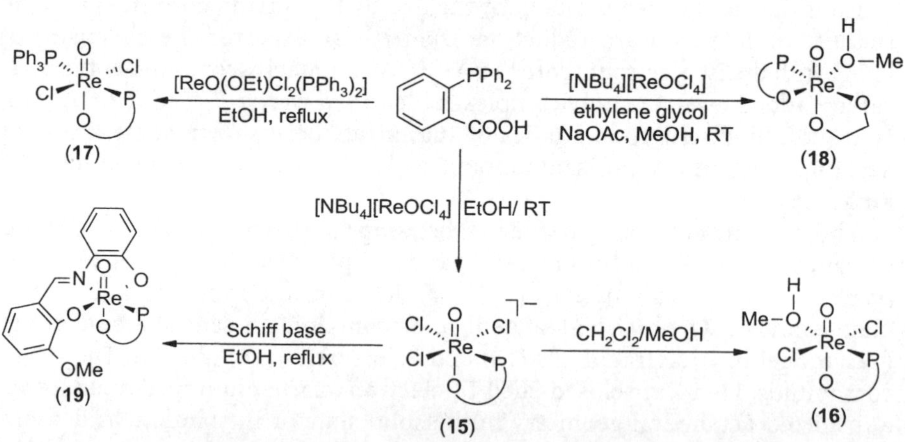

Scheme 3 Oxo complexes anchored by 2-(diphenylphosphanyl)benzoic acid

2.1.1.2
Mixed P, N Ligands

Reactions of the potentially bidentate (*o*-aminophenyl)diphenylphosphine (PNH_2) with $[TcO_4]^-$, $[NBu_4][MOCl_4]$ (*M* is Tc, Re) or $[ReCl_3(MeCN)(PPh_3)_2]$ are quite sensitive to the nature of the solvent, as well as to the ligand-to-metal molar ratio [36–38]. As shown in Scheme 4, starting from $[TcO_4]^-$ in aprotic media the main product formed was $[Tc(PNH)_3]$ (**20**), while in alcohol the isolated complexes were $[TcO(PNH)_2(OR)]$ (**21**). Analogous oxo-alkoxide derivatives with rhenium (**22**) were prepared via ligand-exchange reactions, starting from $[ReOCl_4]^-$ in basic media [36]. In acidic media, the same reaction yields $[ReOCl_3(PNH_2)]$ (**23**) [38]. Interesting imido complexes, $[MCl_2(PN)(PNH)]$ [*M* is Re (**24**), Tc (**25**)], were obtained by reacting $[NBu_4][MOCl_4]$ (*M* is Tc, Re) with an excess of PNH_2 in nonpolar solvents [37]. In all these six-coordinate complexes the PNH_2 ligand always acts as bidentate, although it coordinates as neutral, monoanionic or dianionic, depending on the reaction conditions [36–38].

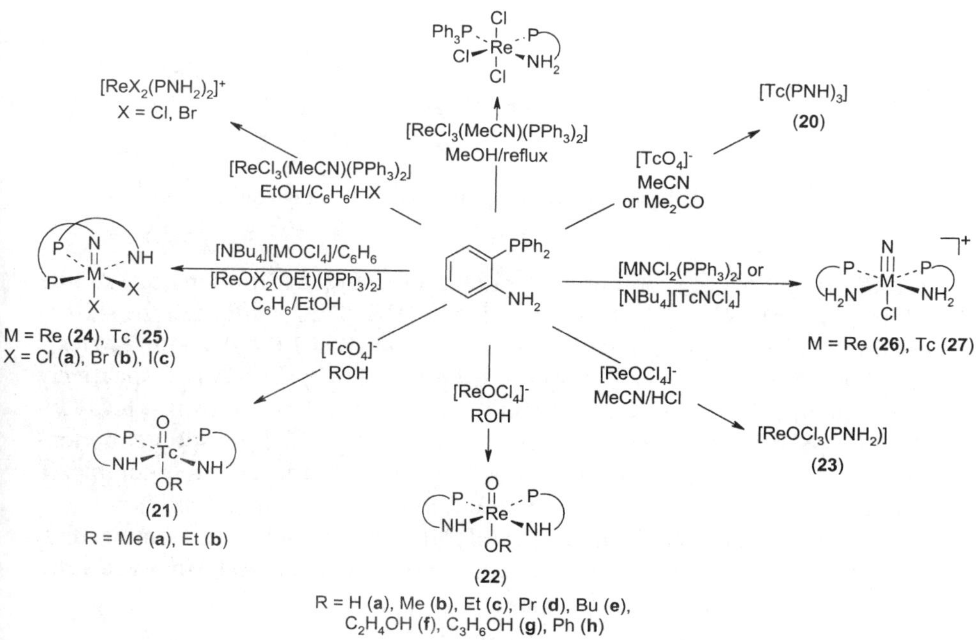

Scheme 4 Complexes with (*o*-aminophenyl)diphenylphosphine

Substitution reactions on $[NBu_4][Tc(N)Cl_4]$ and $[M(N)Cl_2(PPh_3)_2]$ (*M* is Re or Tc) with the PNH_2 ligand yields the cationic nitrido complexes $[M(N)Cl(PNH_2)_2]Cl$ [*M* is Re (**26**), Tc (**27**)] (Scheme 4). The two bidentate and neutral PNH_2 chelators are coordinated with a mutual *cis*-P configuration in the equatorial plane of a distorted octahedron; the site trans to the $M{\equiv}N$ unit is occupied by a Cl ligand [37].

Despite the rich chemistry developed with PNH_2, the high number of species produced precluded the application of this ligand in the development of radiopharmaceuticals. Expecting to overcome this problem, Tisato et al. [38] evaluated the behavior of *o*-(diphenylphosphino)-*N*,*N*′-dimethylaniline ($PNMe_2$) toward different starting materials and using different reaction conditions (Scheme 5). However, reactions of $PNMe_2$ with Re(V) or Re(III) starting materials also provide different neutral or cationic complexes. The resulting complexes are six-coordinate with an approximately octahedral coordination geometry, with the $PNMe_2$ ligand acting as bidentate, through the phosphorus and the nitrogen atoms [38]. In spite of the paramagnetism of the low-spin d^4 ion, all the rhenium(III) complexes with $PNMe_2$ were conveniently characterized in solution by NMR spectroscopy [38].

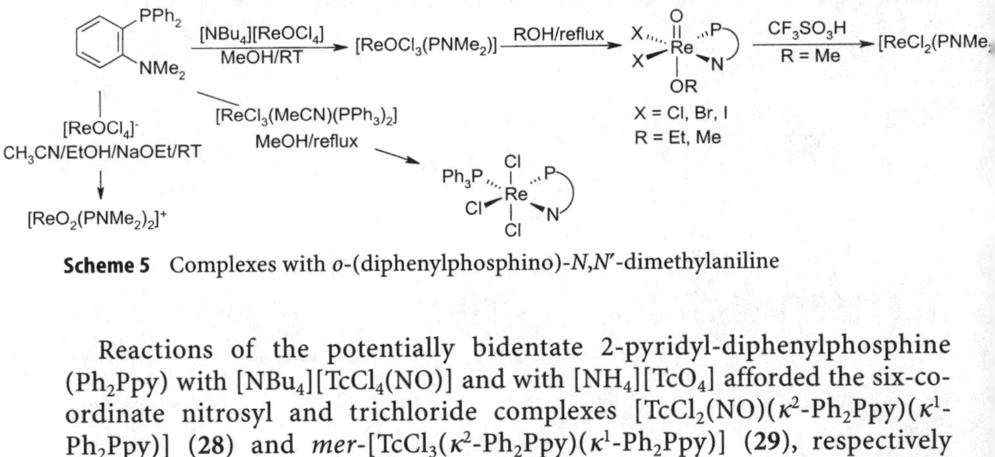

Scheme 5 Complexes with *o*-(diphenylphosphino)-*N*,*N*′-dimethylaniline

Reactions of the potentially bidentate 2-pyridyl-diphenylphosphine (Ph_2Ppy) with $[NBu_4][TcCl_4(NO)]$ and with $[NH_4][TcO_4]$ afforded the six-coordinate nitrosyl and trichloride complexes $[TcCl_2(NO)(\kappa^2\text{-}Ph_2Ppy)(\kappa^1\text{-}Ph_2Ppy)]$ (**28**) and *mer*-$[TcCl_3(\kappa^2\text{-}Ph_2Ppy)(\kappa^1\text{-}Ph_2Ppy)]$ (**29**), respectively (Scheme 6) [39, 40]. During the synthesis of **29**, the cationic *trans*-$[TcCl_2(\kappa^2\text{-}Ph_2P(O)py)_2](PF_6)$ (**30**) [$Ph_2P(O)py$ is diphenyl-2-pyridyl phosphine monoxide] was identified as a by-product. NMR spectroscopic data of these complexes were consistent with the structures found in the solid state [39, 40].

The Tc-6-hydrazinonicotinamide (HYNIC) core for labeling biomolecules, in particular peptides, was introduced by Schwartz et al. [41]. However, since

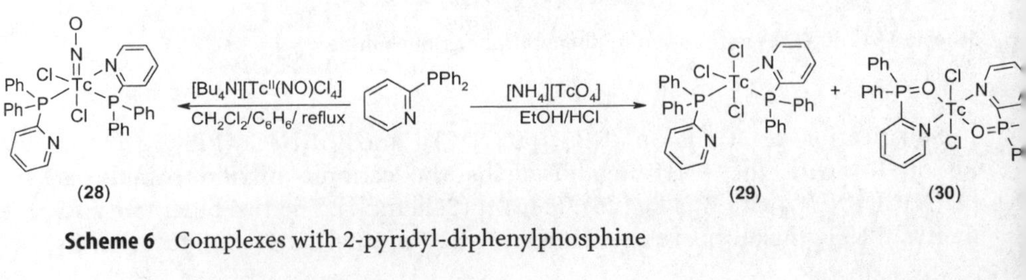

Scheme 6 Complexes with 2-pyridyl-diphenylphosphine

the conjugate HYNIC-biomolecule can only occupy one or two coordination positions, other coligands are needed for preparing stable complexes [42]. The use of the ternary ligand system, HYNIC–biomolecule, tricine, and triphenylphosphine-3,3′,3″-trisulfonate (TPPTS) in the labeling of biologically active substrates has been successfully achieved, with improvement of stability and reduction of the number of isomeric forms [43]. However, a large excess of TPPTS is required, in combination with high-temperature heating, which can be problematic in the labeling of substrates containing disulfide bridges. Trying to overcome this problem, Purohit et al. [44] introduced two novel phosphine-containing HYNIC chelators, HYNIC-Kp-DPPB and HYNIC-Ko-DPPB (Scheme 7). These ligands, with tricine as a coligand, have been proposed to improve stability, reducing most of the drawbacks previously mentioned, and probably, in future, could help in defining unambiguously the real coordination mode of HYNIC.

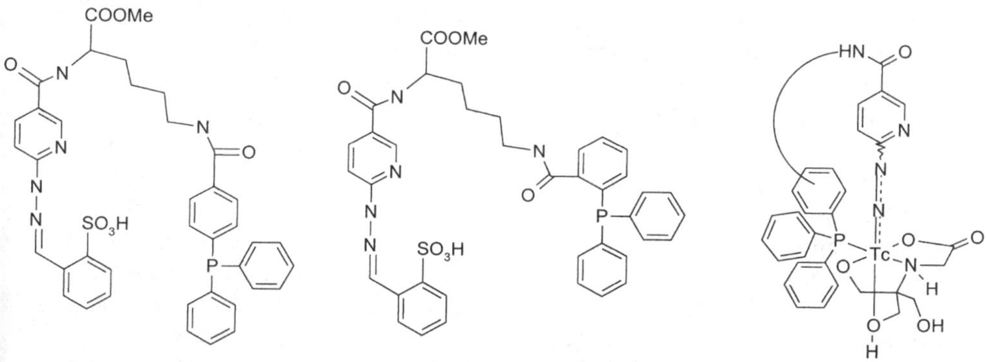

HYNIC-Kp-DPPB **HYNIC-Ko-DPPB**

Scheme 7 Phosphine-containing 6-hydrazinonicotinamide (*HYNIC*) chelators

2.1.1.3 Mixed P, S Ligands

A novel class of trigonal bipyramidal Tc(V) nitrido complexes of the type $[Tc(N)(\kappa^2\text{-PS})_2]$ (**31–33**) were isolated by reacting the precursors $[Tc(N)Cl_2(PPh_3)_2]$ or $[Tc(N)Cl_4]^-$ with bidentate mixed P, S ligands (Scheme 8). Depending on the experimental conditions, the reactions with the starting material $[Tc(N)Cl_4]^-$ could be accompanied by competitive reduction of the metal and abstraction of the Tc≡N core, giving complexes of the type $[Tc(\kappa^2\text{-PS})_2(\kappa^1\text{-SP})]$ or $[Tc(\kappa^2\text{-PS})_2(\kappa^1\text{-SP(O)})]$ [45].

Complexes of the $[Tc(N)(\kappa^2\text{-PS})_2]$ type were also synthesized at noncarrier added level and their biological behavior was studied [46]. These compounds exhibited high initial heart uptake, and elimination through liver and kidneys. The washout kinetics from heart was dependent on the nature of the lateral *R* groups on the phosphine thiol ligands. Extraction of the activity from myo-

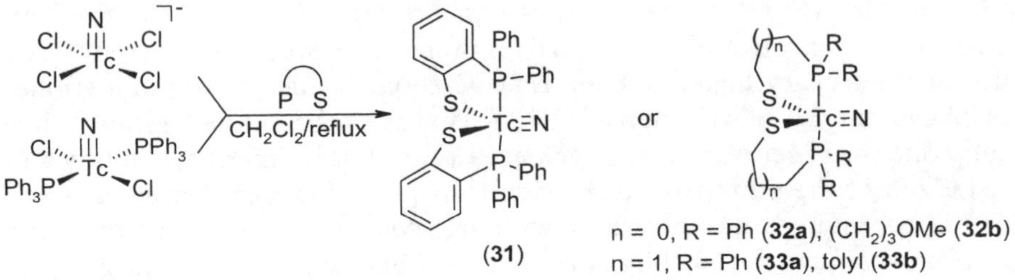

n = 0, R = Ph (**32a**), $(CH_2)_3OMe$ (**32b**)
n = 1, R = Ph (**33a**), tolyl (**33b**)

Scheme 8 Technetium nitrido complexes with bidentate mixed P,S ligands

cardium tissue indicates that no change of the complexes took place, but metabolization to more hydrophilic species was observed in liver and kidneys [46]. The studies performed on complexes **31–33** along with the observation that the PS ligands easily replace the dithiocarbamates in $[Tc(N)(S_2R)_2]$ with a predictable change in the coordination geometry, highlighted that an appropriate mixture of π-donor/π-acceptor coordinating atoms would stabilize the less usual trigonal bipyramidal coordination geometry. These chemical considerations motivated the synthesis of mixed-ligand complexes of the type [Tc(N)(P–*X*–P)(*X*–*Y*)], where P–*X*–P are neutral diphosphines which contain a central heteroatom (*X*), namely $R_2P(CH_2)_2$–O–$(CH_2)_2PR_2$ and $R_2P(CH_2)_2$–$N(CH_2CH_2OCH_3)$–$(CH_2)_2PR_2$ [47]. This elegant and systematic study, supported by theoretical calculations, allowed the introduction of the novel "$[M(N)(P\text{–}N\text{–}P)]^{2+}$" (*M* is Re, Tc) metal fragment, which exhibits selective reactivity toward nucleophilic bidentate ligands (*X*–*Y*) having π donors as coordinating atoms (Scheme 9) [47]. The uni- or dinegative bidentate coligands are easily functionalized with different biomolecules. This approach has been applied to the labeling of small peptides, $5HT_{1A}$ antagonists and benzodiazepine receptor ligands with ^{99m}Tc [48–50].

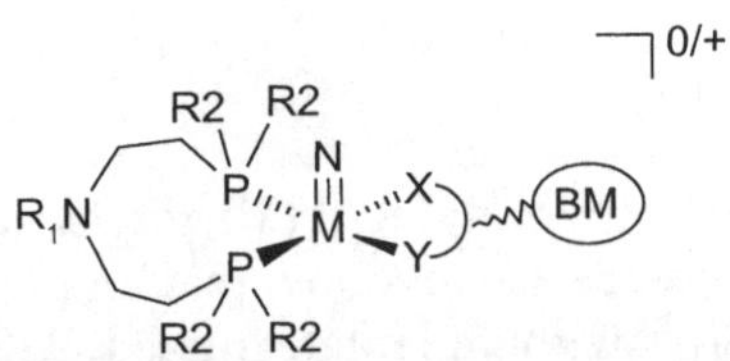

Scheme 9 Metal fragment "$[M(N)(P\text{–}N\text{–}P)]^{2+}$" (*M* is Re, Tc) for labeling biomolecules

Recently, a family of monocationic complexes of the type $[M(N)(P\text{–}N\text{–}P)$-$(S_2CNR_1R_2)]^+$, which exhibit high myocardial uptake in rats, was also described [51, 52]. Some of these complexes possess superior biological properties, compared with Sestamibi and Tetrafosmin. However, because of the possibility of species-dependent effects, further studies of these agents in other animal models are underway, in order to assess their utility as heart perfusion agents [52].

2.1.2
Tridentate Phosphines

2.1.2.1
Mixed P, O, O Ligands

The potentially tridentate and dianionic phosphines bis(*o*-hydroxyphenyl)-phenylphosphine (H_2PO_2) and bis(2-hydroxy-5-methylphenyl)phenylphosphine [$H_2(Me_2PO_2)$] allowed the synthesis of [$MO(\kappa^3\text{-}L)(\kappa^2\text{-}LH)$] {*M* is Tc, *L* is PO_2 (**34**); *M* is Re, *L* is PO_2 (**35**), Me_2PO_2 (**36**)}, and [$ReOCl(PPh_3)(\kappa^3\text{-}PO_2)$] (**37**), either by methathesis reactions with a Re(V) precursor or by reduction/substitution reactions with $[MO_4]^-$ salts (*M* is Re, Tc) (Scheme 10) [22, 53]. The disubstituted complexes **34** and **35** exist as two noninterconvertible diastereomers, as evidenced by $^{31}P\{^1H\}$ NMR, owing to the up or down orientation of the free phenol group relative to the $[Re{=}O]^{3+}$ core [22]. The mixed-ligand complex [$ReO(\kappa^2\text{-}P_1O)(\kappa^3\text{-}PO_2)$] (**38**), obtained by ligand exchange of **37** with P_1OH, was also reported [22]. Compound **38** could not be prepared directly from [$ReOCl_3(PPh_3)_2$] via ligand-exchange reactions, as observed for **34** and **35**.

The characterization of **34–38** demonstrated that these compounds are neutral and octahedral, containing an oxo moiety and tridentate $[Me_2PO_2]^{2-}$ or $[PO_2]^{2-}$ ligands coordinating facially [22, 53].

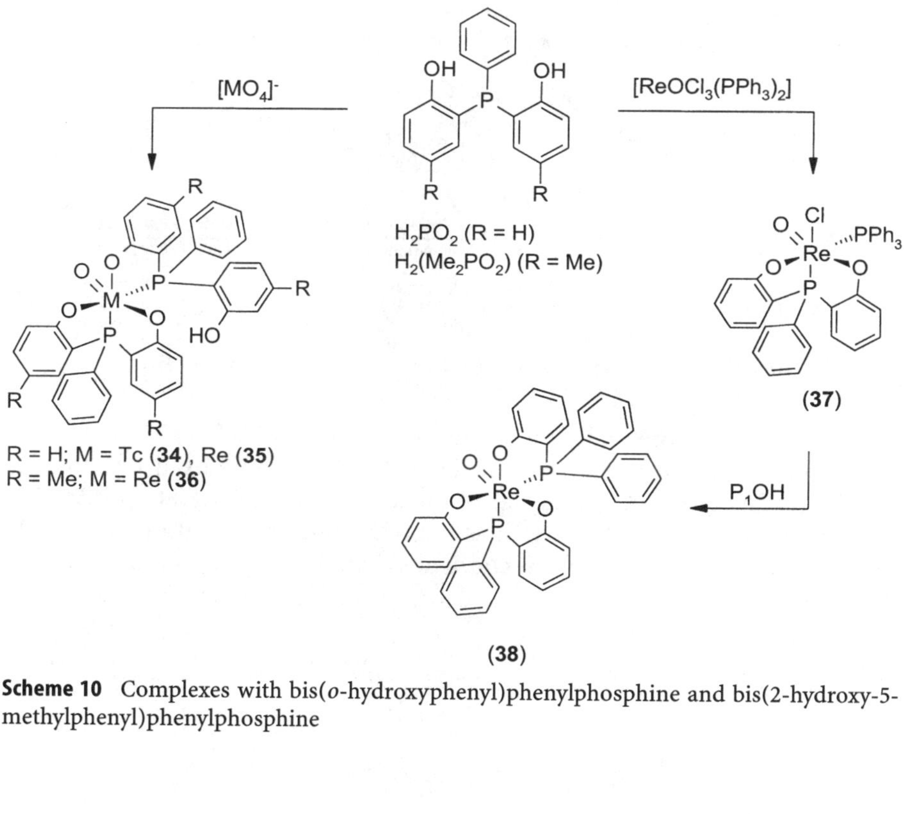

Scheme 10 Complexes with bis(*o*-hydroxyphenyl)phenylphosphine and bis(2-hydroxy-5-methylphenyl)phenylphosphine

Reactions of H_2PO_2 with the isoelectronic Re(V) phenylimido [Re(NPh)-$Cl_3(PPh_3)_2$] afforded the phenylimido complex [Re(NPh)Cl(PPh_3)(κ^3-PO_2)] (**39**), which, depending on the polarity of the reaction media, was isolated in two geometrical isomeric forms [*cis*- or *trans*-(P,P)] [22]. The structural analysis of the *cis*-[Re(NPh)Cl(PPh_3)(κ^3-PO_2)] isomer has shown a facial coordination for the κ^3-PO_2, due to the sterical unavailability of the meridional positions. Other examples of rhenium compounds containing multiple metal-nitrogen bonds and anchored by this type of ligand are *trans*-(P,P)-[Re(NPh)(κ^2-PO)(κ^3-PO_2)] (**40**), [Re(Hhypy)(hypy)(HPO_2)(H_2PO_2)]Cl (**41**) and [Re(Hhypy)(hypy){H(Me_2PO_2){H_2(Me_2PO_2)]Cl (**42**). Complex **40** was obtained from the *trans* isomer of **39**, while the synthesis of **41** and **42** involved reactions of [Re(Hhypy)(hypyH)Cl_3] with H_2PO_2 or H_2(Me_2PO_2) [22, 53].

Complexation reactions of silylated aryloxo proligands with [ReO(OEt)-$Cl_2(PPh_3)_2$] proceed by displacement of one PPh_3 and the subsequent stepwise replacement of the OEt and/or Cl ligands, followed by elimination of Me_3SiOEt or Me_3SiCl (Scheme 11) [54]. The resulting complexes, **43–46**, were fully characterized and the X-ray diffraction analysis of **46** confirmed the κ^3-(P,O,O) coordination of the phosphine ligand, as well as the retention of the trimethylsilyl of the noncoordinated phenyl ring [54].

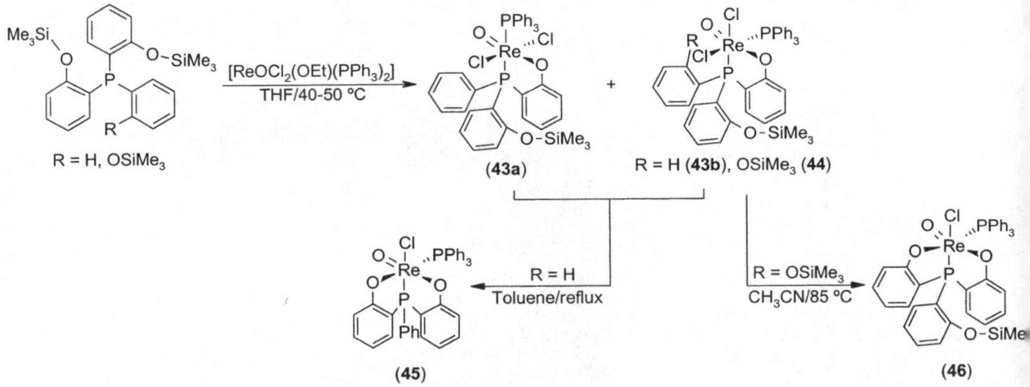

Scheme 11 Reactions of rhenium complexes with silylated phosphine proligands

The softer and potentially tridentate phosphino thiolate PPh(2-C_6H_4SH)$_2$ was used to prepare the mixed-ligand Re(V) complex [ReO{PPh(2-C_6H_6S)$_2$}(S_2CNEt_2)] (**47**) by cleavage of the dimer [$Re_2O_3(S_2CNEt_2)_4$] [55]. The monomeric structure of **47**, as well as the tridentate coordination of the heterofunctionalized phosphine was confirmed by X-ray crystallographic analysis [55].

2.1.2.2
Mixed P, N, *X* (*X* is O, N, S) Ligands

Aiming their application at the design of stable mixed-ligand complexes of the 3+1 type, Correia and coworkers [56–58] have introduced the new potentially tridentate phosphines [(1*R*,2*R*)-*N*-(2-aminocyclohexyl)]-2-(diphenylphosphanyl)benzamide ($HCyPN_2$), *N*-(2-aminoethyl)-2-(diphenylphosphanyl)benzamide (HPN_2), 2-(diphenylphosphanyl)-*N*-(2-hydroxyethyl)benzamide (H_2PNO) and 2-(diphenylphosphanyl)-*N*-(2-thioethyl)-benzamide (H_2PNS). Using different starting materials and reaction conditions, the coordination chemistry of these heterofunctionalized phosphines was extensively explored (Scheme 12).

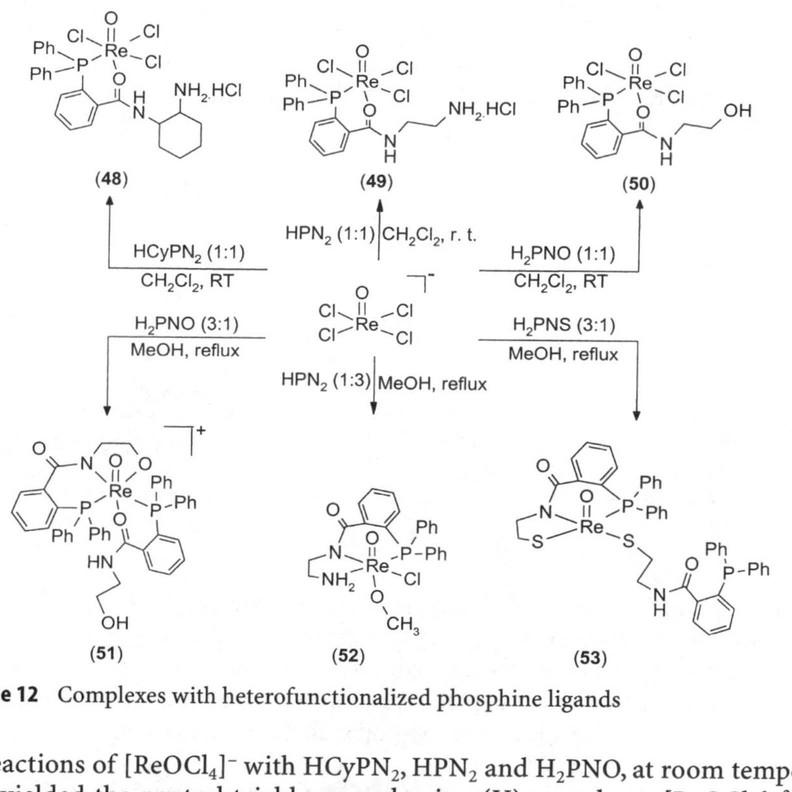

Scheme 12 Complexes with heterofunctionalized phosphine ligands

Reactions of $[ReOCl_4]^-$ with $HCyPN_2$, HPN_2 and H_2PNO, at room temperature, yielded the neutral trichlorooxorhenium(V) complexes $[ReOCl_3\{\kappa^2\text{-}L\}]$ [*L* is $HCyPN_2$ (**48**), HPN_2 (**49**), H_2PNO (**50**)] [56]. By contrast, the similar adduct with the H_2PNS ligand could not be isolated, most probably owing to the higher acidity of the terminal thiol function. By increasing the temperature or by using excess ligand, the deprotonation of the heterofunctionalized phosphines occurs and the reactions with $[Bu_4N][ReOCl_4]$ afforded the monooxorhenium(V) complexes **51–53**, which display coordination environments strongly dependent on the incoming phosphine (Scheme 12) [57, 58]. κ^3- or κ^1-Coordination capabilities were observed for the PNS ligand, in contrast with the HPN_2 or H_2PNO ligands that behave as tridendate or bidentate ligands. The ligand-exchange reaction between H_2PNO and $[ReOCl_3(PPh_3)_2]$ in the presence of NaOAc was also studied. This reaction led to the six-coordinate oxorhenium(V) complex $[ReO(\kappa^3\text{-PNO})(\kappa^2\text{-DPPBA})]$ (**54**), owing to partial hydrolysis of the H_2PNO ligand [57].

The characterization of compounds **48–54** in the solid state or in solution by multinuclear NMR spectroscopy confirmed the coordination mode of the phosphines. In complexes **51–54** the tridentate heterofunctionalized phosphines display a meridional coordination mode, with the remaining coordination sites being occupied by the oxo group and by other bidentate or monodentate coligands [56–58].

The rich and diversified rhenium chemistry developed with the PN*X* (*X* is O, S) phosphines motivated the authors to explore their use in the so-called 3+1 approach [58, 59]. Different model complexes, **55–58**, were synthesized by reacting $[NBu_4][ReOCl_4]$ with H_2PNO or with H_2PNS in the presence of several monodentate thiols as coligands (Scheme 13) [58].

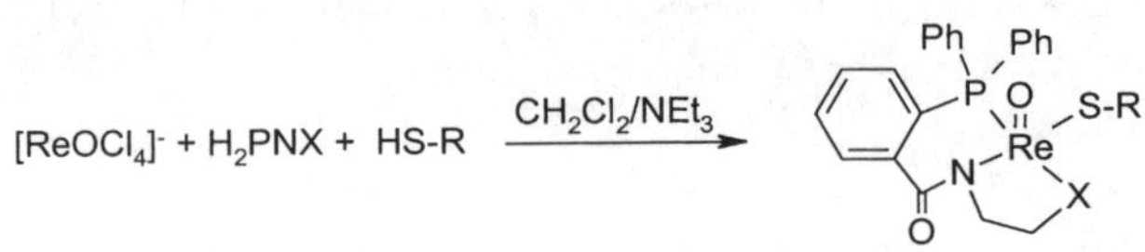

X = O, R = Ph (**55**)

X = S, R = Ph(**56**)
R = *p*-NH$_2$Ph (**57**)
R = CH$_2$CH$_2$COOH (**58**)

Scheme 13 Complexes of the "3+1" type anchored by tridentate heterofunctionalized phosphine ligands

The characterization of complexes **55–58** involved IR, ^{1}H NMR and ^{31}P NMR spectroscopy and X-ray crystallographic analysis in the case of **55** and **56** [58]. The complexes are five-coordinate, adopting a distorted square-pyramidal geometry. The oxo group occupies the axial position and the equatorial plane is defined by the tridentate dianionic chelators and by the sulfur atom of the monothiolate.

An important point in the synthesis of the mixed-ligand complexes, **56–58**, was the $[ReOCl_4]^-$-to-H_2PNS-to-S-*R* molar ratio, which has to be maintained at 1:1:1 to avoid the formation of the previously mentioned complex **53**. This was also an important point when these complexes were prepared at noncarrier added level. The careful optimization of the concentration of the ligands allowed the one-pot synthesis of $[^{99m}TcO(\kappa^3\text{-PN}X)(\kappa^1\text{-SPh})]$ (*X* is O, **55a**; *X* is S, **56a**) in high yield, starting from $[^{99m}TcO_4]^-$ and using stannous chloride as the reducing agent [60]. Stability studies in phosphate-buffered saline (PBS) and in serum, as well as challenge experiments with glutathione, revealed that [^{99m}TcO(κ^3-PNS)(κ^1-SPh)] (**56a**) was the stablest. Biodistribution studies in mice of **56a** have also shown that this complex is able to cross moderately the blood brain barrier (percentage injected dose, ID, per gram of 0.14% at 5 min post injection), with a slow washout (percentage ID per gram of 0.11% at 120 min post injection) [60]. These findings prompted the synthesis of several rhenium and ^{99m}Tc complexes, **59–62**, with monodentate thiol coligands derivatized with a 5-HT$_{1A}$ receptor-binding moiety, 1-(2-methoxyphenyl)piperazine (Scheme 14) [60, 61].

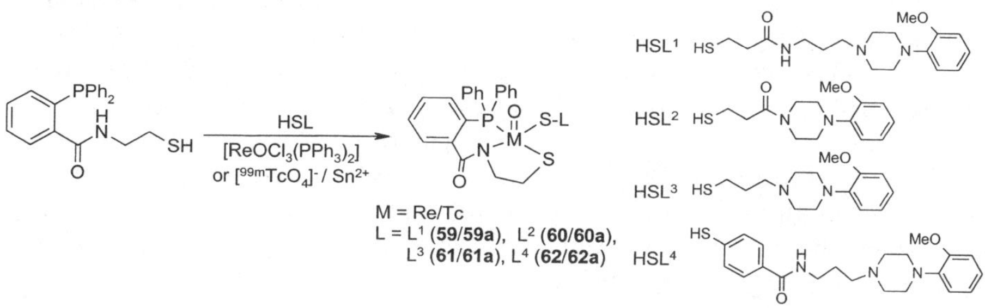

Scheme 14 Complexes of the 3+1 type bearing a monodentate 5-HT$_{1A}$ receptor-binding ligand

Complexes **59–62** have been synthesized by reaction of the precursor [ReOCl$_3$(PPh$_3$)$_2$] with H$_2$PNS and HS*L*, at the molar ratio 1:1:1, in refluxing methanol, using sodium acetate as the base. The analogous ^{99m}Tc complexes (**59a–62a**) were prepared in a one-pot procedure by direct reduction of [^{99m}TcO$_4$]$^-$ with stannous chloride in the presence of the tridentate and monodentate ligands, at 40–50 °C. Labelling yields of 80–95% have been achieved and, after HPLC purification, all the complexes were obtained with a radiochemical purity higher than 95% [60, 61].

In order to evaluate the effect of the chemical modifications introduced in the monodentate receptor-binding ligand, the binding affinity of the rhenium complexes for the 5-HT$_{1A}$ and 5-HT$_{2A}$ receptors was determined. The affinity for 5-HT$_{1A}$ receptors spanned between 2.35 and 1,570 nM. Complex **61** appeared as the one with the highest affinity (IC$_{50}$=2.35 nM) and best selectivity towards the 5-HT$_{1A}$ receptor. Despite the good in vitro and in vivo stabilities and the adequate lipophilicity (log *P*=1.07–2.01) of complexes **59a–62a**, none of them seem capable of crossing the blood brain barrier, as demonstrated by biodistribution studies in mice [60, 61].

Considering that silylation should improve the bioavailability of the complexes [62], especially in terms of passing the blood brain barrier with consequent increase of brain uptake, silylated mixed-ligand rhenium and technetium complexes of the type 3+1 were prepared. Moreover, the accumulation in the target organ would be enhanced if the silyl ethers remain intact in physiological conditions but hydrolysable in the target organ. The first attempts for preparing these types of complexes were performed with the tridentate 3-thiapentane-1,5-dithiol ligand (HSSSH); however, these mixed-ligand complexes were unstable towards trans-chelation with glutathione [63, 64]. The high stability evidenced by the Re/^{99m}Tc mixed-ligand complexes with the H_2PNS chelate prompted us to prepare silylated mixed-ligand complexes in which the tridentate HSSSH ligand was replaced by the heterofunctionalized phosphine H_2PNS.

The new oxorhenium compounds, **63–65**, were prepared by reacting $[NBu_4][ReOCl_4]$ with H_2PNS, and with different hydroxyl silylated monodentate thiols (Scheme 15) [65, 66]. These complexes were fully characterized and used as surrogates of the corresponding $[^{99m}TcO(\kappa^3\text{-PNS})(S(CH_2)_nOSiR_3)]$ compounds (**63a–65a**), obtained by one-pot synthesis, in 60–90% yields (Scheme 15) [65–67]. Considering that eventual radiochemical impurities in the preparation of complexes **63a–65a** could be the hydrolyzed species $[^{99m}TcO(\kappa^3\text{-PNS})(S(CH_2)_nOH)]$ [n=2 (**66a**) n=3 (**67a**)], these complexes and the respective rhenium surrogates were also prepared [67]. All the ^{99m}Tc-silylated complexes, **63a–67a**, were obtained with high radiochemical purities (above 95%), after HPLC purification [65, 67].

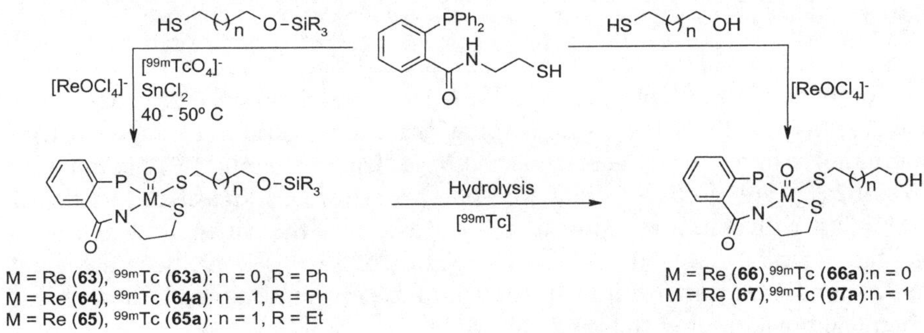

Scheme 15 Silylated mixed-ligand rhenium and technetium complexes of the 3+1 type

The silylated compounds **63a–65a** are stable in pH 7.4 PBS, rat plasma, human serum and whole blood, and do not bind to plasmatic proteins. No exchange with glutathione was observed, even in the presence of concentrated solutions (1 and 10 mM in pH 7.4 PBS, 120 min at 37 °C). There is a marked influence of the silicium substituents on the pH at which the cleavage of the O–Si bond occurs. Under physiological conditions (pH 7.4, 37 °C), rapid cleavage of the O–Si bond was observed for **65a**. However, complexes **63a** and **64a**, with a triphenyl substituent, are stable under the same conditions, requiring a much lower pH to be hydrolyzed [65, 67].

Biodistribution studies in mice demonstrated that the silylated ^{99m}Tc complexes **63a–65a** are stable in vivo and no significant hydrolysis was observed. The complexes are eliminated essentially through the hepatobiliary tract with low urinary elimination. The silylation increased significantly the lipophilicity of the compounds, but no effect was found in their brain uptake or brain retention [67]. In spite of the unsatisfactory brain uptake, compounds **63a–65a**, anchored by the tridentate PNS ligand, are quite stable in vitro and in vivo, contrasting with all the previously described mixed-ligand 3+1 *M*(V) monoxo-complexes. These findings demonstrate the great importance of the tridentate ligand on the in vitro and in vivo stabilities of the 3+1 compounds. The stability promoted by the heterofunctionalized phosphine H_2PNS encourages further studies to improve blood brain barrier penetration, namely by reducing the bulkiness of the PNS framework.

2.1.3
Tetradentate Phosphines

2.1.3.1
Mixed P, N_2, *X* (*X* is P, O, S) Ligands

The potentially tetradentate *N,N'*-bis[2-(diphenylphosphino)phenyl]propane-1,3-diamine ligand ($H_2P_2N_2$) reacts with $[MO_4]^-$ (*M* is Tc, Re) or with $[ReOCl_4]^-$ yielding neutral six-coordinate *M*(V) oxo complexes of the type $[MO(P_2N_2)X]$ [*M* is Tc (**68**), Re (**69**)] (Scheme 16) [68].

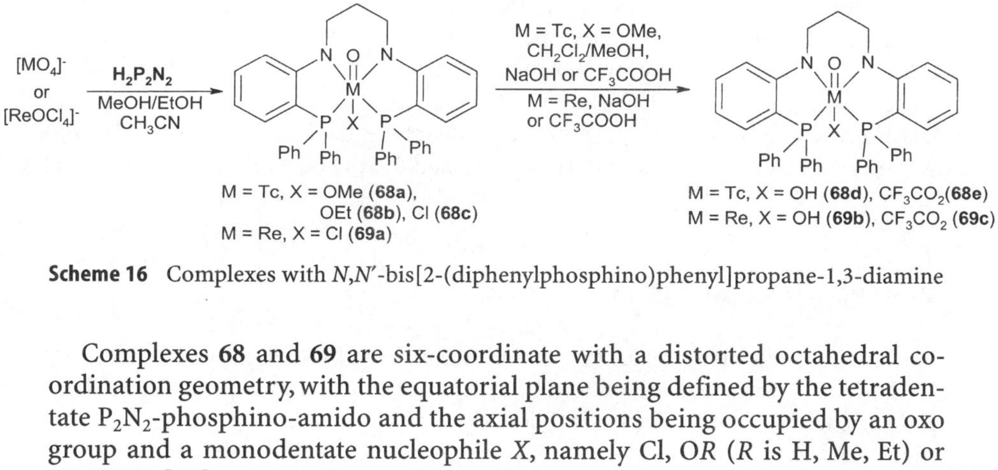

Scheme 16 Complexes with *N,N'*-bis[2-(diphenylphosphino)phenyl]propane-1,3-diamine

Complexes **68** and **69** are six-coordinate with a distorted octahedral coordination geometry, with the equatorial plane being defined by the tetradentate P_2N_2-phosphino-amido and the axial positions being occupied by an oxo group and a monodentate nucleophile *X*, namely Cl, O*R* (*R* is H, Me, Et) or CF_3COO^- [68].

The *N*-{*N*-[3-(Diphenylphosphino)propionyl]glycyl}-L-*S*-benzylcysteine, H_2PN_2S-Bz, and its methyl ester derivative have been used for preparing the oxorhenium(V) complexes **70** and **71**, using $[ReOCl_3(PPh_3)_2]$ or $[ReOCl_2(OEt)(PPh_3)_2]$ as precursors (Scheme 17) [69, 70]. Complexes **70** and **71** are six-coordinate, displaying a distorted octahedral geometry. The equatorial plane is defined by the phosphorus, the two deprotonated nitrogen atoms of the amide groups and by an oxygen or sulfur atom of the derivatized peptide, while

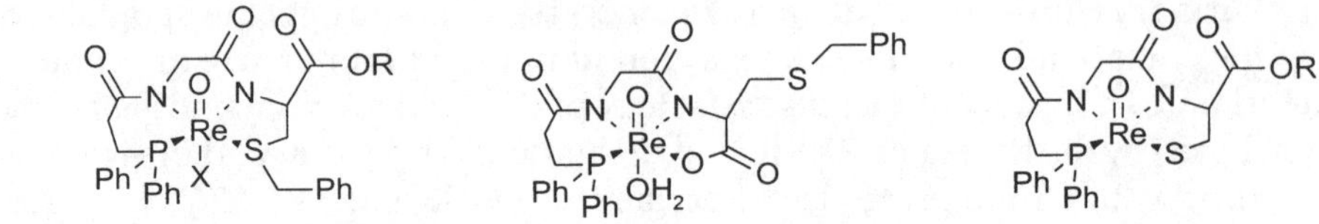

R = Me, X = Cl (**70a**), OH (**70b**)
R = H, X = Cl (**70c**)

71

X = H (**72a**), Me (**72b**)

Scheme 17 Rhenium complexes with phosphorus-containing tripeptides

the axial positions are occupied by an oxo group and by a halogen, a hydroxy or a water molecule [69, 70]. Another phosphorus-containing peptide explored was *N*-[*N*-(3-diphenylphosphinopropionyl)glycyl]-*S*-cysteine (H_3PN_2S) and the corresponding methyl ester. The ligands, protected with an S-trityl group, react with $[ReOCl_3(PPh_3)_2]$ or $Ph_4P[ReOCl_4]$, yielding complexes **72a** and **72b**. Complex **72b** was isolated as two diastereomers, which were separated by reverse-phase HPLC and characterized by NMR, FT-IR and UV–vis spectroscopy and electrospray ionization mass spectrometry (ESI-MS). X-ray analysis was also undertaken and showed a distorted square-pyramidal coordination geometry for the syn isomer of **72b** [71, 72].

The ligand H_2PN_2S-Bz has been coupled to a model peptide (tetragastrin, cholecystokinin fragment) by active ester chemistry, either before or after radiolabelling with ^{99m}Tc. Despite the formation of various isomers, the authors claimed that coordination to the metal involves the PN_2S donor set of the ligands [73]. The H_2PN_2S ligand has also been coupled to the N end of a novel CCK8 derivative, where CCK is cholecystokinin, and the corresponding oxorhenium(V) complex was prepared and characterized by ESI-MS, FT-IR and UV–vis spectroscopy. The structure proposed for the complex is similar to the one reported for **72b** [74].

2.1.3.2
Mixed P_2, X_2 (*X* is S, N) Ligands

Novel water-soluble tetradentate dithiobis(hydroxymethyl)phosphine ligands (P_2S_2) have been synthesized and characterized by the group of Katti. These P_2S_2 phosphines react with $[ReO_2(C_5H_5N)_4]Cl$, in refluxing water, yielding mononuclear or dinuclear Re(V) cationic *trans*-dioxocomplexes (**73, 74**), as established by X-ray crystallography (Scheme 18) [75].

At noncarrier added level, stable mononuclear complexes, $[^{99m}TcO_2(P_2S_2)]^+$, analogous to **74**, were synthesized with high radiochemical purity (above 95%). These compounds, which are stable in vitro over a wide range of pH, were obtained by simply mixing $[^{99m}TcO_4]^-$ with the ligands or by trans-chelation from ^{99m}Tc citrate [76]. The high in vitro and in vivo stability of $[^{99m}TcO_2(P_2S_2)]^+$ prompted the group of Katti to develop a bifunctional chelating agent (P_2S_2-COOH) based on the P_2S_2 framework. Using low concentrations (10^{-5} M) of the

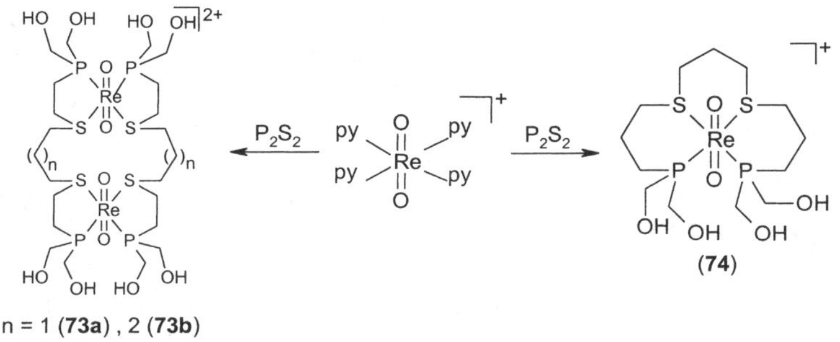

Scheme 18 Complexes with dithiobis(hydroxymethyl)phosphines

P_2S_2-COOH ligand, the complex $[^{99m}TcO_2(P_2S_2\text{-COOH})]^+$ (**75a**) was prepared by direct reduction of $[^{99m}TcO_4]^-$ with Sn^{2+}, or by trans-chelation using ^{99m}Tc gluconate (Scheme 19). This water-soluble complex (**75a**) exhibited fast clearance from the bloodstream and its excretion was through the renal and the hepatobiliary pathways, as revealed by biodistribution studies in mice [77].

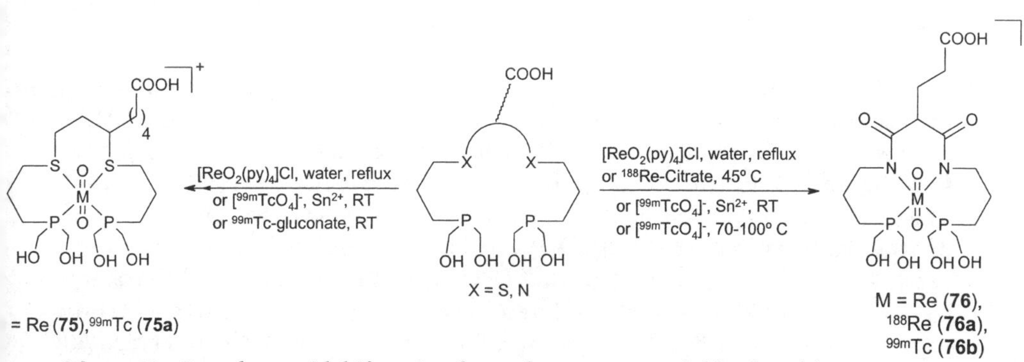

Scheme 19 Complexes with bifunctional tetradentate water-soluble phosphines

With the aim of the preparation of ^{99m}Tc/^{188}Re complexes with improved hydrophilicity relative to **75a**, the ligand 4,4-bis[bis-hydroxymethylphosphanylpropylcarbamoyl)butyric acid (P_2N_2-COOH) was introduced. The coordination properties of P_2N_2-COOH were evaluated with rhenium and technetium, and the complexes formed, **76a** and **76b**, were analogous to **75** and **75a** but negatively charged (Scheme 19) [78, 79].

The in vitro stability of **76a** in aqueous solutions, over a wide pH range, was excellent. In vivo biodistribution studies in normal mice showed that **76a** and **76b** clear almost exclusively via the renal-urinary excretion, and to a lower extent via the hepatobiliary route, reflecting the greater hydrophilic character of the ^{99m}Tc-/^{188}Re-P_2N_2-COOH complexes (**76a, 76b**) if compared with **75a** [77, 79].

Complexes **75** and **75a** were coupled to the N-terminal region of the truncated nine amino acid bombesin (BBN) analogue 5-Ava–Gln–Trp–Ala–Val–Gly–His–Leu–Met–NH_2 [$BBN^{(7-14)}$] by preactivation of the -COOH group with pentafluorophenol. The Re–P_2S_2–$BBN^{(7-14)}$ conjugate retained its bioespecificity (IC_{50}, 0.8±0.4 nM). The biodistribution in normal mice revealed a fast and efficient clearance from the blood pool and a significant uptake in the pancreas, which expresses BBN/gastrin-releasing peptide (GRP) receptors [80].

In order to use the postlabeling approach, a novel and elegant synthetic strategy has been outlined by Gali et al. [81], which allows the efficient synthesis of P_2S_2–peptide conjugates. This strategy enabled the labeling of glycine glycine ethyl ester 5-Ava-$BBN^{(7-14)}NH_2$ and D-Lys^6–luteinizing hormone-releasing hormone (LHRH) with ^{99m}Tc and ^{188}Re. The $^{99m}TcO_2$–P_2S_2–GlyGlyOEt and $^{99m}TcO_2$–P_2S_2–D-Lys^6-LHRH complexes were prepared in yields higher than 95% and 88%, respectively. The identity of the former was confirmed by HPLC comparison with the cold rhenium analogue (ReO_2–P_2S_2–GlyGlyOEt). The ^{188}Re-labelled BBN analogue, obtained with a radiochemical purity of 90% or greater, was stable in vitro and in vivo [82].

2.2 Low-Valent Carbonyl Complexes

The chemistry of the precursor *fac*-$[ReBr_3(CO)_3]^{2-}$ [13, 14] was studied with the water-soluble monophosphine 1,3,5-triaza-7-phosphaadamantane (PTA), and with the bisphosphines 1,2-bis(bis(hydroxymethyl)phosphino)ethane (HMPE) and 1,2-bis(bis(hydroxymethyl)phosphino)benzene (HMPB) [83, 84]. The cyclic monophosphine PTA replaces in a stepwise way one or more of the halides in $[ReBr_3(CO)_3]^{2-}$, yielding $[ReBr_2(CO)_3(PTA)]^-$ (**77**), $[ReBr(CO)_3(PTA)_2]$ (**78**) and $[Re(CO)_3(PTA)_3]^+$ (**79**), which were all characterized by X-ray diffraction analysis. The follow-up of these reactions by ^{31}P NMR techniques revealed that the nature and the concentration of the intermediate species depend very much on the solvent system [83]. The *fac*-$[ReBr(CO)_3L]$ [L is HMPE (**80**), HMPB (**81**)] complexes have also been prepared in water, in good yields, by reaction of the diphosphines with *fac*-$[ReBr_3(CO)_3]^{2-}$. These complexes were characterized by ^1H and ^{31}P NMR spectroscopy, and also by single-crystal X-ray crystallography in the case of **81**. The related cations *fac*-$[Re(OH_2)(CO)_3L]^+$ [L is HMPE (**82**), HMPB (**83**)] were also identified as intermediates in the formation of **80** and **81** [84]. The precursor *fac*-$[^{99m}Tc(OH_2)_3(CO)_3]^+$ reacts with HMPE and HMPB yielding initially the *fac*-$[^{99m}TcCl(CO)_3L]$ [L is HMPE (**80a**), HMPB (**81a**)] complexes, with subsequent lost of the coordinate halide to produce the cationic species *fac*-$[^{99m}Tc(OH_2)(CO)_3L]^+$ [L is HMPE (**82a**), HMPB (**83a**)] (Scheme 20). Complexes **82a** and **83a** are very stable in vitro and experiments performed in normal mice demonstrated a fast clearance from the blood pool and clearance through the hepatobiliary and the urinary pathways [85].

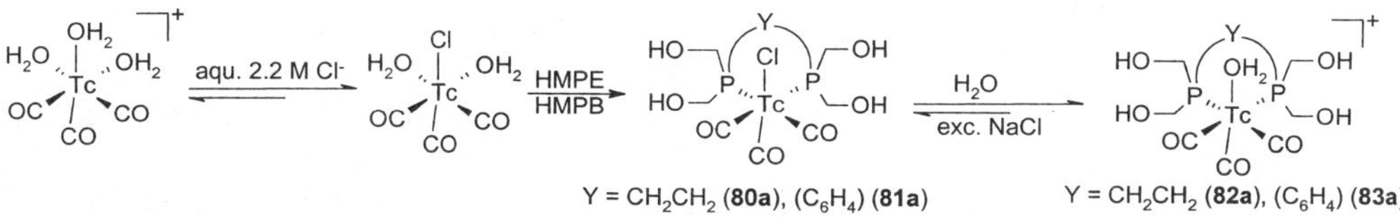

Scheme 20 Tricarbonyl complexes with water-soluble bis-phosphines

The monodentate tris(hydroxymethyl)phosphine (THP) reacts with $[^{99m}Tc(CO)_3(HMPE)(OH_2)]^+$ (**82a**), yielding, albeit slowly, *fac*-$[^{99m}Tc(CO)_3(HMPE)(THP)]^+$ [86]. The high kinetic inertness of the tricarbonyl complexes containing the unidentate THP motivated the use of this phosphine as a coligand in the labeling of a biologically active peptide [87]. Smith et al. have prepared and characterized 2,3-diaminopropionic (Dpr)–BBN conjugates of the type Dpr–Ser–Ser–Ser–Gln–Trp–Ala–Val–Gly–His–Leu–Met–(NH_2), in which the terminal aliphatic diamine behaves as a bidentate ligand toward the organometallic moiety *fac*-$[M(CO)_3]^+$. A metallated BBN conjugate of the type $[Re(OH_2)(CO)_3(\kappa^2$-Dpr–Ser–Ser–Ser–Gln–Trp–Ala–Val–Gly–His–Leu–Met–$(NH_2)]^+$ (**84**), which exhibits high binding affinity to human prostate cancerous PC-3 cells, was prepared and fully characterized. The analogous $[^{99m}Tc(X)(CO)_3(\kappa^2$-Dpr–Ser–Ser–Ser–Gln–Trp–Ala–Val–Gly–His–Leu–Met–$(NH_2)]^+$ [*X* is OH_2 (**84a**), THP (**85a**)] conjugates were also prepared (Scheme 21). These complexes retain high in vitro and in vivo specificity for the

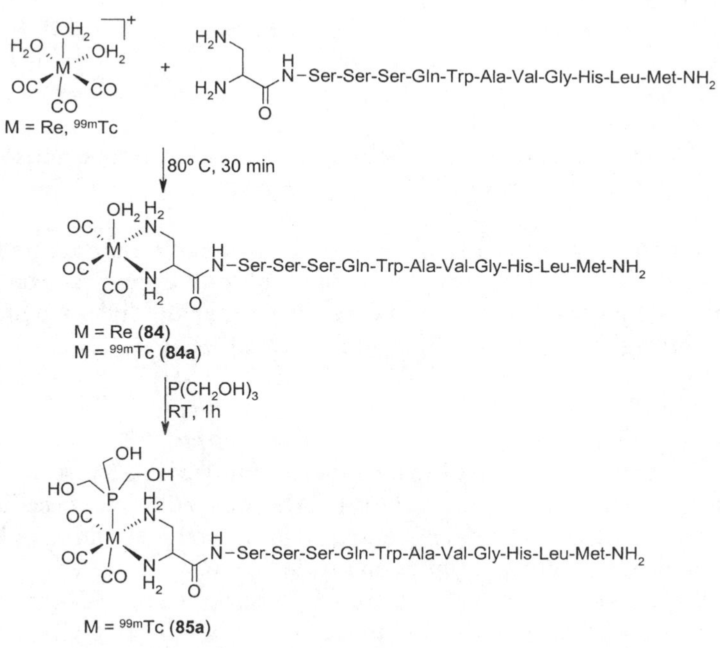

Scheme 21 Tricarbonyl complexes with 2,3-diaminopropionic–bombesin conjugates

targeting of GRPr-expressing cells. The use of the water-soluble phosphine THP, as a coligand, increased noticeably the hydrophilicity of the conjugate, providing an alternative method for tuning the in vivo pharmacokinetics of radiolabeled conjugates [87].

Reactions of $(NEt_4)_2[ReBr_3(CO)_3]$ were also studied with the potentially tri(bi)dentate phosphines HPN_2, H_2PNO, 2-(diphenylphosphanyl)-*N*-(2-hydroxyethyl)benzylamine (H_2CH_2PNO), and *N*-(2-bromo)-2-(diphenylphosphanyl)benzamide ligand (HPNBr). Several neutral complexes, namely $[Re(CO)_3(\kappa^3\text{-}PN_2)]$ (**86**), $[Re(CO)_3(\kappa^2\text{-}H_2PNO)Br]$ (**87**), $[Re(CO)_3(\kappa^2\text{-}H_2CH_2\text{-}PNO)Br]$ (**88**) and $[Re(CO)_3(\kappa^2\text{-}PPh_3oxaz)Br]$ (**89**), [PPh_3oxaz is 2-(2-diphenylphosphinophenyl)oxazoline], were isolated and characterized by ^1H and ^{31}P NMR spectroscopy and by X-ray crystallographic analysis (Scheme 22) [88, 89].

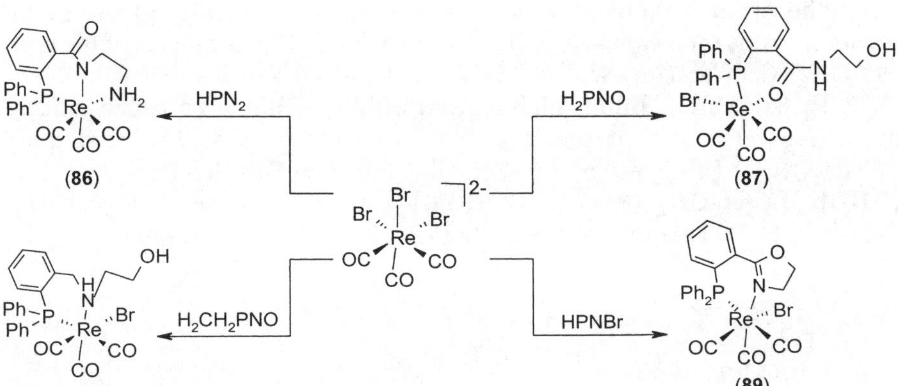

Scheme 22 Tricarbonyl complexes anchored by heterofunctionalized phosphines

These complexes contain an unusual combination of donor atoms for the organometallic *fac*-$[M(CO)_3]^+$ moiety and complex **89** is the first example of a Re(I) tricarbonyl anchored by a phosphorus–oxazoline ligand, which was generated during the course of complex formation (Scheme 22) [88, 89]. The analogous $[^{99m}Tc(CO)_3(\kappa^3\text{-}PN_2)]$ (**86a**), $[^{99m}Tc(CO)_3(\kappa^2\text{-}H_2PNO)Cl]$ (**87a**) and $[^{99m}Tc(CO)_3(\kappa^2\text{-}H_2CH_2PNO)Cl]$ (**88a**) have been prepared by reaction of the corresponding chelating ligands with the aquo ion *fac*-$[^{99m}Tc(OH_2)_3(CO)_3]^+$ [90]. Stability studies indicated higher stability for **87a** and **88a** than for **86a** [90]. Taking advantage of the coordination properties of these ligands (H_2PNO, κ^2-PO; H_2CH_2PNO, κ^2-PN), and also of the stability and lipophilicity of the complexes, a family of arylpiperazine derivatives (L^{1-4}, receptor-ligands for the 5-HT_{1A} subclass of serotonergic receptors) coupled to a P,*X* donor-atom unit (*X* is O, N) were used in the synthesis of $[M(CO)_3Br(\kappa^2\text{-}L^{1-4})]$ (*M* is Re, **90–93**; ^{99m}Tc, **90a–93a**), which were fully characterized [88, 90, 91] (Scheme 23).

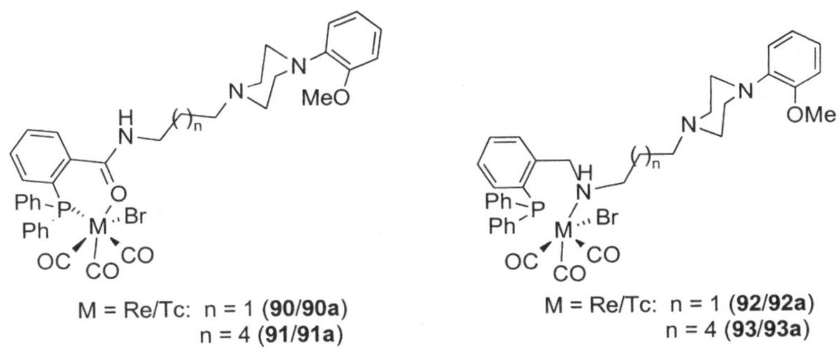

Scheme 23 Tricarbonyl complexes anchored by heterofunctionalized phosphines bearing a 5-HT_{1A} receptor binding unit

The characterization of the complexes at macroscopic level included elemental analysis, IR, ^{1}H and ^{31}P NMR spectroscopy, and X-ray crystallographic analysis [88, 90, 91]. The ^{99m}Tc complexes have been characterized by HPLC comparison with the analytically pure rhenium analogues. The rhenium complexes **90–93** have been used for determination of the receptor-binding affinity and selectivity in rat brain homogenates. The IC_{50} values of these compounds for 5-HT_{1A} receptors were in the range 20–1,100 nM, while for 5-HT_{2A} they were in the range 4,680–1,190 nM. Compound **90** was the one with the best affinity (5-HT_{1A}, IC_{50}=20±0.1 nM) and selectivity (5-HT_{2A}/5-HT_{1A}=234). Besides the effect of the spacer length on the biospecificity of the complexes, the occurrence of intramolecular hydrogen bonding also has a clear influence, as confirmed by structural studies [91].

With the bidentate *o*-(diphenylphosphino)benzaldehyde (PO) and *o*-[(diphenylphosphino)benzylidene]aniline (PN) ligands the complexes [Re*X*(CO)$_3$(PO)] (*X* is halide, **94**) and [Re*X*(CO)$_3$(PN)] (*X* is halide, **95**) have been synthesized and characterized, starting from [Re*X*(CO)$_5$] (*X* is Br, Cl) [92]. Using the same starting material, complexes of the type *fac*-[Re(CO)$_3$(κ^3-*L*) (**96**) were also prepared, where *L* is a phosphorus-containing heterotridentate ligand, obtained by condensation of 2-(diphenylphosphino)aniline with several salicylaldehyde derivatives (Scheme 24) [93].

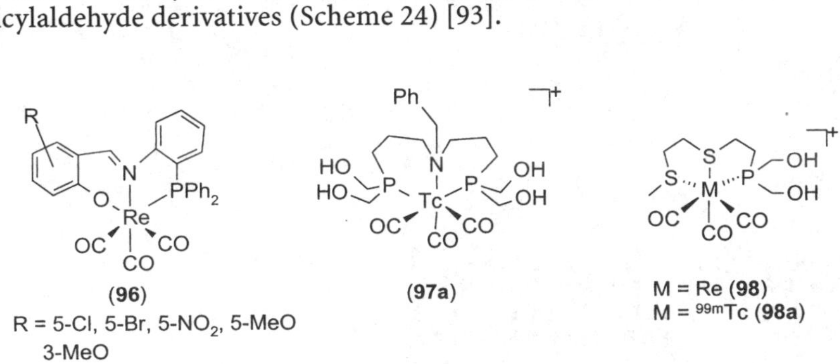

Scheme 24 Tricarbonyl complexes with tridentate phosphines

The novel water-soluble tridentate phosphines, amino-bis(hydroxymethyl)phosphine (P_2N) and a dithioether-phosphine (S_2P) reacted with the organometallic synthon *fac*-$[^{99m}Tc(CO)_3]^+$, yielding single and well-defined complexes which were assumed to be $[^{99m}Tc(CO)_3(\kappa^3\text{-}P_2N)]$ (**97a**) and $[^{99m}Tc(CO)_3(\kappa^3\text{-}PS_2)]^+$ (**98a**), respectively [94, 95]. On the basis of the high in vitro stability of these complexes, the authors considered that P_2N- and PS_2-based bifunctional chelating agents hold potential for labeling biomolecules via the *fac*-$[^{99m}Tc(CO)_3]^+$ precursor.

3
Complexes Anchored by Scorpionates

Since their introduction by Trofimenko, in 1966, a large number of hard scorpionates have been prepared over the years, and have been used in the synthesis of complexes with most metals and metalloids of the periodic table [2, 4–7]. For rhenium, almost all oxidation states (I–VII) have been explored, and a large variety of oxo, carbonyl or polyhydride complexes anchored by hard scorpionates have been synthesized and their reactivity studied [4, 5]. By contrast, the chemistry of technetium with hard scorpionates has been much less explored [5]. Poly(mercaptoimidazolyl)borates, which can be seen as soft congeners of the classical poly(pyrazolyl)borates, have recently been introduced in the chemistry of *d* and *f* elements [96, 97]. These novel sulfur donors also started to be applied in the chemistry of rhenium and technetium. Hard and soft scorpionates, which can be prepared starting from borohydrides, allow the possibility of introducing graded alterations on the chemistry of a given metal, tuning the physicochemical properties of the corresponding complexes, and are easily functionalized with different biomolecules (Structure 2, hard and soft scorpionates).

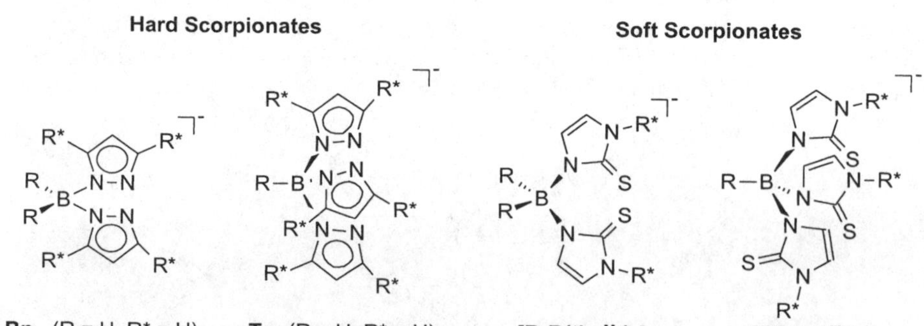

Bp (R = H, R* = H)
Bp* (R = H, R* = Me)

Tp (R = H, R* = H)
Tp* (R = H, R* = Me)
pzTp (R = pz, R* = H)
PhTp (R = Ph, R* = H)

[R$_2$B(tim^{Me})$_2$]$^-$ (R* = Me)

[RB(tim^{Me})$_3$]$^-$ (R* = Me)

Like the pincers of a scorpion, poly(pyrazolyl)- or poly(mercaptoimidazolyl)borates bind to metals by the nitrogen or sulfur heteroatoms from the azole rings attached to the central boron atom. The third azole ring or the hydrogen attached to the boron can rotate forward like a scorpion's tail to "sting" the metal. Nevertheless, the ligands can also be bidentate or monodentate, depending on the nature of the coligands and on the steric congestion around the metal. Concerning hard scorpionates some of the acquired basic chemical knowledge can be considered relevant in the design and development of target-specific radiopharmaceuticals. In this field, the chemistry explored with soft scorpionates, so far, is really unusual and encouraging.

Herein, some of the chemistry with hard and soft scorpionates is reviewed, and for each class of ligands we will take into account the oxidation state of the metal.

3.1 Hard Scorpionates

3.1.1 High- and Intermediate-Valent Oxo Complexes

In the late 1970s, Deutsch and coworkers focused on poly(pyrazolyl)borates in the course of their pioneering studies on the coordination chemistry of technetium. These authors synthesized the oxo-dichloride [$TpTcOCl_2$] (**99**) by reduction of TcO_4^- with excess KTp in aqueous HCl solution (Scheme 25) [98].

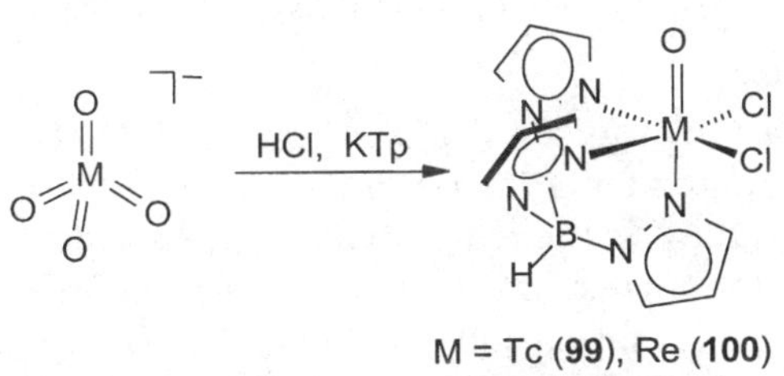

Scheme 25 Hydrotris(pyrazolyl)borate oxo-dichloride complexes

At the time, the synthesis and full characterization of **99**, including X-ray structural analysis, were considered quite relevant in terms of the chemistry of ^{99m}Tc radiopharmaceuticals. Together with [$TcO(SCH_2CH_2S)_2$]$^-$ [99], complex **99** was one of the first examples of Tc(V) oxocomplexes prepared under aqueous conditions and to be fully characterized. These results highlighted the importance of the [Tc=O]$^{3+}$ core in the chemistry of ^{99m}Tc radiopharmaceuticals, afterwards confirmed by the synthesis of several ^{99m}Tc oxocomplexes currently used in diagnostic nuclear medicine [10].

As indicated in Scheme 25, the synthesis of the rhenium congener of **99**, [$TpReOCl_2$] (**100**), was also achieved by reduction of perrhenate with Tp in the presence of hydrochloric acid, but with an isolated yield (40%) lower than the one reported by Deutsch et al. for technetium (60%). Such a difference is certainly due to the different redox properties of rhenium and technetium [100].

Further studies revealed that the high-valent oxides [TpMO$_3$] [M is Re (**101**), Tc (**105**)] could also be obtained starting from [MO$_4$]$^-$ anions, which are the usual starting materials in the synthesis of ^{99m}Tc and $^{186/188}$Re radiopharmaceuticals (Scheme 26). A quite disappointing yield was reported for the synthesis of the Re(VII) oxide (7.5%), while the technetium congener was obtained in a reasonable yield (44%) [101]. Together with compounds anchored by the Klaüi ligand, [(η^5-C$_5$H$_5$)Co{P(OCH$_3$)$_2$(=O)}$_3$$MO_3$] ($M$ is Tc, Re), **101** and **105** are almost unique examples of M(VII) trioxides fully characterized and obtained directly from [MO$_4$]$^-$ as starting materials [102].

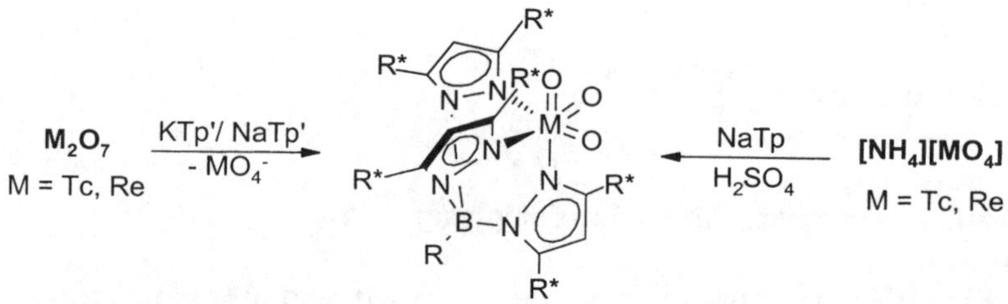

M = Re
R = H; R* = H (**101**), Me (**103**)
R = pz; R* = H (**102**)
R = Ph; R* = H (**104**)

M = Tc
R = H; R* = H (**105**), Me (**106**)

Scheme 26 Hydrotris(pyrazolyl)borate or tetrakis(pyrazolyl)borate trioxides

A more general method for the synthesis of M(VII) oxides of the type [Tp′MO$_3$] [M is Re, Tp′=Tp (**101**), pzTp (**102**), Tp*(**103**), PhTp (**104**); M is Tc, Tp′=Tp (**105**), Tp*(**106**)] is the reaction of M_2O$_7$ (M is Re, Tc) with sodium or potassium salts of the corresponding scorpionates, but these reactions have to be run in organic coordinating solvents (Scheme 26) [103–108]. The trioxides [Tp′MO$_3$] (M is Re, Tc) are remarkably stable toward aerial oxidation or hydrolysis and the metal center is six-coordinate, by three oxygen atoms and by the nitrogen atoms of κ^3-Tp′. The coordination geometry around the metal is pseudo-octahedral and within each compound the Re–N distances are almost identical [range 2.231(3)–2.222(7) Å] as well as the Re–O bond distances [range 1.714(2)–1.710(8) Å], which are consistent with conventional double bonds [103, 105, 106].

As would be expected for complexes with the metal in the highest oxidation state, the chemistry of the trioxides **101–104** is dominated by reductive/deoxygenation processes. This reduction, normally performed with phosphine ligands, yields dinuclear or mononuclear reactive species, which have been formulated as [pzTpReO(μ-O)]$_2$ (**107**) and {Tp′ReO$_2$}, respectively [109–112]. These species, generated in situ, are key compounds to enter into the chemistry of mono or dioxo complexes. As depicted in Scheme 27, several mixed-ligand scorpionate complexes, air- and water-resistant, could be obtained by reacting those fragments with protic or neutral bidentate coligands. Using oxygen, nitrogen or sulfur donor bidentate dianionic coligands, the $[M(\mathrm{O})\mathrm{Tp}']^{2+}$ metal fragment can be easily stabilized, yielding monomeric Re(V) complexes [111–118]. These compounds are six-coordinate, displaying a distorted octahedral coordination geometry. The scorpionate occupies one triangular face of the coordination polyhedron and the donor atoms of the bidentate coligands are in the equatorial plane. The displacement of the rhenium atom toward the oxo ligand and the lengthening of the Re–N bond trans to the oxo group are common features of all the structures [109–112].

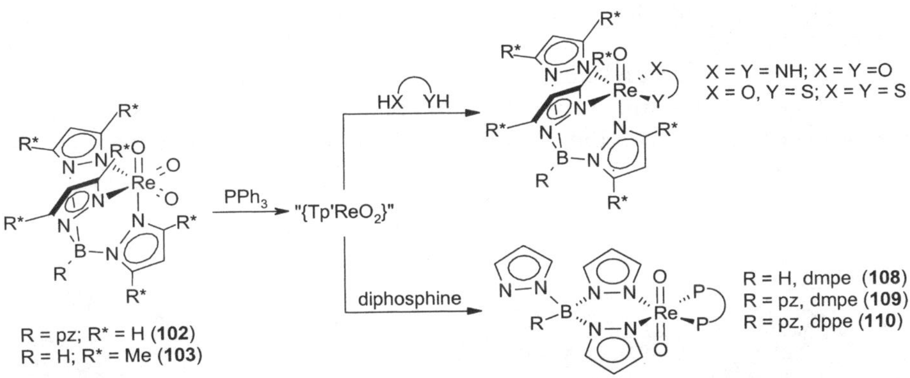

Scheme 27 Monoxo and *trans*-dioxo complexes anchored by hydrotris(pyrazolyl)borates or tetrakis(pyrazolyl)borates

Owing to the $\kappa^3 \Leftrightarrow \kappa^2$ haptotropism of the scorpionates, the trapping of the $\{\mathrm{Tp'ReO_2}\}$ metal fragments can also be achieved in reactions with neutral bidentate ligands, namely diphosphines, yielding *trans*-dioxo derivatives of general formula *trans*-$[(\kappa^2\text{-Tp}')\mathrm{ReO_2(P^\frown P)}]$ **(108–110)** [119, 120]. The structural analysis of these *trans*-dioxo complexes confirmed the κ^2-coordination mode of the scorpionate ligands and the presence of the *trans*-$\mathrm{ReO_2}$ unit. The Re–O bond distances (1.751–1.773 Å) are typical of Re(V) trans-dioxo complexes and are compatible with a certain double character [119, 120]. In solution the structure is maintained, and the abnormally high-field shifted H(3/5) resonances for the uncoordinated pyrazolyl rings rendered ^{1}H NMR spectroscopy a powerful tool for assigning the denticity of the scorpionate ligand [119, 120]. The possibility of generating, at noncarrier added level, units of the type $\{\mathrm{Tp}'M\mathrm{O_2}\}$ would introduce in radiopharmaceutical chemistry an unprecedented and quite reactive synthon that should allow the labeling of a large variety of simple diprotic or neutral bidentate ligands, easily derivatizable with a plethora of biologically active molecules. However, the preparation of the radioactive synthons $\{\mathrm{Tp}'M\mathrm{O_2}\}$ requires the development of new methodologies for the synthesis of trioxides of the type [Tp′*M*O_3] (*M* is ^{99m}Tc, $^{186/188}$Re) under the conditions required for radiopharmaceutical preparations. If these trioxides are available at noncarrier-added level, it is expectable that the synthons [Tp′*M*O_2] (*M* is ^{99m}Tc, $^{186/188}$Re) can be easily obtained by reduction with water-soluble phosphines or by polymer-supported phosphines, avoiding in the latter case eventual competition of the phosphines in the coordination to the naked {Tp′*M*O_2} fragments.

3.1.2
Low-Valent Carbonyl Complexes

Several research groups have studied the chemistry of tris(pyrazolyl)borate tricarbonyl complexes of technetium and rhenium. As depicted in Scheme 28, these studies allowed the synthesis of *fac*-[*M*$(CO)_3$Tp′] [*M* is Tc, Re, Tp′ is Tp (**111/113**), Tp* (**112/114**)] complexes, which were all structurally characterized. Complexes **111–114** were synthesized by reaction of the classical starting materials [*M*$(CO)_5$Br] with the sodium or potassium salts of the correspondent Tp′ ligands [121–123].

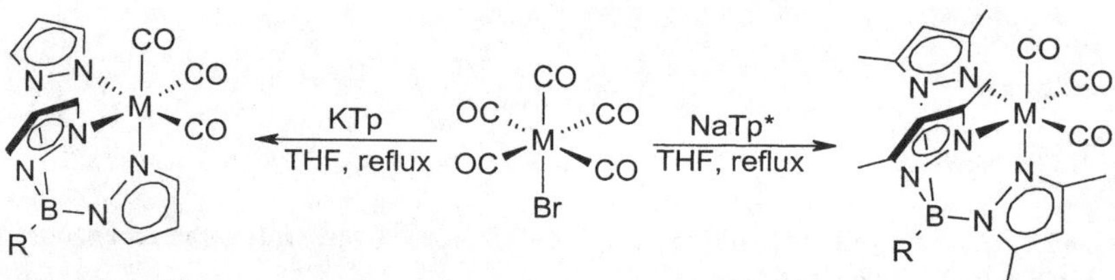

M = Tc (**111**)
M = Re (**113**)

M = Tc (**112**)
M = Re (**114**)

Scheme 28 Tris(pyrazolyl)borate tricarbonyl complexes

Considering that $(NEt_4)_2[Re(CO)_3Br_3]$ in coordinating solvents forms cations of the type *fac*-$[Re(CO)_3(L)_3]^+$ (*L* is a solvent molecule), which are analogous to the *fac*-$[M(CO)_3(OH_2)_3]^+$ (*M* is ^{99m}Tc, $^{186/188}$Re) synthons, our research group evaluated the possibility of preparing complex **114** by reacting $(NEt_4)_2[Re(CO)_3Br_3]$ with NaTp*. During these studies, which were performed in organic (acetonitrile, tetrahydrofuran) or in aqueous media, we found that **114** is hardly formed, and a significant decomposition of the scorpionate ligand was observed. These results probably explain why, in our hands, the labeling of the Tp* ligand with the *fac*-$[^{99m}Tc(CO)_3]^+$ moiety does not proceed with formation of the expected *fac*-$[^{99m}Tc(CO)_3(Tp^*)]$ complex. Instead, another radiochemical species is formed, albeit slowly [124].

At the macroscopic level, and with bis(pyrazolyl)borates, only one complex is fully characterized, and this is *fac*-[Re$(CO)_3$Bp(pzH)] (**115**), obtained by reacting [Re$(CO)_5$Br] with KBp [125]. Reactions of the substituted bis(pyrazolyl)borate Bp* with $(NEt_4)_2[Re(CO)_3Br_3]$ have also been explored in our laboratories [126]. We also found that these reactions are accompanied by degradation of the Bp* ligand, yielding a mixture of species. Complex *fac*-$[Re(CO)_3\{\kappa^3\text{-H}(\mu\text{-OH})B(pz^*)_2\}]$ (**116**) is the only species to have been isolated and fully characterized by IR and NMR spectroscopy and by mass spectrometry. Compound **116**, which was formed when the reaction was run in water, is anchored by a heteroscorpionate, which resulted from the partial hydrolysis of the original ligand. In our hands, the reaction of *fac*-$[^{99m}Tc(CO)_3]^+$ with NaBp* displays rather unfavorable kinetics, yielding a mixture of two radioactive species, but both are different from **116**, as indicated by HPLC analysis [126].

3.2
Soft Scorpionates

3.2.1
Low-Valent Carbonyl Complexes

More recently, great attention has been given to the chemistry of *M*(I) (*M* is Tc, Re) tricarbonyl complexes with bis(mercaptoimidazolyl)borates and tris(mercaptoimidazolyl)borates, searching for target-specific radiopharmaceuticals based on the *fac*-$[M(CO)_3]^+$ moieties (*M* is ^{99m}Tc, $^{186/188}$Re) [127–129].

As depicted in Scheme 29, sodium or lithium salts of tris(mercaptoimidazolyl)borates or bis(mercaptoimidazolyl)borates react, in water or in organic solvents, with $[M(CO)_3X_3]^y$ (*M* is ^{99m}Tc, *X*=H_2O; *M* is Re, *X*=Br), affording the complexes *fac*-$[M\{\kappa^3\text{-}R\text{B}(\text{tim}^{\text{Me}})_3\}(CO)_3]$ [*M* is Re, ^{99m}Tc, *R* is H (**117/117a**) Me (**118/118a**), Ph (**119/119a**)] and *fac*-$[M\{\kappa^3\text{-}R(\mu\text{-H})\text{B}(\text{tim}^{\text{Me}})_2\}(CO)_3]$ [*M* is Re, ^{99m}Tc, *R* is H (**120/120a**), Me (**121/121a**), Ph (**122/122a**)] [127–130]. Based on a similar reaction, the complex *fac*-$[^{99}Tc\{\kappa^3\text{-H}(\mu\text{-H})\text{B}(\text{tim}^{\text{Me}})_2\}(CO)_3]$ (**123**) was also obtained in almost quantitative yield [127].

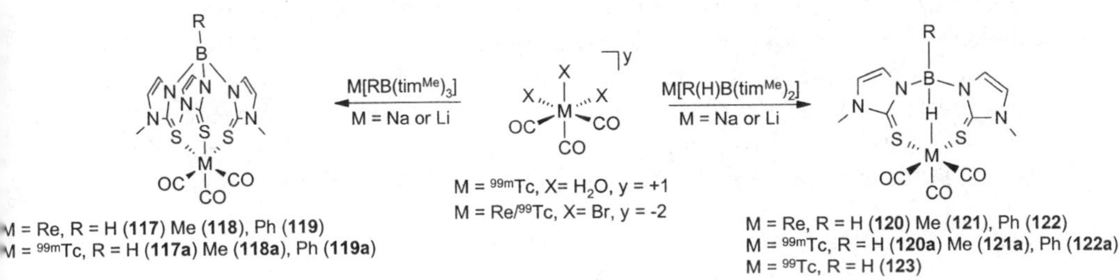

Scheme 29 Bis(mercaptoimidazolyl)borate and tris(mercaptoimidazolyl)borate tricarbonyl complexes

Unlike the harder congeners, the poly(mercaptoimidazolyl)borates react smoothly with the *fac*-$[M(CO)_3]^+$ moieties, yielding, in almost quantitative yield, complexes that are remarkably stable toward aerobic oxidation and hydrolysis. The ^{99m}Tc complexes are obtained with high radiochemical purity (more than 95%), even using low concentrations of the ligands (10^{-5}–10^{-6} M), which confirms the quite favorable kinetics of the soft scorpionates toward the tricarbonyl moieties.

The rhenium and ^{99}Tc complexes (**117–123**) have been characterized by IR, ^{1}H, and ^{11}B NMR spectroscopy, and most of them also by X-ray crystallographic analysis [127–129]. The identification of the ^{99m}Tc complexes has been made by HPLC comparison with the analogous rhenium compounds [128, 130].

The X-ray diffraction analysis of **117** and **118** showed a tridentate coordination of the ligands through the thione sulfur atoms, with the scorpionates adopting a typical propeller-like configuration. Comparing complexes *fac*-[Re{κ^3-HB(tim^{Me})$_3$}(CO)$_3$] (**117**) and *fac*-[Re(CO)$_3$Tp] (**113**), we find that the most striking difference is that in **113** the pyrazolyl rings are almost parallel to the Re–B axis, while in **117** the mercaptoimidazolyl rings lie at an average angle of 60.3° to the B–Re axis. In complex **113** the Tp ligand bite angles are rather less than 90° [average N–Re–N 81.0(7)°], while in compound **117** these angles are close to 90° [average S–Re–S 91.15(7)°] [119, 124]. These structural differences confirm the larger flexibility of {κ^3-HB(tim^{Me})$_3$} compared with Tp, and this most probably explains the fast kinetics of the reactions with [RB(tim^{Me})$_3$]$^-$, as well as the low tendency of this family of ligands to undergo decomposition processes.

In complexes **120–123** the bis(mercaptoimidazolyl)borates are coordinated through the two thione sulfur atoms and through one hydrogen atom attached to boron, defining two six-membered rings and one eight-membered chelate ring. Compounds **120–123** are unique examples of tricarbonyl Re(I) or Tc(I) complexes containing B–H···*M* agostic interactions. For the technetium complex, **123**, this interaction was confirmed by the direct location of the agostic hydride by X-ray structural analysis, while for the other complexes this interaction was based on the short B···Re bond distances [2.832 (12)–2.92(2) Å], from IR and on ^{1}H NMR spectroscopy. Interestingly, the B–H···*M* agostic interactions, found in the solid state, are maintained in solution, even in water or in other coordinating solvents, as clearly indicated by ^{1}H and ^{11}B NMR [127, 128]. Another interesting point is that, during the synthesis of complexes **120a–122a**, the about 10^5 fold excess of Cl$^-$ present in solution does not compete with the agostic ^{99m}Tc–H bond formation. This underlines the excellent properties of these ligands and the versatility of the *fac*-[^{99m}Tc(CO)$_3$]$^+$ moiety, which allowed the unprecedented coordination of a hydrogen atom, the smallest possible ligand, to technetium-99m.

Complexes **117a**, **120a** and **122a** display remarkable in vitro stability, and can be kept in phosphate buffer at 37 °C for at least 6 h without detectable decomposition. Furthermore, challenge experiments with histidine, one of the most powerful chelators for the unit *fac*-[^{99m}Tc(CO)$_3$]$^+$, have shown that, after 6 h, no trans-chelatation reaction occurred for **117a** and **120a**, while for **122a** only negligible displacement of the scorpionate ligand was observed. However, quite different behavior was observed when the complexes were incubated in the presence of a large excess of cysteine. In this case, complex **117a** remained intact, but **120a** and **122a** reacted slowly with cysteine, yielding a novel radiochemical species, which did not result from a trans-chelation process. With the presence of the B–H···Tc agostic interaction in **120a** and **122a** being the most significant difference between these complexes and **117a**, the reaction of cysteine with the agostic hydrides of **120a** and **122a** may explain the reactivity found [130].

Biodistribution studies in mice of complexes **117a**, **120a**, and **122a** demonstrated that these neutral and moderately lipophilic complexes (log P = 1.50–1.64) are able to cross the blood brain barrier. In the case of **117a** there is a fast washout from the brain, but complexes **120a** and **122a** display moderate brain retentions [1 h post injection 0.28% ID/g (**120a**) and 0.38% ID/g (**122a**)] [130]. These biological features have shown that **120a** and **122a** could be adequate building blocks for the labeling of CNS-receptor avid molecules.

The coupling of biologically active substrates to the bis(mercaptoimidazolyl)borate chelator framework has been achieved through the mercaptoimidazolyl rings. Bis(mercaptoimidazolyl)borate anchors have been functionalized, in a symmetric or asymmetric fashion, with 1-(2-methoxy)arylpiperazine fragments, which are known to recognize selectively the 5-HT_{1A} subclass of serotonergic receptors [131, 132]. As depicted in Scheme 30, several symmetric and asymmetric bis(mercaptoimidazolyl)borates have been used in the synthesis of functionalized rhenium tricarbonyl complexes, by reaction with $(NEt_4)_2[ReBr_3(CO)_3]$. The novel compounds **124–128** have been fully characterized, including by X-ray diffraction analysis in the case of **124** and **127** [131, 132].

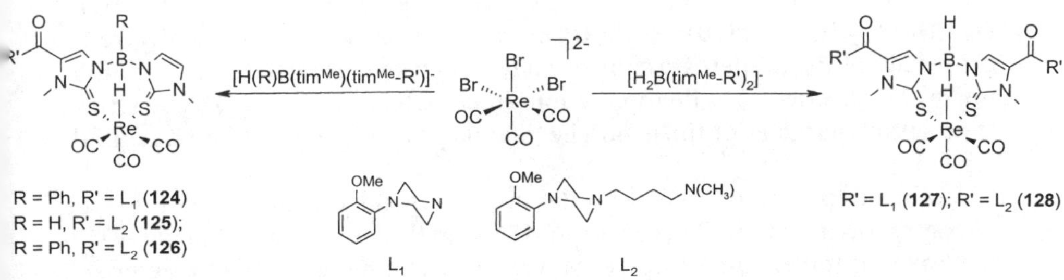

Scheme 30 Tricarbonyl complexes anchored by bis(mercaptoimidazolyl)borates bearing a 5-HT_{1A} receptor ligand unit

The affinity of compounds **124–128** for the 5-HT_{1A} receptors has been measured, and very encouraging IC_{50} values have been found for **125**, **126** and **128** (0.172–0.71 nM). These excellent values contrast with the IC_{50} values found for **124** and **127**, which are higher than 1,000 nM. Interestingly, the symmetric complex, **128**, is the one that shows the best affinity (IC_{50}=0.172 nM), while maintaining excellent selectivity [132]. Hence, the so-called bivalent approach seems to improve the biological properties of this family of compounds [133]. The synthesis of the ^{99m}Tc congeners of complexes **125**, **126** and **128** is currently underway, in order to evaluate if they are able to cross the blood brain barrier.

Another versatile alternative for labeling biologically active substrates, with the building blocks *fac*-[M\{κ^3-R(μ-H)B(tim^{Me})$_2$\}(CO)$_3$] (M is Tc, Re; R is H, alkyl, aryl), relies on the selective and efficient cleavage of the B–H⋯M agostic interaction with ligands carrying a biologically active fragment. Searching for substrates that could cleave efficiently the B–H⋯Re interaction, Garcia et al. [134] studied the reactivity of [Re\{κ^3-H(μ-H)B(tim^{Me})$_2$\}(CO)$_3$] (**120**) toward different unidentate ligands, like imidazoles, pyridines, phosphines or isonitriles (Scheme 31). Thus, novel [Re\{κ^2-H$_2$B(tim^{Me})$_2$\}(CO)$_3$(L)] complexes, **129–133**, have been synthesized and fully characterized, including by X-ray diffraction analysis [134].

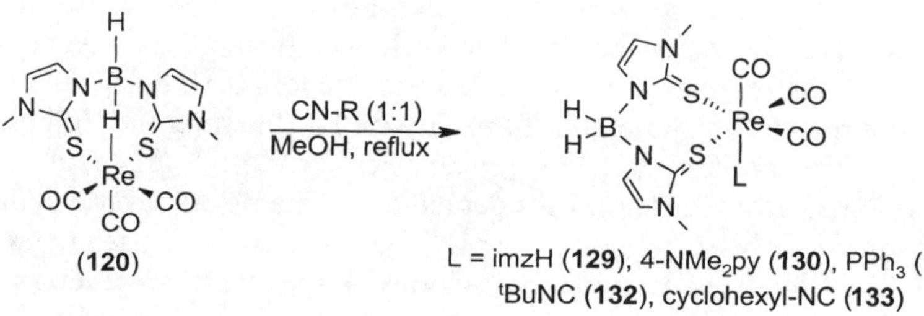

Scheme 31 Bis(mercaptoimidazolyl)borate tricarbonyl complexes of the "2+1" type

Except for the isonitrile derivatives **132** and **133**, all the other adducts (**129–131**) tend to regenerate the starting compound [Re\{κ^3-H(μ-H)B(tim^{Me})$_2$\}(CO)$_3$] (**120**) slowly, if kept in solution. Therefore, the electronic properties of the unidentate coligand have a marked influence on the rebuilding of the agostic B–H⋯Re interaction. In complexes **132** and **133**, the strong π-acceptor character of the isonitrile ligands probably prevents such rebuilding [134].

Taking into account these results, mixed-ligand complexes of the type [Re\{κ^2-H$_2$B(tim^{Me})$_2$\}(CO)$_3$(L)] (**134–136**), with isonitriles bearing 1-(2-methoxy)arylpiperazines fragments, have been synthesized and characterized (Scheme 32) [135]. In vitro receptor binding assays of **134–136** demonstrated that the complexes retained affinity for the 5-HT$_{1A}$ receptors (IC$_{50}$ values between 23.2 and 66.5 nM), although they displayed lower affinities than the complexes containing functionalized bis(mercaptoimidazolyl)borates [132]. The increase of the spacer length between the isonitrile function and the piperazinyl moiety is followed by the improvement of the affinity and selectivity of the complexes. The possibility of preparing the mixed target-specific compounds **134–136** at noncarrier added level (^{99m}Tc) is currently under investigation.

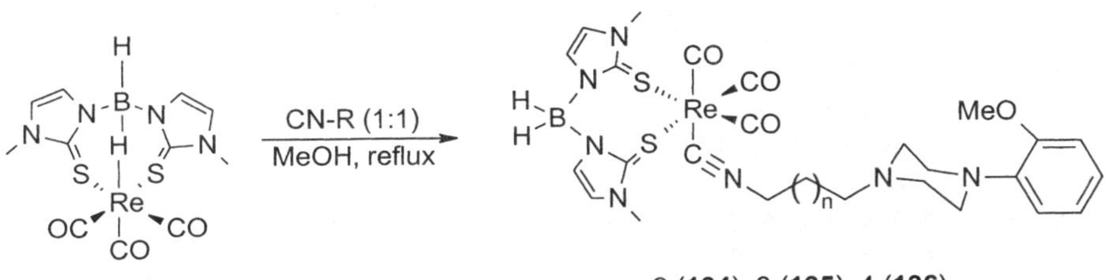

Scheme 32 Bis(mercaptoimidazolyl)borate tricarbonyl complexes of the 2+1 type bearing a 5-HT_{1A} receptor ligand unit

4 Concluding Remarks and Perspectives

The versatility of heterofunctionalized phosphines and scorpionates, two relevant and classical chelating ligands, allowed the synthesis of a large variety of inorganic and organometallic technetium and rhenium complexes, with the metals in the oxidation states +I to +VII. A sustained effort from basic coordination chemistry toward a more applied field confirmed the interest of phosphines and scorpionates in radiopharmaceutical development, although with more mitigated progress in the case of scorpionates.

Heterofunctionalized phosphines, with different denticities and/or donor sets, were successfully applied, at macroscopic and noncarrier added level, to the stabilization of the $[M{=}O]^{3+}$, *trans*-$[MO_2]^+$, $[M{\equiv}N]^{2+}$ and *fac*-$[M(CO)_3]^+$ cores. The unique electronic and/or stereochemical properties of the tridentate PNS phosphine allowed the synthesis of 3+1 mixed-ligand complexes with an unprecedented in vivo robustness. Phosphines as bifunctional chelate ligands have been successfully applied for labeling different biologically active substrates, namely peptides and CNS-receptor avid molecules.

The recent availability of the synthon *fac*-$[Tc(CO)_3(H_2O)_3]^+$ enabled the preparation of unprecedented complexes of the type *fac*-$[Tc\{\kappa^3\text{-}H(\mu\text{-}R)B(tim^{Me})_2\}(CO)_3]$, which contain a coordinated agostic hydrogen, the smallest possible ligand. The coupling of a 5-HT_{1A} receptor-binding piperazinyl moiety to these building blocks yielded functionalized complexes exhibiting excellent subnanomolar affinity for the 5-HT_{1A} receptors. The best performing complex was the one symmetrically functionalized with two biologically active fragments. Apparently, the so-called bivalent approach seems to be a valuable tool to improve the bioespecificity of technetium/rhenium complexes for the targeting of CNS-receptor ligands. The remarkabe water resistance of poly(mercaptoimidazolyl)borates and the successful results obtained with the *fac*-$[^{99m}Tc(CO)_3]^+$ core are strong motivation to extend this research to other oxidation states.

Studies on the basic coordination chemistry of rhenium and technetium with tris(pyrazolyl)borates demonstrated that water-stable Re(VII) trioxides

$[Tp'MO_3]$ (*M* is Re, Tc) can be prepared starting from $[MO_4]^-$, the conventional precursor in radiopharmaceutical synthesis. Although quite challenging, research into the synthesis of radioactive (^{99m}Tc or $^{186/188}$Re) congeners of these trioxides, including that with other tripodal ligands, is of great interest, since it could allow the introduction of innovative labeling procedures. On the basis of those methodologies, robust *M*(VII) (*M* is Re, Tc) complexes would then be achieved without the need for any reducing agent.

References

1. Hierso JC, Amardeil R, Bentabet E, Broussier R, Gautheron B, Meunier P, Kalck P (2003) Coord Chem Rev 236:143
2. Trofimenko S (1999) Scorpionates: the coordination chemistry of polypyrazolylborate ligands. Imperial College Press, London
3. Katti KV, Pillarsetty N, Raghuraman K (2003) Top Curr Chem 229:121
4. Paulo A, Correia JDG, Santos I (1998) Trends Inorg Chem 5:57
5. Paulo A, Correia JDG, Santos I, Campello P (2004) Polyhedron 23:331
6. Santos I, Marques N (1995) New J Chem 19:551
7. Marques N, Sella A, Takats J (2002) Chem Rev 102:2137
8. Hashimoto K, Yoshihara K (1995) Top Curr Chem 176:275
9. (a) Duatti A (1999) In: Nicolini M, Mazzi U (eds) Technetium and rhenium in chemistry and nuclear medicine 5, SGEditoriali, Padova, p 3; (b) Bolzati C, Boschi A, Duatti A, Prakash S, Ucelli L, Refosco E, Tisato F (2000) J Am Chem Soc 122:4510
10. Jurisson SS, Lydon LL (1999) Chem Rev 99:2205
11. Liu S, Edwards DS (1999) Chem Rev 99:2235
12. Volkert WA, Hoffman TJ (1999) Chem Rev 99:2269
13. (a) Alberto R, Schibili R, Egli A, Schubiger AP (1998) J Am Chem Soc 120:7987; (b) Alberto R, Ortner K, Wheatley N, Schibili R, Schubiger AP (2001) J Am Chem Soc 123:3135
14. Alberto R, Schibili, Waibel R, Abram U, Schubiger AP (1999) Coord Chem Rev 190–192:901
15. Schibili R, Schubiger PA (2002) Eur J Nucl Med 29:1529
16. Alberto R (2003) In: McCleverty J, Meyer TJ (eds) Comprehensive coordination chemistry II. Elsevier–Pergamon, Amsterdam, chap 4
17. Tisato F, Refosco F, Bandoli G (1994) Coord Chem Rev 135–136:325
18. Bandoli G, Dolmella A, Porchia M, Refosco F, Tisato F (2001) Coord Chem Rev 214:43
19. Kelly JD, Forster AM, Higley B, Archer CM, Booker FS, Canning LR, Chiu FW, Edwards B, Gill HK, McPartlin M, Nagle KR, Latham IA, Pickett Rd, Storey AE, Webbon PM (1993) J Nucl Med 34:222
20. Rosseti C, Vanoli G, Paganelli G, Kwiatkowski M, Zito F, Colombo F, Bonino C, Carpinelli A, Casati R, Deutsch K, Marmion M, Woulfe SR, Lunghi F, Deutsch E, Fazio F (1994) J Nucl Med 35:1571
21. (a) Treher EN, Francesconi LC, Gougoutas JZ, Malley MF, Nunn AD (1989) Inorg Chem 28:3411; (b) Narra RK, Nunn AD, Kucziynski BL, Feld T, Wedeking P, Eckelman WC (1989) J Nucl Med 30:1830
22. Luo H, Setyawati I, Rettig SJ, Orvig C (1995) Inorg Chem 34:2287
23. Bolzati C, Tisato F, Refosco F, Bandoli G, Dolmella A (1996) Inorg Chem 35:6221
24. Bolzati C, Tisato F, Refosco F, Bandoli G (1996) Inorg Chim Acta 247:125
25. Connac F, Lucchese Y, Gressier M, Dartiguenave, Beauchamp AL (2000) Inorg Chim Acta 304:52

26. Bolzati C, Ucelli L, Refosco F, Tisato F, Duatti A, Giganti M, Piffanelli A (1998) Nucl Med Biol 25:71
27. Bolzati C, Refosco F, Tisato F, Bandoli G (1992) Inorg Chim Acta 201:7
28. Refosco F, Tisato F, Bandoli G, Deutsch E (1993) J Chem Soc Dalton Trans 2901
29. Loiseau F, Connac F, Lucchese Y, Dartiguenave M, Fortin S, Beauchamp AL, Coulais Y (2000) Inorg Chim Acta 306:94
30. Nock B, Maina T, Tisato F, Papadopoulos M, Raptopoulou CP, Terzis A, Chiotellis E (1999) Inorg Chem 38:4197
31. Nock B, Maina T, Tisato F, Papadopoulos M, Raptopoulou CP, Terzis A, Chiotellis E (2000) Inorg Chem 39:2178
32. Nock B, Maina T, Tisato F, Raptopoulou CP, Terzis A, Chiotellis E (2000) Inorg Chem 39:5197
33. Bolzati C, Porchia M, Bandoli G, Boschi A, Malago E, Uccelli L (2001) Inorg Chim Acta 315:205
34. Femia FJ, Chen X, Babich JW, Zubieta J (2001) Inorg Chim Acta 316:145
35. Correia JDG, Domingos A, Santos I, Bolzati C, Refosco F, Tisato F (2001) Inorg Chim Acta 315:213
36. (a) Refosco F, Bolzati C, Moresco A, Bandoli G, Dolmella A, Mazzi U, Nicolini M (1991) J Chem Soc Dalton Trans 3043; (b) Refosco F, Tisato F, Bandoli G, Bolzati C, Dolmella A, Moresco A, Nicolini M (1993) J Chem Soc Dalton Trans 605
37. Refosco F, Tisato F, Moresco A, Bandoli G (1995) J Chem Soc Dalton Trans 3475
38. Tisato F, Refosco F, Bolzati C, Cagnolini A, Gatto S, Bandoli G (1997) J Chem Soc Dalton Trans 1421
39. Nicholson T, Hirsch-Kuchma M, Shellenbarger-Jones A, Davison A, Jones AG (1998) Inorg Chim Acta 267:319
40. Freiberg E, Davis WM, Nicholson T, Davison A, Jones AG (2002) Inorg Chem 41:5667
41. Schwartz DA, Abrams MJ, Hauser MM, Gaul FE, Larsen SK, Rauh D, Zubieta JA (1991) Bioconjugate Chem 2:333
42. Liu S, Edwards DS, Looby RJ, Harris AR, Poirier MJ, Rajopadhye M, Bourque JP (1996) Bioconjugate Chem 7:196
43. Edwards DS, Liu S, Barrett JA, Harris AR, Looby RJ, Ziegler MC, Heminway SJ, Carroll TR (1997) Bioconjugate Chem 8:146
44. Purohit A, Liu S, Casebier D, Edwards DS (2003) Bioconjugate Chem 14:720
45. Bolzati C, Boschi A, Ucelli L, Malago E, Bandoli G, Tisato F, Refosco F, Pasqualini R, Duatti A (1999) Inorg Chem 38:4473
46. Bolzati C, Ucelli L, Boschi A, Malago E, Duatti A, Tisato F, Refosco F, Pasqualini R, Piffanelli A (2000) Nucl Med Biol 27:369
47. Bolzati C, Boschi A, Ucelli L, Tisato F, Refosco F, Cagnolini A, Duatti A, Prakash S, Bandoli G, Vittadini A (2002) J Am Chem Soc 124:11468
48. Boschi A, Bolzati C, Benini E, Malago E, Ucelli L, Duatti A, Piffanelli A, Refosco F, Tisato F (2001) Bioconjugate Chem 12:1035
49. Bolzati C, Mahmood A, Malago E, Ucelli L, Boschi A, Jones AG, Refosco F, Duatti A, Tisato F (2003) Bioconjugate Chem 14:1231
50. Boschi A, Ucelli L, Duatti A, Bolzati C, Refosco F, Tisato F, Romagnoli R, Baraldi PG, Varani K, Borea PA (2003) Bioconjugate Chem 14:1279
51. Boschi A, Bolzati C, Ucelli L, Duatti A, Benini E, Refosco F, Tisato F, Piffanelli A (2002) Nucl Med Commun 23:689
52. Boschi A, Ucelli L, Bolzati C, Duatti A, Sabba N, Moreti E, (2003) J Nucl Med 44:806
53. Kovacs MS, Hein P, Sattarzadeh S, Patrick BO, Emge TJ, Orvig C (2001) J Chem Soc Dalton Trans 3015
54. Cavell RG, Hilts RW, Luo H, Mcdonald R (1999) Inorg Chem 38:897

55. Dilworth JR, Griffiths DV, Parrot SJ, Zheng Y (1997) J Chem Soc Dalton Trans 2931
56. Correia JDG, Domingos Â, Santos I (2000) Eur J Inorg Chem 7:1523
57. Correia JDG, Domingos Â, Paulo A, Santos I (2000) J Chem Soc Dalton Trans 14:2477
58. Correia JDG, Domingos Â, Santos I, Spies H (2001) J Chem Soc Dalton Trans 15:2245
59. Palma E, Correia JDG, Domingos A, Santos I (2002) Eur J Inorg Chem 2402
60. Fernandes C, Correia JDG, Gano L, Santos I, Seifert S, Shyre R, Spies H (2001) J Labelled Compd Radiopharm 44 (Suppl 1):518
61. Fernandes C, Seifert S, Spies H, Syhre R, Bergman R, Gano L, Santos I (2004) 12th European symposium on radiopharmacy and radiopharmaceuticals, September, Gdansk, Poland
62. (a) Beckett AH, Taylor DC, Garrod JW (1975) J Pharm Pharmacol 27:588; (b) Millership JS, Shanks ML (1988) J Pharm Sci 77:116; (c) Przuntek H, Westarp ME, Vohl ML, Gerlach M, Jutzi P, Wekerle H (1987) Neuropharmacology 26:255; (d) Tsareva TA, Kramarova EP, Zharkovskii AM (1982) Farmakol Toksikol 45:20
63. (a) Kniess T, Spies H, Santos I, Zablotskaya A (2002) J Labelled Compd Radiopharm 45:629; (b) Spies H, Fietz, Th, Zablotska A, Belyakov S, Lukevics E (1999) Chem Heterocycl Compd 35:112
64. (a) Gupta A, Seifert S, Syhre R, Scheunemann P, Brust P, Johannsen B (2001) Radiochim Acta 89:43; (b) Syhre R, Seifert S, Spies H, Gupta A, Johannsen B (1998) Eur J Nucl Med 25:793
65. Fernandes C, Kniess T, Santos I, Seifert S, Spies H, Zablotskaya A (2002) In: Nicolini M, Bandoli G, Mazzi U (eds) Technetium, rhenium and other metals in chemistry and nuclear medicine 6. SGEditoriali, Padova, p 211
66. Kniess T, Fernandes C, Santos I, Kraus W, Spies H (2003) Inorg Chim Acta 348:237
67. Fernandes C, Kniess T, Gano L, Seifert S, Spies H, Santos I (2004) Nucl Med Biol 31:785
68. Tisato F, Refosco F, Moresco A, Bandoli G, Dolmella A, Bolzati C (1995) Inorg Chem 34:1779
69. Santimaria M, Maina T, Mazzi U, Nicolini M (1995) Inorg Chim Acta 240:291
70. Santimaria M, Mazzi U, Gatto S, Dolmella A, Bandoli G, Nicolini M (1997) J Chem Soc Dalton Trans 1765
71. Visentin, Santimaria M, Giron MC, Mazzi U, Nicolini M (1999) In: Nicolini M, Mazzi U (eds) Technetium, rhenium and other metals in chemistry and nuclear medicine 5. SGEditoriali, Padova, p 173
72. Visentini R, Rossin R, Giron MC, Dolmella A, Bandoli G, Mazzi U (2003) Inorg Chem 42:950
73. Santimaria M, Blok D, Feitsma RIJ, Mazzi U, Pauwels EKJ (1999) Nucl Med Biol 26:251
74. Morelli G, De Luca S, Tesauro D, Saviano M, Pedone C, Dolmella A, Visentin R, Mazzi U (2002) J Pept Sci 8:373
75. Smith CJ, Kati KV, Volkert WA, Barbour LJ (1997) Inorg Chem 36:3928
76. Smith CJ, Li N, Kati KV, Higginbotham, Volkert WA (1997) Nucl Med Biol 24:685
77. Schibli R, Karra SR, Gali H, Katti KV, Higginbotham C, Volkert WA (1998) Radiochim Acta 83:211
78. Prabhu KR, Pillarsetty N, Gali H, Katti KV (2000) J Am Chem Soc 122:1554
79. Kothary KK, Gali H, Prabhu KR, Pillarsetty N, Owen NK, Katti KV, Hoffman TJ, Volkert WA (2002) Nucl Med Biol 29:83
80. Karra SR, Schibli R, Gali H, Katti KV, Hoffman TJ, Higginbotham C, Sieckman GL, Volkert WA (1999) Bioconjugate Chem 10:254
81. Gali H, Karra SR, Reddy VS, Katti KV (1999) Angew Chem Int Ed Engl 38:2020
82. Gali H, Hoffman TJ, Sieckman GL, Owen NK, Katti KV, Volkert WA (2001) Bioconjugate Chem 12:354
83. Schibli R, Katti KV, Volkert WA, Barnes CL (1998) Inorg Chem 37:5306

84. Schibli R, Katti KV, Volkert, WA, Barnes CL (2001) Inorg Chem 40:2358
85. Schibli R, Katti KV, Higginbotham C, Volkert WA, Alberto R (1999) Nucl Med Biol 26:711
86. Kothari KK, Pillarsety NK, Katti KV, Volkert WA (2003) Radiochim Acta 91:53
87. Smith CJ, Sieckman GL, Owen NK, Hayes DL, Mazuru DG, Kannan R, Volkert WA, Hoffman TJ (2003) Cancer Res 63:4082
88. Correia JDG, Domingos Â, Santos I, Alberto R, Ortner K (2001) Inorg Chem 40:5147
89. Kniess T, Correia JDG, Domingos Â, Palma E, Santos I (2003) Inorg Chem 42:6130
90. Correia JDG, Santos I, Alberto R, Ortner K, Spies H, Drews A (2001) J Labelled Compd Radiopharm 44(Suppl 1):507
91. Palma E, Correia JDG, Domingos Â, Santos I, Alberto R, Spies H (2004) J Organomet Chem, in press
92. Chen X, Femia FJ, Babich JW, Zubieta J (2001) Inorg Chim Acta 315:147
93. Faller JW, Mason G, Parr J (2001) J Organomet Chem 626:181
94. Kothari KK, Raghuraman K, Pillarsetty NK, Hoffman, Owen NK, Katti KV, Volkert WA (2003) Appl Radiat Isot 58:543
95. Kothari KK, Raghuraman K, Pillarsetty NK, Volkert WA, Jurisson SS, Hoffman TJ (2002) In: Nicolini M, Bandoli G, Mazzi U (eds) Technetium, rhenium and other metals in chemistry and nuclear medicine 6. SGEditoriali, Padova, p 69
96. Garner M, Reglinski J, Cassidy I, Spicer MD, Kennedy AR (1996) Chem Commun 1975
97. (a) Maria L, Domingos A, Santos I (2001) Inorg Chem 40:6863; (b) Maria L (2003) PhD Thesis. IST, Lisbon
98. Thomas RW, Estes GW, Elder RC, Deutsch E (1979) J Am Chem Soc 101:4581
99. Smith JE, Byrne EF, Cotton FA, Sekutowski JC (1978) J Am Chem Soc 100:5571
100. Abrams MJ, Davison A, Jones AG (1984) Inorg Chim Acta 82:125
101. Thomas JA, Davison A (1991) Inorg Chim Acta 190:231
102. Banbery HJ, Hussain W, Evans IG, Hamor TA, Jones CJ, McCleverty JA, Schulte H-J, Engles B, Klaüi W (1990) Polyhedron 9:2549
103. Degnan IA, Herrmann WA, Herdtweck E (1990) Chem Ber 123:1347
104. Coe BJ (1992) Polyhedron 9:1085
105. Domingos A, Marçalo J, Paulo A, Pires de Matos A, Santos I (1993) Inorg Chem 32:5114
106. Garcia R, Xing YH, Domingos A, Paulo A, Santos I, Inorg Chim Acta (2003) 343:27
107. Joachim JE, Apostolidis C, Kanellakopulos B, Maier R, Ziegler ML (1993) Z Naturforsch 48b:227
108. Matano Y, Northcutt TO, Brugman J, Bennet BK, Lovell S, Mayer JM (2000) Organometallics 19:2781
109. Paulo A, Domingos A, Marçalo J, Pires de Matos A, Santos I (1995) Inorg Chem 34:2113
110. Paulo A, Marçalo J, Pires de Matos A, Santos I (1995) In: Nicolini M, Bandoli G, Mazzi U (eds) Technetium and rhenium in chemistry and nuclear medicine. SGEditoriali, Padova, p 227
111. Gable KP, Abubaker A, Zientara K, Wainwright AM (1999) Organometallics 18:173
112. Gable KP, Chuawong P, Yokochy AFT (2002) Organometallics 21:929
113. Tisato F, Bolzati C, Duatti A, Bandoli G, Refosco F (1993) Inorg Chem 32:2042
114. Nunes D, Domingos A, Paulo A, Patrício L, Santos I, Carvalho MFNN, Pombeiro AJL (1998) Inorg Chim Acta 271:65
115. Brown SN, Mayer JM (1992) Inorg Chem 31:4091
116. Paulo A, Domingos A, Pires de Matos A, Santos I, Carvalho MFNN, Pombeiro AJL (1994) Inorg Chem 33:4729
117. Faller JW, Lavoie AR (2000) Organometallics 19:3957
118. Kettler PB, Chang Y, Chen Q, Zubieta J, Abrams MJ, Larsen SK (1995) Inorg Chim Acta 231:13
119. Paulo A, Reddy KR, Domingos A, Santos I (1998) Inorg Chem 37:6807

120. Paulo A, Domingos A, Santos I (1999) In: Nicolini M, Bandoli G, Mazzi U (eds) Technetium and rhenium in chemistry and nuclear medicine. SGEditoriali, Padova, p 245
121. McCleverty JA, Wolochowicz I (1979) J Organomet Chem 169:289
122. Angaroni M, Ardizzoia GA, D'Alfonso G, La Monica G (1990) J Chem Soc Dalton Trans 1895
123. Joachim JE, Apostolidis C, Kanellakopulos B, Maier R, Marques N, Meyer D, Müller J, Matos AP, Nuber B, Rebizant J, Ziegler M (1993) J Organomet Chem 448:119
124. Garcia R, Paulo A, Santos I (2003) Faculty of Sciences, University of Lisbon, May, Portugal, private communication
125. Bond A, Green M (1971) J Chem Soc A 682
126. Garcia R, Paulo A, Santos I (2003) Annual report, Instituto Tecnológico e Nuclear, Portugal
127. Garcia R, Paulo A, Domingos A, Santos I, Ortner K, Alberto R (2000) J Am Chem Soc 122:11240
128. Garcia R, Paulo A, Domingos A, Santos I (2001) J Organomet Chem 632:41
129. Garcia R, Paulo A, Domingos A, Santos I (2003) J Chem Soc Dalton Trans 2757
130. Garcia R, Paulo A, Gano L, Santos I (2003) 1st congress of the Portuguese Society of Pharmaceutical Sciences, April, Lisbon, Portugal
131. Garcia R, Xing YH, Paulo A, Domingos A, Santos I (2002) J Chem Soc Dalton Trans 4236
132. Garcia R, Paulo A, Santos I, Spies H, Pietzsch H-J, Bergmann R (2004) 2nd international symposium on biorganometallic chemistry, July, Zurich, Switzerland
133. Tamiz AP, Zhang J, Zhang M, Wang CZ, Johnson KM, Kozikowski AP (2000) J Am Chem Soc 122:5393
134. Garcia R, Paulo A, Domingos A, Santos I, Alberto R (2002) Inorg Chem 41:2422
135. Garcia R, Paulo A, Domingos A, Santos I, Spies H, Pietzsch HJ, Bergmann R, Alberto R (2002) In: Nicolini M, Mazzi U (eds) Technetium and rhenium in chemistry and nuclear medicine 6. SGEditoriali, Padova, p 143

Top Curr Chem (2005) 252: 85–115
DOI 10.1007/b101225

Development of Technetium-99m and Rhenium-188 Radiopharmaceuticals Containing a Terminal Metal–Nitrido Multiple Bond for Diagnosis and Therapy

Alessandra Boschi · Adriano Duatti (✉) · Licia Uccelli

Laboratory of Nuclear Medicine, Department of Clinical and Experimental Medicine, University of Ferrara, Via L. Borsari, 46, 44100 Ferrara, Italy
dta@unife.it

Abstract The chemistry of nitrido complexes of technetium has played an important role in the search of new Tc-99m radiopharmaceuticals and in providing efficient labeling procedures of biologically active molecules. A number of potentially useful ^{99m}Tc-labeled nitrido compounds have been found for imaging the most important organs and, in particular, for heart imaging. The relevance of nitrido complexes is also currently expanding to the field of Re-188 radiopharmaceuticals utilized as therapeutic agents in the treatment of degenerative diseases. Presumably, the most useful characteristic of nitrido complexes of technetium and rhenium lies in their unique structural features that are strictly controlled by the electronic structure of the Tc≡N multiple bond. Five-coordination is the most common arrangement, and structures may range between the two ideal limits of square pyramidal and trigonal bipyramidal geometry. It was found that geometrical changes can be efficiently manipulated through the careful selection of the type of coordinating atoms bound to the Tc≡N group. A set of four π-donor atoms favors square pyramidal geometry, while a combination of two π donors and two π acceptors causes the structure to switch toward trigonal bipyramidal geometry. The precise geometrical control conjoined with the high stability of the Tc≡N group easily allows a rational design of different categories of radiopharmaceuticals and the determination of their structure–activity relationships. In turn, this makes possible a fine tuning of the biological properties of selected lead compounds. In this review, a summary of the more recent results obtained in the application of nitrido complexes of technetium and rhenium to nuclear medicine and molecular imaging will be given with specific emphasis to the underlying chemical principles.

Keywords Technetium-99m · Rhenium-188 · Nitrido complexes · Radiopharmaceuticals · Imaging · Therapy

Abbreviations

DBODC	Diethoxyethyldithiocarbamato
DEDC	Diethyldithiocarbamato
DTCZ	Dithiocarbazic acid
HCC	Hepatocellular carcinoma
HcysD	Cysteine–distamycin
H_2DMSA	*meso*-2,3-Dimercaptosuccinic acid
H_2SB	*S*-Methyl-3-(2′-hydroxybenzylidene)dithiocarbazate
MO	Molecular orbital
NOET	(*N*-Ethoxy, *N*-ethyl)dithiocarbamato
PNP5	Bis[(dimethoxypropylphosphino)ethyl]ethoxyethylamine)
PPh_3	Triphenylphosphine
R	Organic functional group
sp	Square pyramidal
tbp	Trigonal bipyramidal

1 Introduction

1.1 Preparation of Nitrido Technetium-99m Radiopharmaceuticals

The first attempt to prepare a Tc-99m complex containing a terminal technetium–nitrogen multiple bond (Tc≡N) in physiological solution with concentrations as low as 10^{-6} mol dm^{-3} (micromolar concentration scale or tracer level), and under sterile and pyrogen-free conditions, for example, in strictly controlled conditions suitable for intravenous injection, was first reported by Baldas and coworkers [1–3]. The proposed method was based on the reaction of $[^{99m}TcO_4]^-$ with sodium azide (NaN_3) in the presence of HCl, and required a complicated, multistep procedure that made it difficult to ensure the sterility and apyrogenicity of the resulting solution. Besides, the observed yield of formation of the final product was far below the limit of 90% radiochemical purity required to allow administration to a patient.

A few years later, an alternative route to the preparation of this category of complexes came out unexpectedly from results obtained in a study of a new reaction procedure carried out at tracer level [4, 5]. This reaction is depicted in Fig. 1. In this procedure, $[^{99m}TcO_4]^-$ is reacted with triphenylphosphine (PPh_3) and the Schiff base derivative of dithiocarbazic acid (DTCZ), *S*-methyl-3-(2′-hydroxybenzylidene)dithiocarbazate (H_2SB), under acidic conditions. Surprisingly, this preparation gave rise to the high-yield (above 95%) formation of the nitrido Tc-99m complex, $[^{99m}Tc(N)(SB)(PPh_3]$, shown in Fig. 1. $[^{99m}Tc(N)(SB)(PPh_3]$ is a five-coordinated Tc(V) complex composed of a terminal Tc≡N multiple bond bound to a PPh_3 group and a doubly deprotonated SB^{2-} ligand [6]. This latter compound is coordinated in its thiol form, and acts as a tridentate ligand through the negative, hydroxylic oxygen atom, the neutral, aldiminic nitrogen atom and the negative, thiol sulfur atom. The unexpected result was the discovery that H_2SB could play the role of both a coordinating ligand and a donor of nitrido nitrogen atoms (N^{3-}) to afford the Tc≡N group. This N^{3-}-donor property of derivatives of DTCZ was confirmed by employing *N*-methyl *S*-methyl dithiocarbazate $[H_2N–N(CH_3)–C(=S)SCH_3]$ in the reaction illustrated in Fig. 1 [4]. In these conditions, addition of $[^{99m}TcO_4]^-$ yielded a mixture of different products. Though the exact chemical nature of these intermediate complexes has not yet been established, it was easily demonstrated that all these species incorporate a terminal Tc≡N group. This result was simply achieved by reacting the intermediate mixture with some suitable coordinating ligand. For instance, quantitative conversion to a single product could be rapidly accomplished through the reaction with a dithiocarbamate ligand of the type $[R_2N$-$C(=S)S]Na$ (*R* is an organic

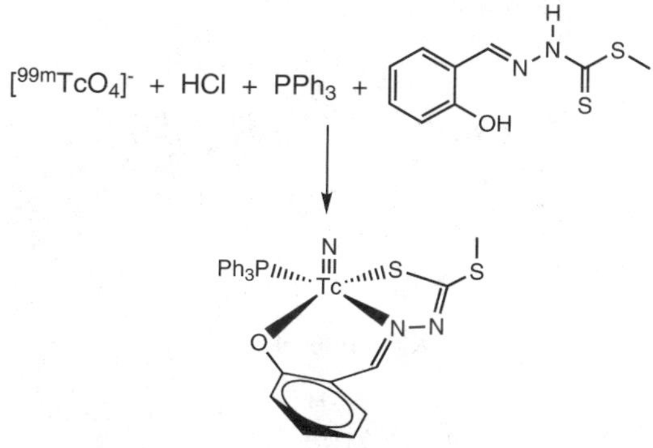

Fig. 1 Reaction diagram for the synthesis of the complex $[Tc(N)(PPh_3)(L)]$, and schematic drawing of its chemical structure showing the resulting square pyramidal arrangement with an apical Tc≡N group bound to a tridentate *L* ligand and a monodentate PPh_3 ligand positioned on the basal plane. In this reaction, the Schiff base ligand *L* also plays the role of a donor of nitrido nitrogen atoms

functional group) yielding the bis-substituted, nitrido Tc(V) complex $[^{99m}Tc(N)(R_2NCS_2)_2]$.

It soon became apparent that almost all derivatives containing the basic hydrazine-like motif >N–N < were able to behave as donor of N^{3-} groups in reactions with $[^{99m}TcO_4]^-$. The further requirement to formulate the preparation of the Tc≡N core as a freeze-dried, pharmaceutical kit led to the selection of the compound succinic dihydrazide (Fig. 2) as a convenient source of N^{3-} atoms owing to its high solubility in water and low toxicity. Another step toward the development of a suitable lyophilized kit formulation was brought about by the discovery that PPh_3 could be conveniently replaced by $SnCl_2$. As this reagent is widely utilized in radiopharmaceutical preparations, demonstration that it could be successfully employed for preparing the Tc≡N group in physiological solution was particularly useful. The composition of the currently available kit formulation for preparing nitrido Tc-99m radiopharmaceuticals is illustrated in Fig. 2. The procedure involves introducing generator-eluted $[^{99m}TcO_4]^-$ into a vial containing $SnCl_2$ and succinic dihydrazide, in the presence of a phosphate buffer and of the Sn^{2+}-complexing agents 1,2-diaminopropane-*N,N,N′,N′*-tetraacetic acid or ethylenediaminetetraacetic acid. Upon addition of $[^{99m}TcO_4]^-$, the formation of the Tc≡N group takes place within 15 min, at room temperature [7–12].

In summary, early investigations into the chemistry of Tc-99m radiopharmaceuticals containing a terminal Tc≡N bond clearly showed that this chemistry is perfectly feasible at tracer level and in physiological solution, thus opening the door for the exploration of the biological behavior of a completely new class of potential diagnostic agents. In the next sections, a survey of current developments of the chemistry of nitrido Tc-99m radiopharmaceuticals and their relevance for nuclear medicine applications will be presented. Particular emphasis will be given to the relations existing between the structural design of a nitrido Tc-99m radiopharmaceutical, required to match those essential fea-

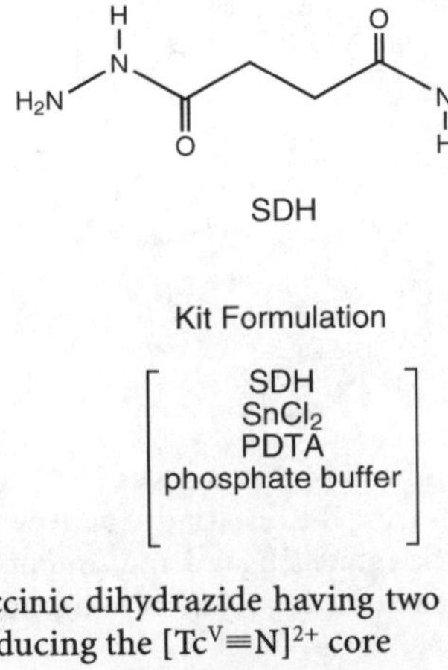

Fig. 2 The N^{3-} donor succinic dihydrazide having two terminal hydrazine-like (>N–N<) moieties essential for producing the $[Tc^V{\equiv}N]^{2+}$ core

tures for targeting a specific biological function, and the electronic properties of the Tc≡N core.

1.2 The Electronic Structure of the Tc≡N Group

Fragment theory [13] offers a simple and convenient tool for describing the electronic structure of the Tc≡N group. The remarkable stability of this core allows it to be classified as a true inorganic functional group exhibiting peculiar chemical characteristics. Therefore, it is convenient to describe it as a single molecular entity. It turns out that the main structural properties of complexes containing the Tc≡N core can be easily accounted for on the basis of the electronic structure of the technetium nitrido fragment.

A simplified molecular orbital (MO) diagram of the Tc≡N group is illustrated in Fig. 3. This was drawn by assuming that the technetium atom has a +5 oxidation number, which is the most common metal oxidation state for technetium nitrido complexes. This corresponds to a d^2 electronic configuration for the metal. As a consequence, the nitrido nitrogen atom was given a –3 oxidation state corresponding to a filled $2s^2 2p^6$ electronic configuration. Assuming a C_{4v} symmetry associated with a five-coordinated, square pyramidal (sp) complex with an apical Tc≡N multiple bond and four identical basal ligands, the symmetry type of the atomic orbitals gives the combinations shown in Fig. 3.

Filling the resulting MOs with the eight valence electrons gives the corresponding highest occupied MO (HOMO), which is a *xy*-type nonbonding orbital,

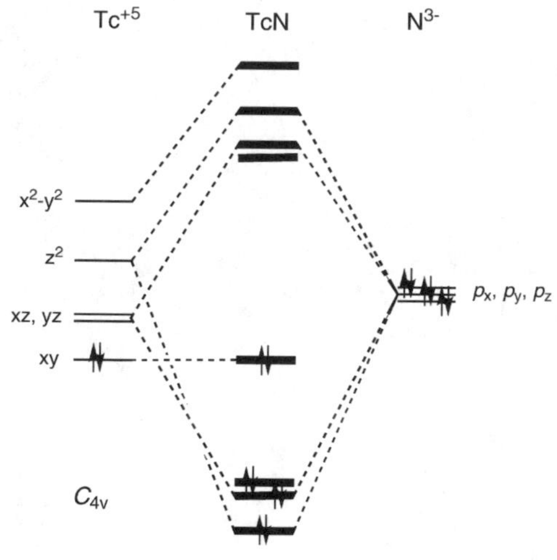

Fig. 3 Simplified molecular orbital (*MO*) diagram of the fragment $[Tc^V{\equiv}N]^{2+}$ in C_{4v} symmetry (energy scale is arbitrary). The *xy* highest occupied MO (*HOMO*) and the *xz*, *yz* pair of lowest unoccupied MOs (*LUMOs*) are the frontier orbitals regulating the reactivity of the fragment

and the corresponding lowest unoccupied MOs (LUMOs), which are two degenerate π^* antibonding orbitals. The distribution of the electronic density of these frontier orbitals is able to account for all structural features of sp Tc(V) nitrido complexes. It should be noted that similar arguments can be applied to the description of the electronic structure of the isoelectronic $[Tc^{V}{\equiv}O]^{3+}$ fragment.

The alternative geometry for a five-coordinate nitrido Tc(V) complex is a trigonal bipyramidal (tbp) structure. Though replacement of one of the five identical ligands of an ideal tbp geometry by a nitrido group determines the lowering of the symmetry from D_{3h} to C_{2v}, for practical reasons it is convenient to attribute a full D_{3h} symmetry to this class of complexes. The transition between sp and tbp geometries, therefore, can be simply accomplished through a Berry pseudorotation as shown in Fig. 4. In the final tbp arrangement the

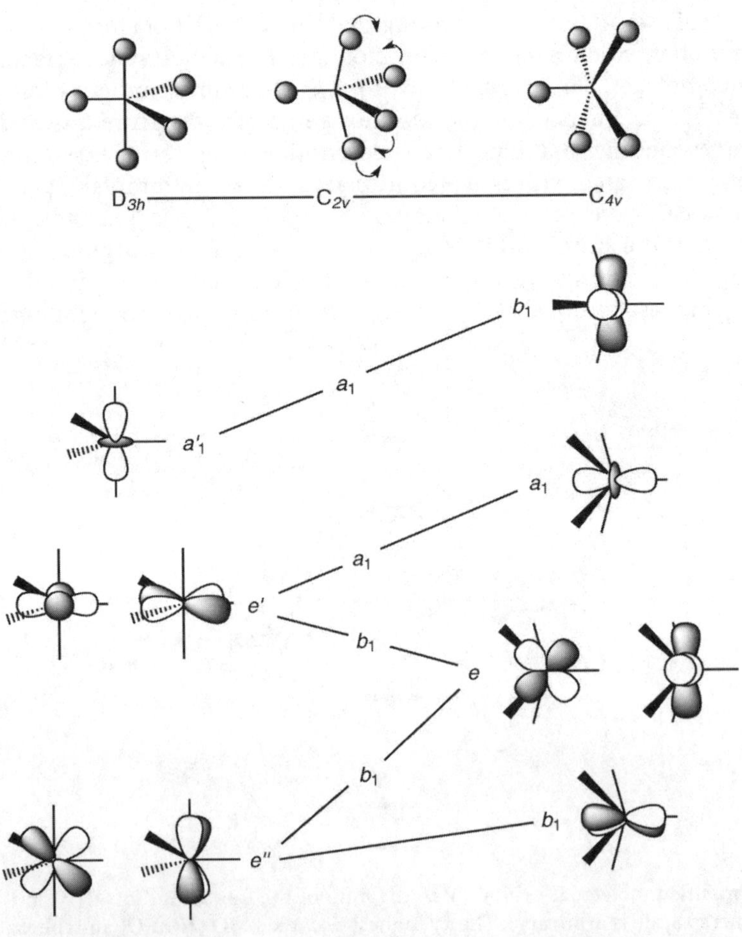

Fig. 4 Berry pseudorotation for the transition from C_{4v} to D_{3h} symmetry transforming a square pyramidal geometry into a trigonal bipyramidal geometry

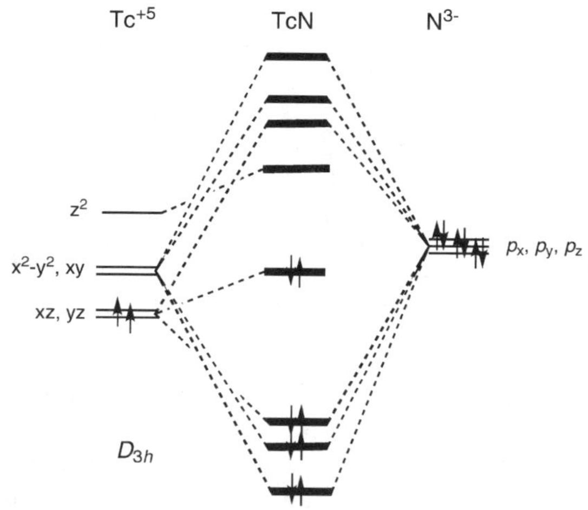

Fig. 5 Schematic MO diagram of the fragment $[Tc^V{\equiv}N]^{2+}$ in D_{3h} symmetry (energy scale is arbitrary). The *yz* HOMO and the z^2 LUMO are the frontier orbitals regulating the primary reactivity of the fragment toward substitution by π-acceptor ligands along the axial positions. This set of orbitals behaves independently from the *xy*, x^2-y^2 pair of antibonding orbitals regulating the reactivity of the fragment toward substitution by π-donor ligands on the trigonal plane

Tc≡N group lies along the *x*-axis on the trigonal plane perpendicular to the *z*-axis. The simplified MO diagram of the fragment $[Tc^V{\equiv}N]^{2+}$ in D_{3h} symmetry is illustrated in Fig. 5. The set of the resulting frontier orbitals is composed of a *yz*-type HOMO nonbonding orbital and a z^2-type LUMO nonbonding orbital.

2 Structural Properties of Five-Coordinate Nitrido Technetium(V) Complexes

2.1 Square Pyramidal Complexes

Five-coordinated Tc(V) nitrido complexes may assume two distinct structural arrangements, namely a sp geometry and a tbp geometry. The choice between the two limiting geometries is always dictated by the nature of the four coordinating atoms bound to the Tc≡N group. A survey of crystal structure determinations reported to date [14, 15] easily reveals that a sp configuration is highly preferred when the four coordinated atoms are π-donor Lewis bases. A π donor is defined here as a coordinating atom having filled atomic orbitals with appropriate symmetry to overlap with the two empty π^* LUMOs of the Tc≡N group.

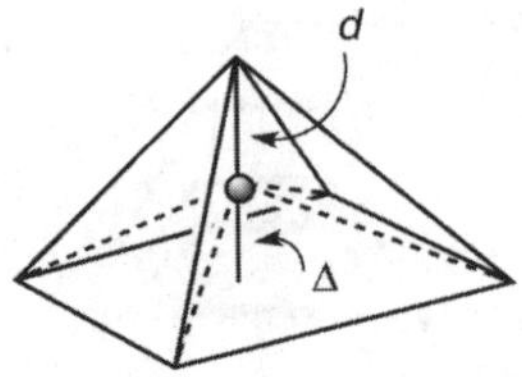

Fig. 6 Representation of the structural parameters Δ and *d*. In a square pyramidal geometry, the parameters Δ and *d* measure the displacement of the central ion from the basal plane and the length of the terminal multiple bond, respectively

Two structural parameters, derived from crystallographic studies, are particularly useful in the description of the structural features of sp Tc(V) nitrido complexes. These are the displacement (Δ) of the technetium atom from the basal plane of the square pyramid, and the Tc≡N bond length (*d*) (Fig. 6). When *d* is plotted as a function of Δ the curve represented in Fig. 7 is obtained. Inspection of this diagram reveals that there exists a close relationship between the two parameters Δ and *d*. Specifically, points on the right side of the curve represent complexes where the lengthening of the Tc≡N bond length is always associated with a corresponding increase of the parameter Δ. This effect leads to a progressive distortion of the sp geometry toward an "umbrella-shaped" structure, and is determined by the rise of the "soft" basic character of the four donor atoms coordinated on the basal plane. Thus, the right portion of the diagram corresponds to complexes containing sulfur ligands such as thiols or dithiocarbamates. Conversely, replacement of "soft" π donors with "hard" π donors causes the joint decrease of the parameters Δ and *d*, and the transition of the coordination geometry toward a more ideal sp structure. As a result, the minimum of the curve corresponds to the Tc(VI) complex $[Tc(N)Cl_4]^-$ where

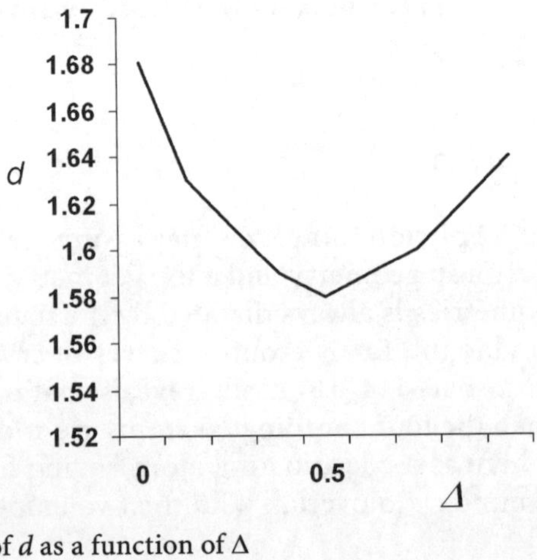

Fig. 7 The variation of *d* as a function of Δ

the Tc≡N group is fully surrounded by four "hard" chloride atoms. Points on the left part of the curve, beyond the minimum point, correspond to octahedral complexes. This category includes complexes with ligands which do not possess suitable π orbitals, but are mostly σ donors such as polyamines and tetraazamacrocycles. They are characterized by small values of the parameter Δ, but always exhibit a concomitant increase of the Tc≡N bond length. Usually, a weakly bound sixth ligand is coordinated in the trans position to the Tc≡N group.

The observed structural behavior of sp Tc(V) nitrido complexes is easily accounted for by considering the corresponding MO diagram of the $[Tc{\equiv}N]^{2+}$ core. Actually, soft π-donor ligands can also be viewed as possessing high-energy filled orbitals, which correctly overlap with π^* antibonding orbitals of the Tc≡N fragment. Owing to the specific spatial distribution of metallic LUMO π^* orbitals, the coordinated atoms are bent away from the Tc≡N multiple bond and the metal ion is forced to lie over the sp plane. Therefore, the resulting geometry is characterized by high values of the Δ parameter. However, filling up of the metal fragment π^* orbitals determines, as a consequence, the corresponding weakening of the Tc≡N bond length. The highly distorted, umbrella-shaped structure of these complexes also prevents a sixth ligand from entering in the trans position to the metal–nitrogen multiple bond.

By decreasing the soft character of the coordinating atoms, the electronic density on the π^* orbitals is consequently reduced. Thus, the lengthening of the Tc≡N bond becomes less pronounced and the effect of geometrical distortions is actually reversed. Furthermore, the gradual decrease of the Δ parameter allows a sixth coordinating ligand to overlap with the high-energy, σ^* orbital of the metal fragment, giving rise to octahedral structures. This latter interaction accounts for the concurrent increase of the Tc≡N bond length on the left side of the curve.

2.2 Trigonal Bipyramidal Complexes

The number of nitrido Tc(V) complexes characterized by a tbp structure is relatively small compared with the number of sp complexes. However, available structural data [14–16] clearly indicate that a tbp arrangement is preferred when the set of four coordinating atoms around the Tc≡N group is formed by a combination of two π-donor and two π-acceptor atoms. A π acceptor is defined here as a coordinating atom possessing empty orbitals with a symmetry appropriate to overlap with filled π orbitals of the metal. Despite the fact that sterical distortions from ideal geometry frequently make it difficult to distinguish sharply between sp and tbp structures, a key structural preference observed in mixed π-donor–π-acceptor nitrido Tc(V) complexes is always apparent. This can be expressed by the remark that the two π-acceptor atoms usually manifest a strong tendency to occupy a reciprocal trans position [16, 17]. Since a tbp arrangement can only accommodate two trans ligands along the C_3 axis perpendicular to the trigonal plane, this suggests that the arrangement with two axial π-acceptor and two equatorial π-donor ligands lying in the same plane of the nitrido group is the most preferred configuration.

The MO diagram of the fragment $[Tc^V{\equiv}N]^{2+}$ (Fig. 5) provides a simple description of the axial interaction between a π-acceptor ligand and the Tc≡N group in tbp complexes. The HOMO possesses the correct symmetry to overlap with empty π orbitals of the two axial π-acceptor atoms. Similarly, a σ bond can be formed by overlapping the fragment LUMO with filled, axial atomic orbitals having the same symmetry. The remaining two coordination positions on the trigonal plane can only accommodate donor atoms having filled π orbitals that overlap with π^* antibonding orbitals of the metal fragment.

This picture of the bonding in tbp nitrido Tc(V) complexes clearly suggests that there exist two different sets of metal fragment orbitals which can be used independently. Specifically, the yz and z^2 orbitals control the axial interaction with the two π-acceptor ligands and, conversely, the π^* orbitals account for the equatorial bonding with two π-donor ligands. This representation will be particularly useful for describing the interaction of five-coordinate tbp nitrido Tc(V) complexes with bidentate ligands discussed next.

3 Five-Coordinate Nitrido Technetium(V) Complexes with Bidentate Ligands

3.1 Symmetrical Complexes

An important class of five-coordinate nitrido Tc(V) complexes is represented by compounds containing two bidentate ligands bound to the same Tc≡N group. Owing to the strong steric hindrance exerted by the electron-rich Tc≡N multiple bond, it is usually found that coordination by two bidentate ligands is more favored than that of a single tetradentate ligand. Generally, this difference determines, at tracer level, faster reaction kinetics for bidentate ligands in comparison with tetradentate species.

Nitrido Tc(V) complexes with a bidentate ligand may occur both as symmetrical and as asymmetrical compounds. A symmetrical complex is represented by the general formula $[Tc(N)(L)_2]^{n+/0/n-}$, where L is a bidentate ligand and n is an integer. Asymmetrical complexes can be written as $[Tc(N)(L)(L')]^{n+/0/n-}$, where L and L' are different bidentate ligands. Symmetrical complexes may assume both sp and tbp geometries depending on the set of coordinating atoms. An example of sp compounds is given by the class of bis(dithiocarbamato) nitrido Tc(V) complexes of the type $\{Tc(N)[R(R')–CS_2]_2\}$ [18, 19]. Symmetrical complexes having a tbp structure are represented by the class of bis(phosphinothiol) nitrido Tc(V) compounds, $\{Tc(N)[R_2P(CH_2)_nS]_2\}$, where n=2, 3 [16, 20].

3.2 Asymmetrical Complexes

Asymmetrical nitrido Tc(V) complexes (simply defined as heterocomplexes) are defined as coordination compounds in which two different bidentate ligands are bound to the same Tc≡N group, and are represented by the general formula $[Tc(N)(L)(L')]^{n+/0/n-}$. The attempt to develop a high-yield synthesis of these types of complexes may first appear to be prevented by basic chemical considerations. Actually, it is reasonable to expect that the reaction of two different bidentate ligands, A and B, with the same Tc≡N group would always yield a statistical mixture of symmetrical and asymmetrical complexes, namely $[Tc(N)(A)_2]$, $[Tc(N)(B)_2]$ and [Tc(N)(A)(B)]. However, the peculiar properties of mixed π-acceptor–π-donor ligands offered the route to the solution of this synthetic problem. The key approach can be outlined as follows.

As previously discussed, the description of the electronic structure of tbp, nitrido Tc(V) complexes revealed the existence of two separated bonding interactions for π-acceptor and π-donor ligands. In particular, π-acceptor ligands interact with an axial set of metal fragment orbitals, thus assuming a reciprocal trans position. Conversely, π-donor ligands overlap with another set of metal orbitals positioned on the equatorial plane containing the Tc≡N group. This picture suggests that the stability of the resulting complex should not be influenced by the way in which the two π-acceptor and two π-donor atoms are connected to each other. More precisely, if in symmetrical $\{Tc(N)[R_2P(CH_2)_nS]_2\}$ complexes the links between the P and S atoms in the two phosphino-thiol ligands were removed and replaced by two separate bridges, one connecting the two P atoms and the other the two S atoms as illustrated in Fig. 8, the resulting complex would be composed by a Tc≡N group coordinated to one diphosphine P~P ligand and one disulfide S–S ligand. It is reasonable to expect that this simple change in the connectivity between atoms would not alter significantly the electronic distribution of the nitrido group. This conversion, therefore, would allow the formation of an asymmetrical complex, but without affecting the starting arrangement of coordinating atoms around the metal center and, consequently, the whole electronic stability.

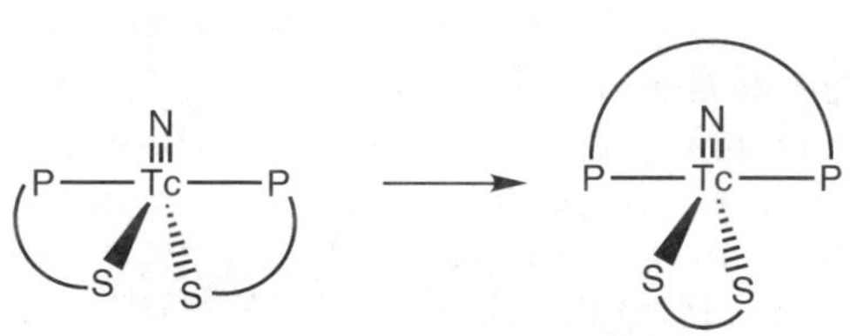

Fig. 8 Conversion of a symmetrical bipyramidal nitrido Tc(V) complex containing two identical mixed π-acceptor–π-donor bidentate ligands into an asymmetrical nitrido Tc(V) complex containing one π-acceptor and one π-donor bidentate ligand. Conversion is simply obtained by changing the connectivity between coordinating atoms without altering the first coordination sphere

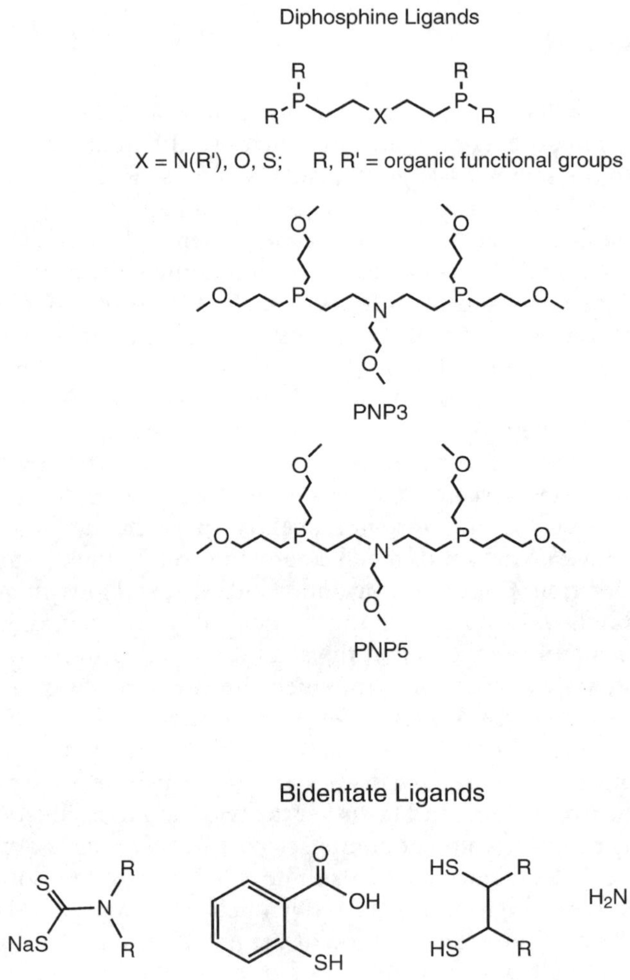

Fig. 9 Representative diphosphine and bidentate ligands utilized in the synthesis of asymmetrical nitrido Tc(V) complexes

Experimental results completely confirmed these theoretical predictions [21, 22]. In a series of reactions, a diphosphine ligand (P~P) and a bidentate ligand (Y~Z), selected from the list reported in Fig. 9, were simultaneously mixed with the precursor nitrido complex $[Tc(N)Cl_2(PPh_3)_2]$ to yield asymmetrical heterocomplexes of general formula $[Tc(N)(P{\sim}P)(Y{\sim}Z)]^{+/0}$. It is important to note that no formation of the corresponding symmetrical complexes was detected in these preparations.

Another strong piece of evidence supporting the predicted theoretical behavior was obtained from the study of monosubstituted complexes of the type $[Tc(N)Cl_2(P{\sim}P)]$. These species were easily prepared by substitution reactions

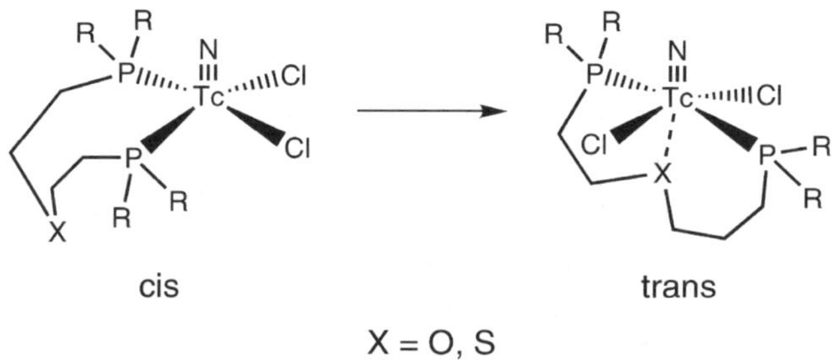

Fig. 10 Schematic drawings of the structures of cis and trans isomers of the complexes $[Tc(N)Cl_2(POP)]$ and $[Tc(N)Cl_2(PSP)]$. The cis isomer exhibits a distorted square pyramidal structure with the two P atoms in a reciprocal cis position, whereas the trans isomer has a distorted trigonal bipyramidal structure with the two P atoms occupying the axial positions

of a diphosphine ligand (P~P)onto the precursor complex $[Tc(N)Cl_2(PPh_3)_2]$. With P~P ligands containing either an oxygen or a sulfur heteroatom within the carbon chain connecting the two terminal P atoms (POP and PSP diphosphines, respectively), the resulting complexes $[Tc(N)Cl_2(PXP)]$ (*X* is O, S) were found to occur in two distinct isomeric forms (Fig. 10). One complex corresponds to a cis isomer with the two P atoms in a reciprocal cis position, while the other complex represents a trans isomer where the same atoms assume a reciprocal trans position. Interestingly, in solution, the cis isomer transformed into the trans isomer, which, therefore, constitutes the thermodynamically stable species.

Different behavior was observed with diphosphine ligands containing an amino nitrogen within the carbon bridge between the two P atoms [PN(*R*)P diphosphines]. The configuration of the resulting complexes, $\{Tc(N)Cl_2[PN(R)P]\}$, was found to be strongly influenced by the nature of the *R* group on the N atom. When *R* was composed by a chain of carbon atoms, only the cis isomer was isolated [22, 23]. In contrast, using H for *R*, only the trans isomer was obtained. Surprisingly, however, this latter species exhibits an octahedral structure in which the PN(*R*)P ligand behaves as a tridentate ligand assuming a meridional configuration, in which the two P atoms are coordinated in a reciprocal trans position, and the N(*R*) atom is bound in a cis position relative to the Tc≡N group (Fig. 11) (G. Bandoli, unpublished data). These findings suggest that, in $\{Tc(N)Cl_2[PN(R)P]\}$ complexes, achievement of the stable trans configuration is strongly hindered by sterical constrains brought about by the pendant *R* group on the nitrogen atom. Removal of this group causes the rapid rearrangement of the structure to acquire the stablest trans arrangement. Evidently, these observations give further support to the general statement concerning the preference of the two π-acceptor atoms for a reciprocal trans position in nitrido Tc(V) complexes.

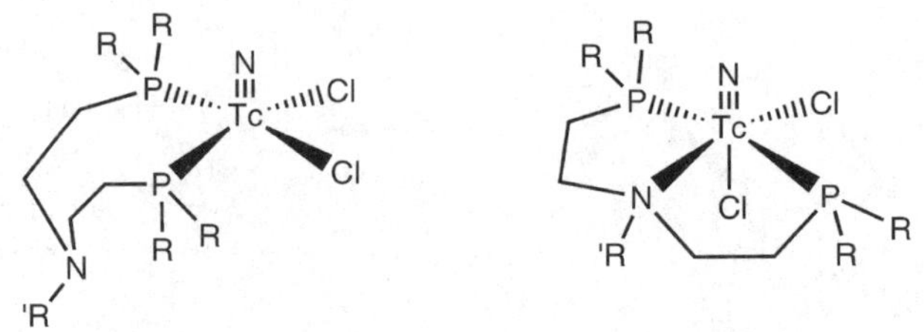

cis trans

R' = organic functional group R' = H

Fig. 11 Schematic drawings of the structures of cis and trans isomers of the complexes {$Tc(N)Cl_2[PN(R)P]$}. The cis isomer exhibits a distorted square pyramidal structure with the two P atoms in a reciprocal cis position, whereas the trans isomer has an octahedral structure with the two P atoms in a reciprocal trans position

4 Nitrido Technetium(V) Radiopharmaceuticals

4.1 Heart Imaging

A cardiac perfusion tracer is a compound that is selectively captured by the myocardium, and its accumulation is proportional to the blood flow passing through the heart. An ideal heart perfusion imaging agent should possess a few important properties, such as a high first-pass extraction, i.e., it should be quickly extracted by myocardial tissue immediately after intravenous injection. Moreover, it should be rapidly washed out from critical, nontarget organs such as lungs, liver, and from the blood to allow a clear visualization of the cardiac region. A further requirement is that cardiac uptake should be linear over the whole range of blood flows passing through the organ [24].

The first Tc-99m radiopharmaceuticals for heart perfusion imaging were introduced more than a decade ago [25, 26]. However, the properties of these tracers are still far from ideal. Both categories of symmetrical and asymmetrical nitrido Tc(V) complexes were found to have potential application for cardiac imaging. In particular, the complexes {$^{99m}Tc(N)[Et(OEt)N{-}CS_2]_2$} (^{99m}TcN-NOET) and $[^{99m}Tc(N)(PNP5)(DBODC)]^+$ (^{99m}TcN-DBODC), where PNP5 is bis[(dimethoxypropylphosphino)ethyl]ethoxyethylamine) and DBODC is diethoxyethyldithiocarbamate, showed the most interesting biological properties, which, in some respect, appear to approach more closely the ideal behavior. These agents are described in the following paragraphs.

4.2
^{99m}TcN-NOET

The first class of potentially useful radiopharmaceuticals characterized by the presence of a $[Tc{\equiv}N]^{2+}$ core was that of neutral bis(dithiocarbamato) nitrido Tc(V) complexes of general formula $\{^{99m}Tc(N)[R(R')N{-}CS_2]_2\}$ [19, 26]. These complexes have a sp geometry with an apical Tc≡N group and two identical dithiocarbamate ligands symmetrically arranged around the metal ion. It was found that these complexes accumulate in the myocardium both in animal models and in humans. Depending on the chemical nature of the lateral groups, retention of activity in the heart region was highly variable, ranging from a few minutes to hours. ^{99m}TcN-NOET (Fig. 12) was the member of this class showing the most promising imaging properties and, therefore, was selected for further clinical evaluation in patients.

Biodistribution studies in healthy volunteers demonstrated that heart uptake of ^{99m}TcN-NOET is remarkably high, approaching 4% of injected activity. Washout is slow with a half-life of approximately 3 h [27]. Heart uptake was found to be almost proportional to the blood flow at low and medium flux rates, becoming approximately 80% at higher fluxes. Accumulation of activity after the first cardiac transit (first-pass extraction) is in the range 80–90% [28, 29]. The lungs are another target organ and they are visualized immediately after injection, but successively undergo fast clearance with a half-life of approximately 10 min. Elimination occurs mostly through the liver, where a large portion of activity is retained and only slowly eliminated. Lung uptake and persistent liver accumulation may become an important drawback in patients with a significant impairment of heart function causing abnormal blood circulation from the lung compartment.

A relevant feature of ^{99m}TcN-NOET derives from its in vivo kinetic behavior, which gives rise to the phenomenon called "'redistribution". This term is commonly referred to as the ability of a tracer to move across viable heart regions filled with a different concentration of activity, a situation that usually occurs when some areas within the heart tissue receive a lower blood supply owing to the partial occlusion of coronary vessels (ischemia). When an ischemic patient is injected with a cardiac tracer under stress conditions, a sharp contrast is observed between normally perfused tissue and ischemic areas (defects). In contrast, when the patient is at rest, the activity becomes more uniformly dis-

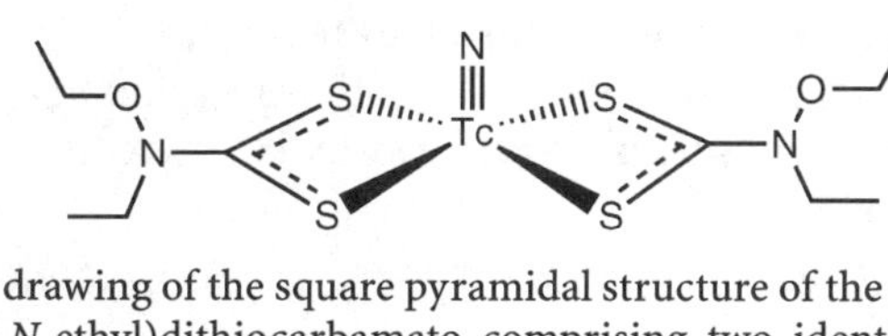

Fig. 12 Schematic drawing of the square pyramidal structure of the cardiac imaging agent ^{99m}TcN-(*N*-ethoxy, *N*-ethyl)dithiocarbamato comprising two identical bidentate ligands coordinated to the same metal center

tributed as a result of the lower demand for blood supply by normal tissues, and defect areas are usually not evidenced. However, in the presence of a necrotic lesion (infarct), in which no viable myocites survive, the defect remains persistent also under rest conditions. Therefore, a tracer which is capable of differentiating between ischemic, but still viable, and infarcted myocardium would be of great clinical utility. This characteristic can be observed only for those tracers which are able to diffuse slowly from high-activity regions toward low-activity regions (redistribution) when the patient is allowed to rest. In these conditions defects usually disappear when repeating a patient scan after a few hours, but obviously this does not occur when necrotic tissues are already present. Thus, with a single injection it would be possible to evaluate if the patient is suffering from a reversible ischemic attack or if tissue has been irreversibly damaged. Only the monocation Tl-201 was previously known to undergo redistribution in going from stress to rest conditions. Monocationic Tc-99m complexes do not possess this feature, and two injections, under stress and rest conditions, are usually required to discriminate between ischemia and infarct. ^{99m}TcN-NOET was the first Tc-99m agent exhibiting this useful, thallium-like behavior that, at present, remains the most unique trait of this agent.

It turns out that ^{99m}TcN-NOET was the first neutral agent showing a prolonged uptake within the heart. Previously, only Tc-99m complexes carrying a monopositive charge were found to possess a stable myocardial localization, and this feature was mostly determined by their specific localization mechanism [30]. Subcellular studies conducted on isolated rat's cardiac tissue demonstrated that, after entering a myocite through passive diffusion, a monocationic heart Tc-99m agent is captured by the mitochondria as a result of its interaction with the negative charge gradient existing across the mitochondrial membrane. It was thought, therefore, that only this category of positively charged complexes would be able to accumulate in the heart tissue, neutral complexes necessarily being rapidly washed out [31]. Thus, the high heart uptake and slow clearance of ^{99m}TcN-NOET were somewhat unexpected. Subcellular distribution studies revealed that the localization site of this tracer is not the mitochondrial membrane, but rather the external cell membrane [32]. In vitro experiments conducted with calcium blockers suggested that uptake of ^{99m}TcN-NOET would be related to the affinity of this complex for channel proteins involved in calcium transport [33]. Recent studies have pursued another possibility that the primary uptake site of ^{99m}TcN-NOET is cardiac endothelial cells and not the myocites [33]. If confirmed, this result would have a significant impact on the assessment of the diagnostic usefulness of this agent, mostly because cardiac endothelium always receives the first injury at the onset of an ischemic process. Thus, the possibility to visualize this damage before it spreads to the surrounding myocites would dramatically improve the early detection of the cardiac infarct [34].

4.3 ^{99m}TcN-DBODC

Conversion of the symmetrical, bis-substituted complexes {^{99m}Tc(N)-[*R*(*R*′)N–CS_2]$_2$} to asymmetrical species containing two different bidentate ligands coordinated to the same Tc≡N group had a dramatic impact on the observed biological properties. The preparation of the monocationic nitrido Tc(V) heterocomplexes {Tc(N)[PN(*R*)P][*R*(*R*′)N–CS_2]}$^+$ was accomplished through the simultaneous reaction of a suitable nitrido Tc(V) precursor with the ligands PN(*R*)P and [*R*(*R*′)N–CS_2]Na. In particular, using the diphosphines bis[(dimethoxypropylphosphino)ethyl]methoxyethylamine) and PNP5 (Fig. 9), a novel class of myocardial tracers exhibiting superior imaging qualities was obtained [35]. Within this class, the compound [^{99m}Tc(N)(PNP5)(DBODC)]$^+$ (^{99m}TcN-DBODC), where DBODC is diethoxyethyldithiocarbamato, was selected as a promising candidate for clinical applications [36]. The structure of this complex is illustrated in Fig. 13. It is composed of one PNP5 ligand and one DBODC ligand bound to the same [Tc≡N]$^{2+}$ core through the two P atoms, and the two S atoms of the CS_2^- group, respectively. The resulting geometry is highly distorted sp.

The first biological evaluation of ^{99m}TcN-DBODC was carried out in rats. Biodistribution data showed that the heart is the most important target organ for this complex. Lung clearance is remarkably fast and elimination occurs both through kidneys and liver. However, liver activity is rapidly washed out into the intestine, thus leaving the cardiac region completely emptied of background activity. This result appears particularly evident when heart/lung and heart/liver ratios are considered. The variation in time of heart/lung and heart/liver ratios for ^{99m}TcN-DBODC as compared with two commercial tracers, ^{99m}Tc-Sestamibi and ^{99m}Tc-Tetrofosmin, is reported in Fig. 14.

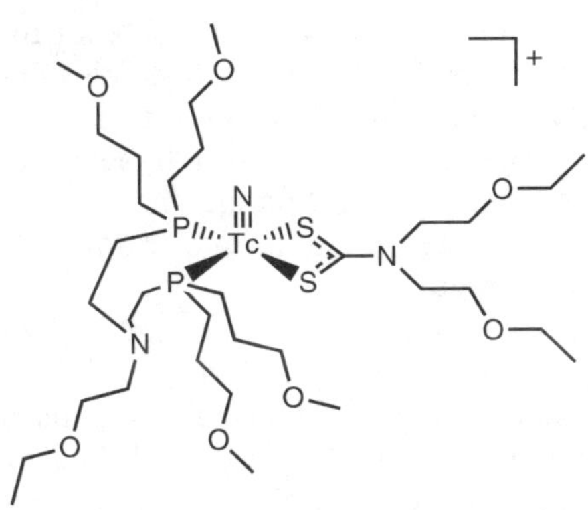

Fig. 13 Schematic drawing of the distorted square pyramidal structure of the cardiac imaging agent ^{99m}TcN-diethoxyethyldithiocarbamato (*DBODC*) comprising two different bidentate ligands coordinated to the same metal center

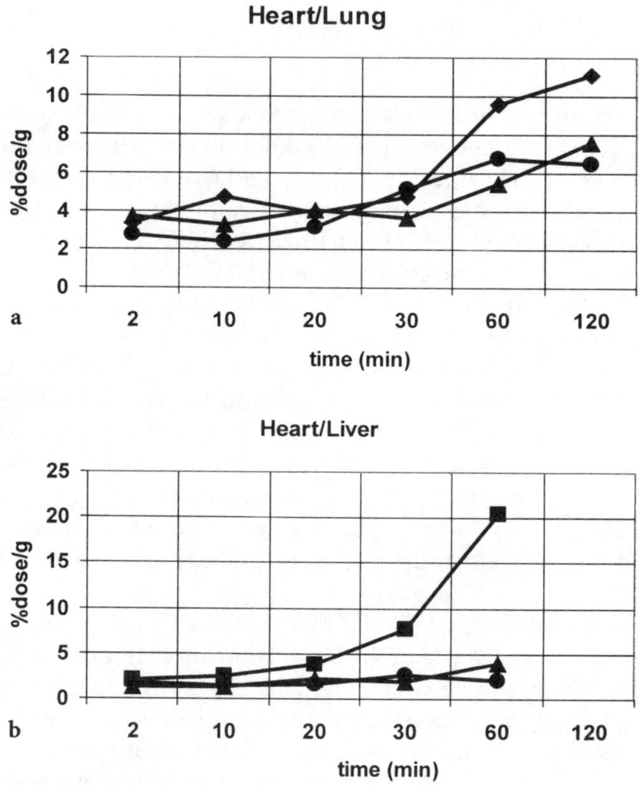

Fig. 14 Time variation of **a** heart/lung and **b** heart/liver ratios for the complexes ^{99m}TcN-DBODC (*squares*), ^{99m}Tc-Sestamibi (*circles*) and ^{99m}Tc-Tetrofosmin (*triangles*) measured at a specific time after injection in rats. The values are expressed as the ratio between the activity localized in the heart and the activity localized in the lungs and the liver, respectively

Though heart/lung ratios are nearly comparable for the three compounds, the heart/liver ratio of ^{99m}TcN-DBODC rises dramatically over time becoming much greater, at 1 h post injection, than the ratios of the other two agents, which remain almost constant. This biological behavior may provide a useful tool for improving the quality of heart images. A tomographic view of the rat heart, obtained with a dedicated small-animal scanner [37] at 1 h after injection of ^{99m}TcN-DBODC, is reported in Fig. 15. The images clearly show the cardiac region without interference from surrounding activity. Liver is not visualized, activity being localized mostly in kidneys and intestine. Dog studies nicely confirmed the results obtained in rats and demonstrated that the kinetic behavior of ^{99m}TcN-DBODC parallels that of ^{99m}Tc-Sestamibi [37]. The unprecedented biological properties of ^{99m}TcN-DBODC appear to approach more closely those of an ideal perfusion tracer and, therefore, this new agent is currently under further investigation for assessing the feasibility in humans of myocardial perfusion imaging with superior quality.

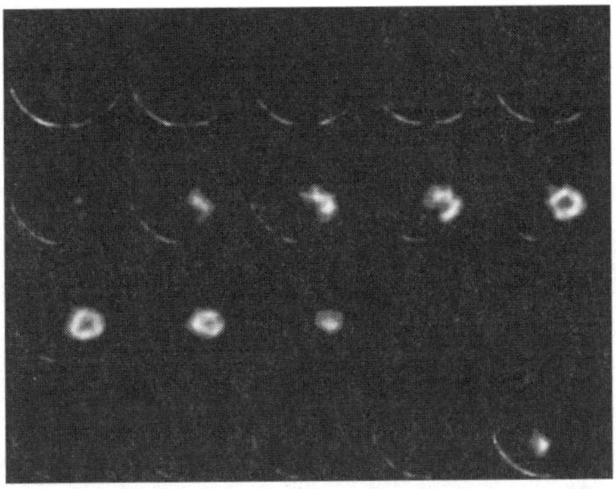

Fig. 15 Tomographic images of the rat heart obtained with a small-animal scanner at 1 h after injection of ^{99m}TcN-DBODC. The images represent heart horizontal slices moving from the top of the heart (*upper left*) to the heart apex (*middle*). The last image corresponds to kidney uptake (*bottom right*)

4.4 Labeling of Biologically Active Molecules

In recent years, the labeling of biologically active molecules has become the most effective tool to generate affinity of a Tc-99m complex for a specific biological target. The most common design of this type of diagnostic agent is based on the application of two main methods called bifunctional (or pendant) and integrated approaches [38]. The first step in both strategies always requires the selection of a convenient biologically active molecule or a pharmaceutical drug having affinity for a definite biological substrate. After this common onset, the two methods follow separate routes. The bifunctional approach literally suggests putting together the bioactive group and the metal complex through a suitable linkage, a result that can be conveniently achieved using a bifunctional ligand. This species could be represented as a composite molecule combining two different molecular blocks such as a strong chelating group for the metal and a biomolecule. Thus, in the final ligand, these two moieties become tightly connected by means of a chemical linker. Clinging the chelating system to the metal results in the formation of a conjugate complex, which will incorporate the bioactive group within its structure as an appended side chain. In contrast, the integrated approach focuses on the selected biomolecule itself. In fact, the structure of this substance may serve as a template for shaping the structure of the final Tc-99m radiopharmaceutical. The key step is to identify a region of the biomolecule that can be easily replaced without affecting its biological properties. The final radiopharmaceutical, therefore, will be assembled simply by removing this nonessential part and filling the resulting empty position with a metal-containing fragment having structural similarity with the substituted portion. In principle, if the size and the geometry of the metallic block fit correctly within the edges of the molecular mould, the final integrated complex should display the same biological properties of the primitive biomolecule. Obviously, the ultimate success of both designs lies in their ability to keep unaltered the intrinsic biological behavior of the starting biomolecule.

The previously mentioned approaches provide a theoretically useful picture of the molecular structure of a radioactive tracer as composed of different pieces or fragments, which can be conveniently put together to obtain the final radiopharmaceutical. However, the combination of the various fragments to yield a stable product is not always simple to perform. In recent years, an alternative solution to this problem has emerged. The method exploits the chemical properties of certain types of substitution-labile technetium complexes showing a marked reactivity only toward ligands bearing some specific sets of coordinating atoms. The characteristic feature of these complexes resides in the presence of two different groups of ligands in the coordination environment around the metal center. One set is composed by ligands that are tightly bound to the metal ion. The resulting strong ligand field allows the relevant stabilization of the metal oxidation state and prevents the occurrence of oxidation–reduction reactions. The residual positions of the coordination arrangement are usually occupied by another set of ligands forming weak bonds with the metal. This weakly bound group can be easily removed when some incoming ligand having selective bonding affinity for the robust part of the complex (metal fragment) undergoes strong interaction with the metal center. The reaction is expected to be highly selective, thus affording the final substituted complex in great yield.

Metal fragments can be efficiently employed for tethering a biomolecule with a Tc-99m complex providing that the bioactive group includes the appropriate set of coordinating atoms. The high affinity of the precursor metal fragment for the specific bonding sites on the bioactive ligand would ensure the careful fitting of these two molecular building blocks to form the final, conjugate complex. It was found that the novel mixed-ligand complexes of the type $[Tc(N)Cl_2(P{\sim}P)]$ discussed in the previous sections provide a definite example of the metal-fragment approach. In these complexes, the arrangement of atoms $[Tc(N)(P{\sim}P)]^{2+}$, formed by a Tc≡N group coordinated to a chelating diphosphine ligand P~P, plays the role of a robust metallic block. In contrast, the two chlorine atoms can be easily displaced by an incoming bidentate ligand (L) to afford the asymmetrical complexes $[^{99m}Tc(N)(P{\sim}P)(L)]^{0/+}$ [22, 23]. A salient property of this fragment comes from the fact that its structure can switch between sp and tbp geometries depending on the relative cis or trans position of the two π-acceptor P atoms. A tbp arrangement is more favored as a result of the strong tendency of the two P atoms to span a reciprocal trans position. This constitutes the key factor controlling the selective reactivity of this fragment toward nucleophilic ligands. Actually, the tendency toward a tbp structure also brings about the complete rearrangement of the electronic density, which appears to concentrate only on the two π-acceptor atoms [23]. This leaves the other two coordination positions depleted of electrons and, thus, activated toward reaction with electron-rich ligands surrounded by a set of filled atomic orbitals having the appropriate symmetry to overlap with the empty antibonding π orbitals of the metal fragment. Owing to its peculiar electronic structure, the fragment $[Tc(N)(P\sim P)]^{2+}$ exhibits strong electrophilic character, and reacts selectively only toward soft π-donor bases such as S^- or O^-. When a biomolecule carrying an appropriate set of two π-donor atoms is selected, it easily combines with the fragment $[Tc(N)(P\sim P)]^{2+}$ to yield the corresponding conjugate complex. Thus, the metal synthon $[^{99m}Tc(N)(P\sim P)]^{2+}$ could be conveniently utilized to obtain a very broad class of asymmetrical nitrido Tc(V) complexes with a variety of bidentate bioactive ligands.

The finding that the amino acid cysteine displays excellent coordinating properties toward the fragment $[Tc(N)(P\sim P)]^{2+}$ provided an almost natural way for incorporating short peptide chains into a nitrido Tc-99m asymmetrical complex [39]. Cysteine binds the fragment $[Tc(N)(P\sim P)]^{2+}$ either through the $[NH_2, S^-]$ pair or, alternatively, the $[O^-, S^-]$ pair (Fig. 16). These reactions are highly specific and quantitative. Thus, peptide sequences having a cysteine group sufficiently accessible to bind the technetium atom of the basic motif $[Tc(N)(P\sim P)]^{2+}$ would be the most obvious candidates for labeling studies. The application of the metal-fragment approach can be outlined as follows. A selected peptide is combined with a cysteine residue either through the terminal amino group to give N-functionalized *O,S*-cysteine ligands or, alternatively, through the terminal carboxylic group to yield the corresponding COO-functionalized *N,S*-cysteine ligands. The resulting bifunctional ligand (cys~) is then reacted with the metal fragment $[^{99m}Tc(N)(P\sim P)_2]^{2+}$ to afford the mixed, asymmetrical complex $[^{99m}Tc(N)(P\sim P)(cys\sim)]^{0/+}$. As a result, the peptide sequence

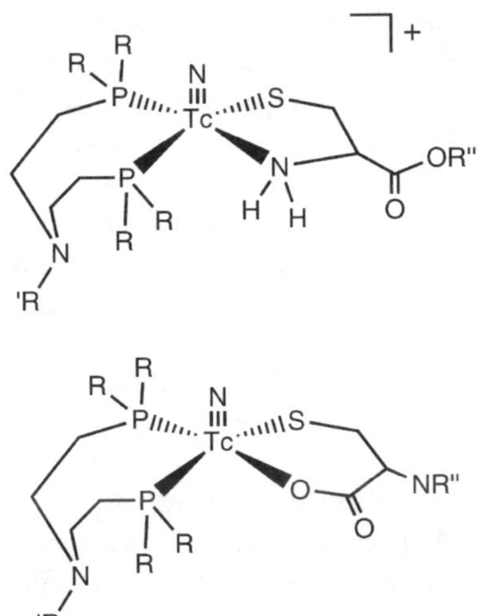

Fig. 16 Schematic drawing of the possible structures of nitrido Tc-99m heterocomplexes with cysteine. This compound can bind to the metal center either as $[NH_2, S^-]$ monoanionic bidentate ligand or as $[O^-, S^-]$ dianionic bidentate ligand

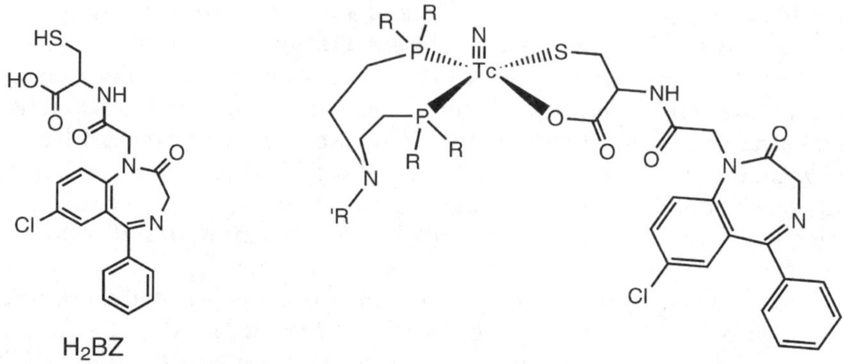

Fig. 17 Schematic drawing of the structures of the bifunctional ligand H_2BZ and of the corresponding nitrido Tc-99m heterocomplexes for imaging benzodiazepine receptors

remains stably incorporated into the structure of the final complex. The overall charge of the resulting asymmetrical complex is dependent on the pair of π-donating atoms of cysteine. A $[NH_2, S^-]$ chelating system yields monocationic complexes, while neutral complexes arise when cysteine coordinates through the $[O^-, S^-]$ pair.

Attempts to develop Tc-99m imaging agents for the central nervous system using the metal-fragment approach with the $[Tc(N)(P{\sim}P)]^{2+}$ synthon have recently been undertaken [40]. A receptor-specific complex for imaging benzodiazepine receptors was designed by first selecting the drug desmethyldiazepam as the starting bioactive molecule. This substance was successively modified by appending a cysteine moiety through its terminal amino group to the central seven-membered ring of the benzodiazepine moiety. The resulting bifunctional ligand H_2Bz (Fig. 17) was in turn reacted with the fragment $[Tc(N)(P{\sim}P)]^{2+}$ to yield the neutral asymmetric complexes, $[Tc(N)(P{\sim}P)(Bz)]$ (Fig. 17). These complexes occur in two distinct isomeric forms depending on the syn or anti orientation of the pendant benzodiazepine group relative to the $Tc{\equiv}N$ multiple bond. A similar approach was utilized in the investigation of receptor-specific tracers for $5HT_{1A}$ receptors [41]. The 2-methoxyphenylpiperazine pharmacophore, which displays preferential affinity for $5HT_{1A}$ receptors, was conjugated to the amino group of cysteine to obtain the N-functionalized *O,S*-cysteine ligand H_2HT (Fig. 18). The strong electrophilic $[Tc(N)(P{\sim}P)]^{2+}$ metal fragment efficiently reacts with this bifunctional chelating ligand affording the neutral nitrido Tc(V) heterocomplexes, $[Tc(N)(P{\sim}P)(HT)]$ (Fig. 18). As observed before, a mixture of syn and anti isomers was obtained, the latter being the thermodynamically favored species. Biodistribution studies were carried out in rats, and affinity for benzodiazepine and $5HT_{1A}$ receptors was assessed through in vitro binding experiments on isolated rat cerebral membranes. The results showed that the labeling procedure determined the almost complete loss of receptor affinity of the conjugate com-

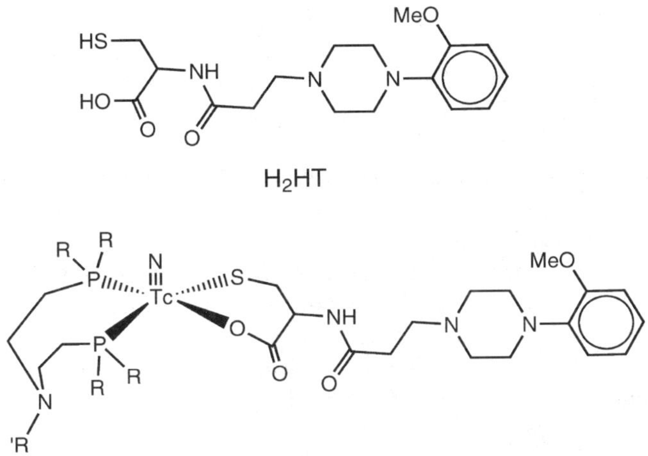

Fig. 18 Schematic drawing of the structures of the bifunctional ligand H_2HT and of the corresponding nitrido Tc-99m heterocomplexes for imaging $5HT_{1A}$ receptors

plexes presumably because of the strong perturbation caused by the proximity of the metal-containing block to the bioactive group.

A novel category of Tc-99m radiopharmaceuticals for diagnostic imaging applications in oncology was developed by selecting the compound distamycin as a bioactive substrate for labeling with the metal fragment $[Tc(N)(P\sim P)]^{2+}$ [42]. The tripyrrole peptide distamycin A is a naturally occurring antibiotic which binds to the minor groove of DNA by hydrogen bonds, van der Waals contacts and electrostatic interactions. This oligopeptide exhibits a high selec-

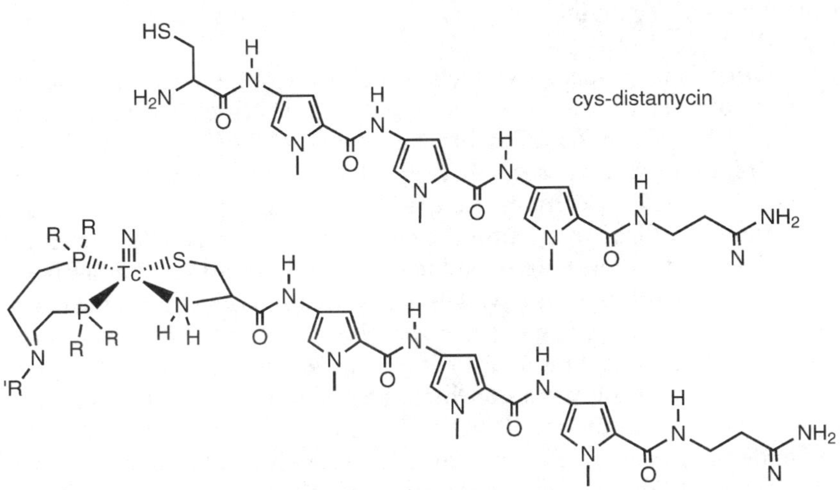

Fig. 19 Schematic drawing of the structures of the bifunctional ligand cysteine–distamycin and of the corresponding nitrido Tc-99m heterocomplexes

tivity for AT rich sequences and sticks preferentially to 5′-AAATT-3′ sequences. Introduction of L-cysteine on the N-terminus of deformyl distamycin gave the bifunctional ligand shown in Fig. 19. The resulting hybrid cysteine–distamycin (HcysD) retained an efficiency in arrested polymerase-chain reactions experiments comparable to that of the parent distamycin A. The ligand HcysD was efficiently labeled with the metal synthon $[Tc(N)(P{\sim}P)]^{2+}$ to give the new class of monocationic heterocomplexes reported in Fig. 19. In these complexes, HcysD binds the metal center through the deprotonated thiol sulfur atom and the neutral amino nitrogen atom. The complexes were found to be highly stable in physiological solution and in plasma. In vitro experiments showed that they display both DNA binding activity and sequence-selectivity and, therefore, can be used as suitable models for the design of potentially useful Tc-99m imaging agents for the diagnosis of various types of cancerous diseases.

5 Nitrido Rhenium-188 Radiopharmaceuticals

5.1 Reduction of the Tetraoxo Rhenium-188 Anion

The use of radiolabeled compounds for the treatment of degenerative diseases currently constitutes a rapidly expanding field in nuclear medicine [43]. The basic principle of this therapeutic method involves administration of a radioactive molecule that should incorporate at least one radionuclide emitting high-energy, massive particles such as β-electrons or α-particles. After reaching the target tissue, the radiopharmaceutical will deliver its radiation in situ, and the interaction with the decay particle may eventually kill the diseased cell, thus producing a therapeutic effect.

Rhenium-188 is a β-emitting nuclide that is currently attracting much interest as a potential candidate for therapeutic applications because of its useful nuclear properties and availability. Rhenium-188 decays through the emission of a high-energy β-particle ($E_{\beta max}$=2.1 MeV) with a half-life of 16.9 h. Associated with this decay mode there is also a γ-emission of 152 keV that can be conveniently utilized to monitor the course of therapy using a conventional γ camera. Another important advantage of employing Re-188 radiopharmaceuticals comes from the easy availability of this radionuclide, which is produced through a transportable generator system under the chemical form of the tetraoxo perrhenate anion $[^{188}ReO_4]^-$. This generator is, essentially, composed of an alumina column onto which the father nuclide W-188 is absorbed as tetraoxo wolframate anion, $[^{188}WO_4]^{2-}$. Radioactive β-decay of the W-188 nuclide yields the parent $[^{188}ReO_4]^-$, which is successively eluted with a physiological solution. This situation, therefore, parallels completely that of the nuclide Tc-99m, which is obtained through the $^{99}Mo/^{99m}Tc$ generator system in the form of the tetraoxo pertechnetate anion $[^{99m}TcO_4]^-$, which always constitutes the starting compound for preparing Tc-99m radiopharmaceuticals. Likewise, $[^{188}ReO_4]^-$ is the ubiquitous starting material for the preparation of Re-188 radiopharmaceuticals.

However, since technetium and rhenium belong to the same group 7 of the transition series, the similarities between Tc-99m and Re-188 radiopharmaceuticals are even more pronounced. In fact, owing to lanthanide contraction, technetium and rhenium have almost identical ionic radii. This indicates that, when these two elements form analogous complexes having exactly the same chemical structure and stability, and differing only in the metal center, these species should exhibit the same in vivo biological behavior. Despite this, there exists a fundamental difference between the values of the standard reduction potentials of the redox reactions involving technetium and rhenium compounds. On average, $E°$ of a technetium process is 200 mV higher than that of the corresponding rhenium process. This implies that reduction of $[^{188}ReO_4]^-$ should be much more difficult than that of $[^{99m}TcO_4]^-$. As a consequence, the methods utilized for the preparation of Re-188 radiopharmaceuticals cannot simply follow routes employed for obtaining Tc-99m complexes, and usually more drastic conditions are required [44]. This fact always constitutes a fundamental obstacle for the development of new Re-188 radiopharmaceuticals.

A solution to this problem has recently been proposed [45]. This approach was inspired by a basic principle of inorganic chemistry sometimes called expansion of the coordination sphere [46]. Briefly, the principle states that in electron-transfer reactions involving coordination complexes, a redox process is more favored when the geometries of the oxidized and reduced pair are very similar. Since $[^{188}ReO_4]^-$ constitutes the ubiquitous starting material in all radiopharmaceutical preparations, conversion of this species to a final complex containing the metal in a lower oxidation state (less than +7) is always an inevitable step. However, the tetraoxo anion has a four-coordinate, tetrahedral geometry, while, commonly, rhenium complexes in lower oxidation states possess a more expanded five- or six-coordinate geometry. This suggests that the geometrical change always occurring in redox reactions involving $[^{188}ReO_4]^-$ should have a detrimental influence on the standard reduction potential of the whole process. To avoid this negative effect it would be necessary to literally expand the coordination arrangement of $[^{188}ReO_4]^-$ before effecting the electron transfer. This result could be simply achieved by forming an intermediate Re(VII) complex, thus having an expanded coordination arrangement, through the reaction of $[^{188}ReO_4]^-$ with some suitable ligand. In this way, the transfer of electrons will occur between two species having similar geometries, a situation that should bring about a significant increase of the standard reduction potential.

It was found that oxalate ions, $(C_2O_4)^{2-}$, are excellent ligands for rhenium in the +7 oxidation state. This simple dicarboxylic acid is able to form intermediate Re(VII) complexes that are sufficiently stable to allow the expansion of the coordination sphere, but also sufficiently substitution-labile to permit the re-

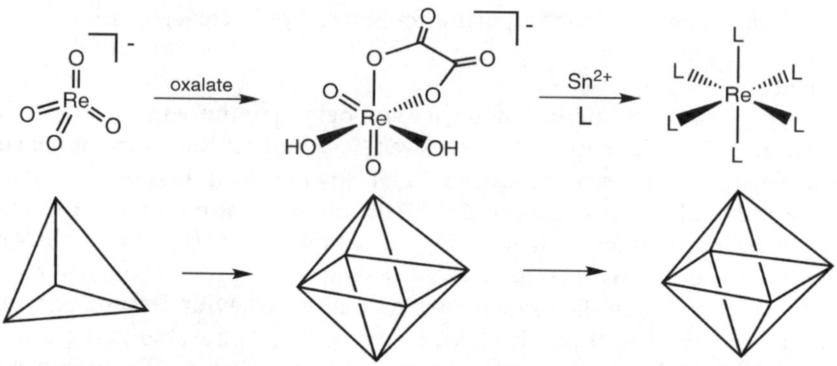

Fig. 20 Pictorial illustration of the hypothetical mechanism of the action of oxalate on the reduction of $[^{188}ReO_4]^-$. Oxalate ions react first with the teraoxo anion forming an intermediate Re(VII) complex and causing the concomitant expansion of the coordination sphere of the metal from tetrahedral to octahedral. Successively, electron transfer takes place from Sn^{2+} ions to the octahedral metal center

placement of the ancillary ligands by some stronger ligand of biological interest. The results demonstrated that addition of $Na_2[C_2O_4]$ to radiopharmaceutical preparations starting from generator-eluted $[^{188}ReO_4]^-$ dramatically improved the yield of the final product. Moreover, no drastic conditions such as high temperature or high concentrations of the reducing agent were required. A pictorial representation of the action of oxalate is reported in Fig. 20.

The first application of the new, oxalate-based reduction method was carried out for preparing the complex $[^{188}Re(O)(DMSA)_2]^-$, where H_2DMSA is *meso*-2,3-dimercaptosuccinic acid. This complex had been previously obtained, under strong acidic conditions, by heating $[^{188}ReO_4]^-$ at high temperature for a prolonged time, and in the presence of a large amount of $SnCl_2$ as a reductant. Such conditions are completely unsuitable for any in vivo clinical study in humans. However, addition of oxalate ions changed dramatically the course of this reaction [45]. The high-yield production of the complex $[^{188}Re(O)(DMSA)_2]^-$ was easily obtained at pH>5, by keeping the reaction solution at room temperature for a few minutes and using a small amount of $SnCl_2$ (less than 0.5 mg).

5.2 Bis-Dithiocarbamato Nitrido Rhenium-188 Complexes

The oxalate-based approach has been subsequently utilized to develop the first efficient procedure for producing the $[^{188}Re{\equiv}N]^{2+}$ core from $[^{188}ReO_4]^-$, at tracer level and under physiological conditions [47]. For this purpose, the method originally developed for the tracer-level preparation of the analogous $[^{99m}Tc{\equiv}N]^{2+}$ core was employed [4]. As mentioned before, this procedure involves the reaction of $[^{99m}TcO_4]^-$ with the N^{3-}-donor DTCZ and $SnCl_2$ to afford a mixture of intermediate nitrido Tc-99m complexes. This intermediate mixture is suddenly converted into a single product upon addition of a strong coordinating ligand. In the first attempt at transferring the procedure to the preparation of nitrido Re-188 radiopharmaceuticals, dithiocarbamate ligands of the type $[R_2N\text{-}C(=S)S]Na$ were employed and, thus, the formation of bis-dithiocarbamato nitrido Re-188 complexes was investigated. However, it was found that the plain application of the experimental conditions used with Tc-99m completely failed to give the corresponding Re-188 complexes, $[^{188}Re(N)(R_2NCS_2)_2]$. In contrast, addition of sodium oxalate allowed these complexes to be obtained in very high yield. The observed results clearly suggest that the rate-determining step in these reactions is always associated with the reduction of the tetraoxo perrhenate anion. The synthesis of the complexes $[^{188}Re(N)(R_2NCS_2)_2]$ was carried out using a two-step procedure. In the first step, $[^{188}ReO_4]^-$ was reacted with DTCZ, $SnCl_2$ and oxalate. Chromatographic analysis showed that this reaction yielded two intermediate complexes. Though the exact chemical nature of these species was not fully determined, it was found that, upon addition of the appropriate dithiocarbamate ligand, the intermediate mixture was rapidly converted into the final nitrido complex $[^{188}Re(N)(R_2NCS_2)_2]$. This strongly supports the hypothesis that the molecular structure of the two intermediate complexes should comprise a terminal $Re{\equiv}N$ group.

Biological experiments carried out in rats showed that the biodistribution of $[^{188}Re(N)(R_2NCS_2)_2]$ complexes parallels exactly that observed for the analogous Tc-99m complexes $[^{99m}Tc(N)(R_2NCS_2)_2]$ described previously [19]. In particular, heart was one of the most important target organs. This fact clearly demonstrates that, inside a matching pair of technetium and rhenium complexes possessing identical molecular structure and stability toward in-vivo redox reactions, the corresponding radiocompounds always exhibit the same biological behavior.

5.3 Rhenium-188 Lipiodol

The complexes $[^{188}Re(N)(R_2NCS_2)_2]$ have been successively utilized for developing the first nitrido Re-188 therapeutic agent for the treatment of hepatocellular carcinoma (HCC) [48, 49]. This pathological state is one of the most common cancerous diseases in the world and is particularly frequent in Asiatic and South American areas. It often appears late, and successful surgical resection is difficult or impossible in most patients. Carriers of therapeutic agents that treat hepatoma effectively, without damage to normal tissues, are actively investigated. An example of this type of carrier is lipiodol, an iodinated ethyl ester of the poppy-seed fatty oil containing 38% iodine by weight, and is usually employed as a radiological contrast material. Administration through the hepatic artery in HCC patients is followed by selective and prolonged retention of lipiodol within the HCC. This unique characteristic suggests its potential as a carrier for radiotherapeutic agents. Iodine-131 and yttrium-90 labeled lipiodol have been developed for the treatment of hepatoma, but these agents also accumulate in nontarget organs, especially in the lungs, gut and bone marrow, a fact that strongly limits their clinical usefulness. Various attempts at labeling lipiodol with Re-188 have been proposed, but most of them are exceedingly complicated and difficult to apply under controlled conditions [50]. A simple and elegant approach involves dissolution of a strongly lipophilic Re-188 compound into lipiodol, which constitutes a highly hydrophobic material [51]. Using this strategy, Re-188 would be tightly retained into the fatty oil as a consequence of its strong hydrophobic interaction with the lipophilic metal complex. The successful application of this approach always requires the high-yield production of a Re-188 complex that possess enough stability and lipophilicity to remain firmly trapped in lipiodol. An attempt to employ this labeling method has recently been reported [51–53]. A series of oxo complexes of Re-188 were prepared by reacting $[^{188}ReO_4]^-$ with derivatives of the tetradentate ligand 3,3,10,10-tetramethyl-1,2-dithia-5,8-diazacyclodecane, and then mixed with lipiodol; however, the final labeling yield was low and the Re-188 complexes were not stably retained in hepatoma. This result reflects the difficulty in obtaining Re-188 complexes in satisfactory yield and the intrinsic instability of oxo rhenium complexes.

The high lipophilic character of the complexes $[^{188}Re(N)(R_2NCS_2)_2]$ was predicted on the basis of the measured water-to-octanol partition coefficient (P) of the corresponding Tc-99m counterparts [19]. Owing to their identical chemical structure, the analogous Re-188 complexes exhibit exactly the same bulk properties. Therefore, the complex $[^{188}Re(N)(DEDC)_2]$, where DEDC is diethyldithiocarbamato, was selected as a suitable candidate for labeling lipiodol through the previously mentioned method. The preparation of the nitrido Re-188 complex was accomplished in two steps using a two-vial, lyophilized kit formulation. The first vial contained DTCZ, as a donor of nitrido nitrogen atoms, $SnCl_2$ and sodium oxalate to allow the formation of the Re≡N group. The pH was kept approximately in the range 4.5–5.0 by addition of glacial acetic acid. The second vial, filled with the sodium salt of the ligand DEDC and a carbonate buffer, was utilized to accomplish the facile substitution reaction onto the $[^{188}Re{\equiv}N]^{2+}$ to give the bis-substituted product $[^{188}Re(N)(DEDC)_2]$, in high yields (95–99%). This reaction was carried out under basic conditions to favor the coordination of the CS_2^- group of the ligand.

After preparation, the reaction solution containing the complex $[^{188}Re(N)(DEDC)_2]$ was mixed with a sample of lipiodol, thus allowing the activity to be partitioned between the aqueous and hydrophobic phases. Approximately 98% of the initial activity was incorporated into the lipiodol phase as a consequence of the high lipophilicity of $[^{188}Re(N)(DEDC)_2]$. Subsequent separation of the two phases led to the recovery of ^{188}Re-labeled lipiodol for in vivo studies. A schematic representation of the labeling procedure is reported in Fig. 21.

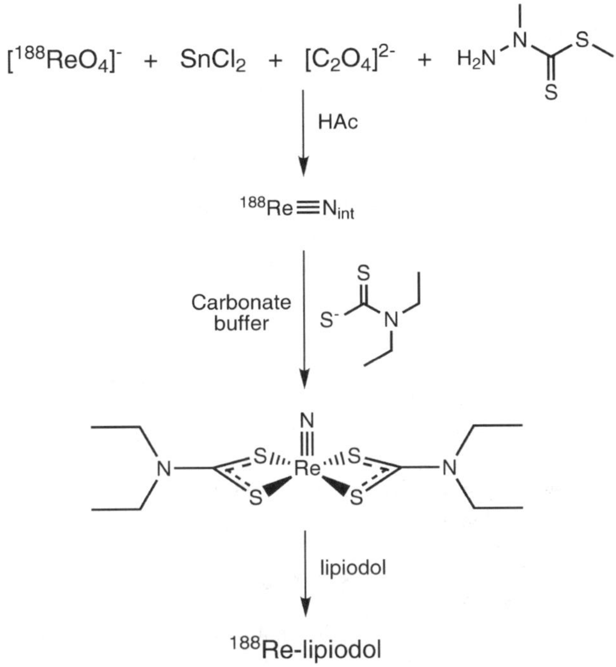

Fig. 21 Schematic representation of the procedure for preparing ^{188}Re-labeled lipiodol. Mixing of the lipophilic complex [^{188}Re(N)(diethyldithiocarbamato)$_2$] with lipiodol causes the quantitative extraction and retention of the activity in the hydrophobic phase

Whole-body γ-imaging of HCC patients within 1–4 h of intrahepatic arterial administration of ^{188}Re-labeled lipiodol demonstrated excellent uptake in the lesion without significant activity in the gut and lungs. Stable retention of activity in hepatoma was revealed at 20 h post administration with minimal increase in colonic activity and some uptake in the spleen. In particular, no lung activity was observed in any patient as opposed to treatment of hepatocellular carcinoma with ^{131}I-lipiodol where lung uptake approaches 35% of administered activity. The excellent biological properties and selective uptake in the target tissue indicates that nitrido ^{188}Re-lipiodol may become an effective therapeutic agent for the treatment of this type of neoplastic lesion.

References

1. Baldas J, Bonnyman J, Williams GA (1986) Inorg Chem 25:150
2. Baldas J, Bonnyman J (1985) Int J Appl Radiat Isot 36:133
3. Baldas J, Bonnyman J (1988) Nucl Med Biol 15:451
4. Pasqualini R, Comazzi V, Bellande E, Duatti A, Marchi A (1992) Appl Radiat Isot 43:1329
5. Duatti A, Marchi A, Pasqualini R (1990) J Chem Soc Dalton Trans 3729
6. Marchi A, Duatti A, Rossi R, Magon L, Pasqualini R, Bertolasi V, Ferretti V (1988) J Chem Soc Dalton Trans 1743

7. Marchi A, Rossi R, Magon L, Duatti A, Pasqualini R, Ferretti V, Bertolasi V (1990) J Chem Soc Dalton Trans 1411
8. Uccelli L, Bolzati C, Boschi A, Duatti A, Morin C, Pasqualini R, Giganti M, Piffanelli A (1999) Nucl Med Biol 26:63
9. Bolzati C, Uccelli L, Duatti A, Venturini M, Morin C, Chéradame S, Refosco F, Ossola F, Tisato F (1997) Inorg Chem 36:3582
10. Duatti A, Marchi A, Bertolasi V, Ferretti V (1991) J Am Chem Soc 113:9680
11. Marchi A, Garuti P, Duatti A, Magon L, Rossi R, Bertolasi V, Ferretti V (1990) Inorg Chem 29:2091
12. Bellande E, Comazzi V, Laine J, Lecayon M, Pasqualini R, Duatti A, Hoffschir D (1995) Nucl Med Biol 22:315
13. Hoffman R (1981) Science 211:995
14. Bandoli G, Dolmella A, Porchia M, Tisato F, Refosco F (2001) Coord Chem Rev 214:43
15. Schwochau K (2000) Technetium chemistry and radiopharmaceutical applications. Wiley-VCH, Weinheim, Germany
16. Bolzati C, Boschi A, Uccelli L, Malagò E, Bandoli G, Tisato F, Refosco F, Pasqualini R, Duatti A (1999) Inorg Chem 38:4473
17. Abrams MJ, Larsen SK, Shaikh SN, Zubieta J (1991) Inorg Chim Acta 185:7
18. Baldas J, Bonnyman J, Pojer PM, Williams GA, Mackay MF (1981) J Chem Soc Dalton Trans 1798
19. Pasqualini R, Duatti A, Bellande E, Comazzi V, Brucato V, Hoffschir D, Fagret D, Comet C (1994) J Nucl Med 35:334
20. Bolzati C, Uccelli L, Boschi A, Malagò E, Duatti A, Tisato F, Refosco F, Pasqualini R, Piffanelli A (2000) Nucl Med Biol 27:369
21. Bolzati C, Boschi A, Duatti A, Prakash S, Uccelli L, Refosco F, Tisato F, Bandoli G (2000) J Am Chem Soc 122:4510
22. Bolzati C, Boschi A, Uccelli L, Tisato F, Refosco F, Cagnolini A, Duatti A, Prakash S, Bandoli G, Cittadini A (2002) J Am Chem Soc 124:11468
23. Refosco F, Bolzati C, Duatti A, Tisato F, Uccelli L (2000) Recent Res Devel Inorganic Chem 2:89
24. Berman DS, Hayes SW, Shaw LJ, Germano G (2001) Curr Probl Cardiol 16:1
25. Jurisson S, Lindoy JD (1999) Chem Rev 99:2205
26. Duatti A, Uccelli L (1996) Trends Inorg Chem 4:27
27. Vanzetto G, Fagret D, Pasqualini R, Mathieu JP, Chossat F, Machecourt J (2000) J Nucl Med 41:141
28. Vanzetto G, Glover DK, Ruiz M, Calnon DA, Pasqualini R, Watson DD, Beller GA (2000) Circulation 101:2424
29. Takehana K, Beller GA, Ruiz M, Petruzella FD, Watson DD, Glover DK (2001) J Nucl Med 42:1388
30. Dahlberg ST, Leppo JA (1994) Cardiol Clin 12:169
31. Johnson LL, Seldin DW (1990) Am J Cardiol 66:63E
32. Uccelli L, Giganti M, Duatti A, Bolzati C, Pasqualini R, Cittanti C, Colamussi P, Piffanelli A (1995) J Nucl Med 36:2075
33. Riou L, Ghezzi C, Pasqualini R, Fagret D (2000) J Nucl Cardiol 7:365
34. Riou LM, Unger S, Toufektsian MC, Ruiz M, Watson DD, Beller GA, Glover DK (2003) J Nucl Med 44:1092
35. Boschi A, Uccelli L, Bolzati C, Duatti A, Sabba N, Moretti E, Di Domenico G, Zavattini G, Refosco F, Giganti M (2003) J Nucl Med 44:806
36. Boschi A, Bolzati C, Uccelli L, Duatti A, Benini E, Refosco F, Tisato F, Piffanelli A (2002) Nucl Med Commun 23:689

37. Hatada K, Riou LH, Ruiz M, Lima R, Goode AR, Watson DD, Beller GA, Glover DK (2002) Circulation 106:386
38. Hom RK, Katzenellebogen (1997) Nucl Med Biol 24:485
39. Boschi A, Bolzati C, Benini E, Malagò E, Uccelli L, Duatti A, Piffanelli A, Refosco F, Tisato F (2001) Bioconjugate Chem 12:1035
40. Boschi A, Uccelli L, Duatti A, Bolzati C, Refosco F, Tisato F, Romagnoli R, Baraldi PG, Varani K, Borea PA (2003) Bioconjugate Chem 14:1279
41. Bolzati C, Mahmood A, Malagò E, Uccelli L, Boschi A, Jones AG, Refosco F, Duatti A, Tisato F (2003) Bioconjugate Chem 14:1231
42. Baraldi PG, Romagnoli R, Duatti A, Bolzati C, Piffanelli A, Bianchi N, Mischiati C, Gambari R (2000) Bioorg Med Chem Lett 10:1397
43. Ercan MT, Caglar M (2000) Curr Pharm Des 6:1085
44. Deutsch E, Libson K, Vanderheyden J-L, Ketring A, Maxon HR (1986) Nucl Med Biol 13:465
45. Bolzati C, Boschi A, Uccelli L, Duatti A, Franceschini R, Piffanelli A (2000) Nucl Med Biol 27:309
46. Vajo JJ, Aikens DA, Ashley L, Poeltl DE, Bailey RA, Clark HM, Bunce SC (1981) Inorg Chem 20:3328
47. Boschi A, Bolzati C, Uccelli L, Duatti A (2003) Nucl Med Biol 30:381
48. Duatti A, Martindale AA, Turner JH, Boschi A, Giganti M, Bolzati C, Uccelli L (2002) Alasbimn J 5(17):September
49. Boschi A, Uccelli L, Duatti A, Colamussi P, Cittanti C, Filice A, Rose AH, Martindale AA, Claringbold PG, Kearney D, Galeotti R, Turner JH, Giganti M (2004) Nucl Med Commum 25:691
50. Wang SJ, Lin WY, Chen MN, Hsieh BT, Shen LH, Tsai ZT, Ting G, Knapp FF Jr (1996) Eur J Nucl Med 23:13
51. Keng GH, Sundram FX (2003) Ann Acad Med Singapore 32:518
52. Sundram FX, Yu SW, Jeong JM, Somanesan S, Premaraj J, Saw MM, Tan BS (2001) Ann Acad Med Singapore 30:542
53. Jeong JM, Kim YJ, Lee YS, Ko JI, Son M, Lee DS, Chung JK, Park JH, Lee MC (2001) Nucl Med Biol 28:197

Top Curr Chem (2005) 252: 117–153
DOI 10.1007/b101226

6-Hydrazinonicotinamide Derivatives as Bifunctional Coupling Agents for ^{99m}Tc-Labeling of Small Biomolecules

Shuang Liu (✉)

Department of Industrial and Physical Pharmacy, School of Pharmacy and Pharmaceutical Sciences, Purdue University, 575 Stadium Drive, West Lafayette, IN47907-2051, USA
lius@pharmacy.purdue.edu

Abstract ^{99m}Tc-labeled small biomolecules are a class of receptor-based, target specific radiopharmaceuticals for the detection of various diseases, such as cancer, thrombosis, infection and inflammation. To label a small biomolecule with ^{99m}Tc, a bifunctional coupling agent (BFC) is needed so that the ^{99m}Tc can be tightly attached to the biomolecule and can localize at the site of the diseased tissue. For the last several years, 6-hydrazinonicotinamide (HYNIC) has been used as the BFC for the ^{99m}Tc-labeling of small biomolecules, including chemotactic peptides and leukotriene B_4 receptor antagonists for imaging infection or inflammation, integrin $\alpha_v\beta_3$ receptor antagonists and somatostatin analogs for tumor imaging, and a glycoprotein IIb/IIIa receptor antagonist for imaging thrombosis. This review will focus on some important issues associated with the HYNIC technology and recent developments in the use of HYNIC derivatives as BFCs for the ^{99m}Tc-labeling of small biomolecules. It will also discuss some new techniques for the characterization of ^{99m}Tc radiopharmaceuticals at the tracer level. The topics of this chapter include ^{99m}Tc-labeling of HYNIC–biomolecule conjugates, protection of the hydrazine moiety, formulation development, ^{99m}Tc-labeling efficiency, effect of the coligand on physical and biological properties of ^{99m}Tc radiopharmaceuticals, characterization of ^{99m}Tc radiopharmaceuticals at the tracer level, structural characterization of ternary ligand technetium complexes, and examples of ^{99m}Tc-labeled HYNIC–biomolecules as new diagnostic radiopharmaceuticals.

Keywords 6-Hydrazinonicotinamide · 6-Hydrazinonicotinamide–biomolecule conjugates · Target-specific radiopharmaceuticals

1 Introduction

Abrams and coworkers [1, 2] first reported the use of aromatic hydrazines, including 6-hydrazinonicotinamide (HYNIC), as bifunctional coupling agents (BFCs) for the ^{99m}Tc-labeling of polyclonal IgG for imaging inflammation and infection diseases. Since then, the HYNIC technology has successfully been used for the ^{99m}Tc-labeling of antibodies [3–8] and small biomolecules (BMs), including chemotactic peptides [9–18], somatostatin analogs [19–25], "stealth" liposomes [26, 27], antisense oligonucleotides [28–31], a folate receptor ligand [32], and polypeptides [32–34]. A ternary ligand system (HYNIC, tricine, and trisodium triphenylphosphine-3,3′,3″-trisulfonate, TPPTS) has been used for the ^{99m}Tc-labeling of chemotactic peptides [35] and leukotriene B_4 (LTB_4) receptor antagonists [36–39] for imaging infection and inflammation, integrin $\alpha_v\beta_3$ receptor antagonists for tumor imaging [40], and a glycoprotein IIb/IIIa (GPIIb/IIIa) receptor antagonist for imaging thrombosis [41–48]. The advantage of HYNIC is its high labeling efficiency (rapid and high yield radiolabeling) and the high in vivo stability of the resulting complexes [35–46, 49–52]. Since HYNIC can only occupy one or two coordination sites, a coligand such as tricine is needed to complete the coordination sphere of technetium. The choice of coligands allows easy modification of hydrophilicity and pharmacokinetics of ^{99m}Tc-labeled small BMs [35–46, 49–52].

2
^{99m}Tc-Labeling of HYNIC–Biomolecule Conjugates: Problems and Solutions

2.1
Binary Ligand Technetium Complexes

Babich and coworkers [9–14, 18] have studied the ^{99m}Tc labeled HYNIC-conjugated chemotactic peptides (such as fMLFK–HYNIC) extensively for the development of radiopharmaceuticals for imaging infection. It was found that the ^{99m}Tc labeling of fMLFK–HYNIC could be performed using glucoheptonate (GH), mannitol, and glucamine as coligands [18]. Very high specific activity (20,000 mCi/μmol and greater) was achieved using GH as the coligand. Different biodistributions were observed for binary ligand ^{99m}Tc complexes [^{99m}Tc(XV120)(L)$_2$] [Fig. 1: L is GH, mannitol, and glucamine and XV120 is cyclo(D-Val–NMeArg–Gly–Asp–Mamb(5-(6-(6-hydrazinonicotinamido)hexanamide)))]. Animal studies have shown evidence of binding of the ^{99m}Tc-labeled fMLFK–HYNIC to leukocytes in vivo and localization at the infection site [9–18, 35].

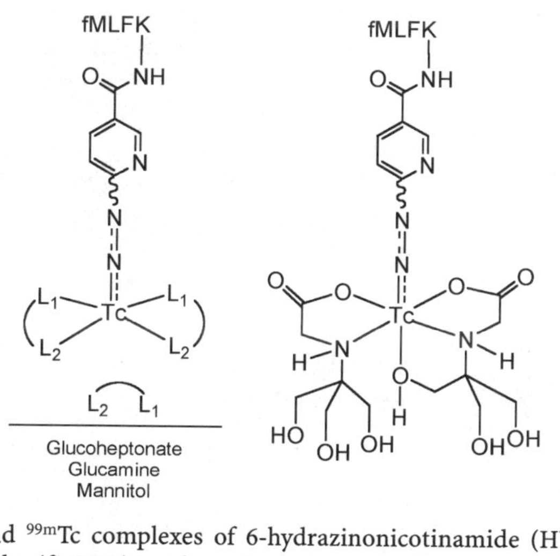

Fig. 1 Binary ligand ^{99m}Tc complexes of 6-hydrazinonicotinamide (HYNIC)-conjugated chemotactic peptide (fMLFK). Coligands such as tricine are used to stabilize the ^{99m}Tc–HYNIC core

Liu et al. [41] reported the ^{99m}Tc-labeling of an HYNIC-derivatized platelet GPIIb/IIIa receptor antagonist (Scheme 1) using tricine as the coligand. ^{99m}Tc-labeling was performed by reacting XV120 with $^{99m}TcO_4^-$ in the presence of excess tricine and stannous chloride at pH 4–5. Very high specific activity (20,000 mCi/μmol and greater) was achieved for the binary ligand technetium complex [^{99m}Tc(XV120)(tricine)$_2$] (Scheme 1: RP431). However, it was found that RP431 is not stable, particularly under dilute conditions, and exists as multiple species in solution. Similar characteristics were also observed for ^{99m}Tc complexes, [^{99m}Tc(fMLFK–HYNIC)(L)$_2$] (Fig. 1: L is tricine and GH) [35, 49]. Although the biological data from animal studies show that these agents are able to localize at the disease site, it would still be difficult to develop them for

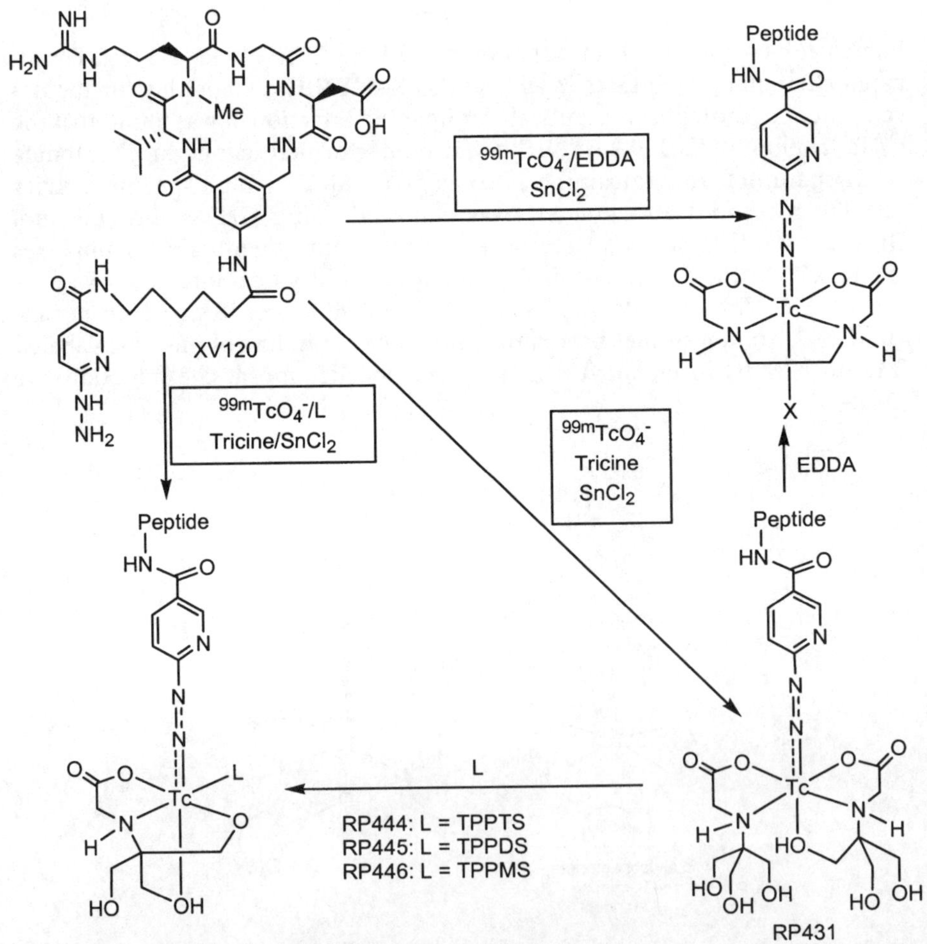

Scheme 1 Synthesis of ^{99m}TC complexes of cyclo(D-Val–NMeArg–Gly–Asp–Mamb(5-(6-(6-hydrazinonicotinamido)hexanamide))) (XV120)

routine clinical use mainly owing to their solution instability and the presence of many isomers.

In order to prepare [^{99m}Tc]HYNIC complexes with higher solution stability and less isomerism, ethylenediamine-*N*, *N'*-diacetic acid (EDDA) has been used for ^{99m}Tc-labeling of XV120 (Scheme 1). It was found that the complex [^{99m}Tc(XV120)(EDDA)$_n$] was stable for at least 12 h in the reaction mixture; but it exists in at least three isomeric forms in solution [41]. It is clear that the replacement of tricine by EDDA produces [^{99m}Tc]HYNIC complexes with higher stability and fewer isomers. The composition of the complex [^{99m}Tc(XV120)(EDDA)$_n$] remains unknown; but recent liquid chromatography–mass spectrometry (LC–MS) data seem to indicate that there are two EDDA coligands in the complex [^{99m}Tc(HYNIC–TOC)(EDDA)$_n$], where TOC is D-Phe[1]-Tyr[3]-octreotide [53].

2.2 Ternary Ligand Technetium Complexes

Another approach to increase solution stability and decrease the number of isomers in [^{99m}Tc]HYNIC complexes involves the use of a water-soluble phosphine (Fig. 2: TPPTS, disodium triphenylphosphine-3,3′-disulfonate, TPPDS, and sodium triphenylphosphine-3-monosulfonate, TPPMS) as a second coligand [42]. It was found that the combination of XV120 with tricine and a phosphine coligand results in a versatile ternary ligand system that forms ^{99m}Tc

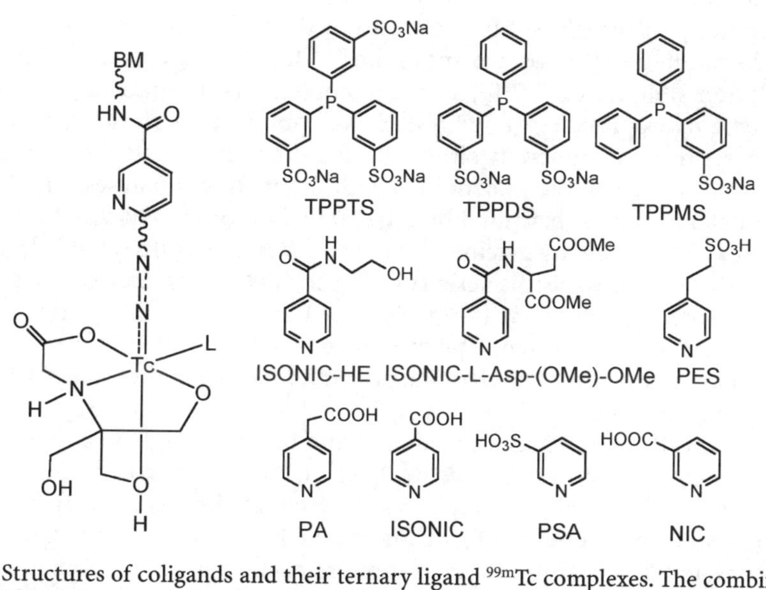

Fig. 2 Structures of coligands and their ternary ligand ^{99m}Tc complexes. The combination of HYNIC, tricine, and a phosphine or pyridine analog results in a versatile ternary ligand system that forms ternary ligand ^{99m}Tc complexes with extremely high stability

complexes, [^{99m}Tc(XV120)(tricine)(L)] (Scheme 1: RP444, L is TPPTS; RP445, L is TPPDS; RP446, L is TPPMS) in high yield and high specific activity 20,000 Ci/mmol and greater). These complexes can be readily prepared by reacting XV120 directly with $^{99m}TcO_4^-$ in the presence of excess tricine, water-soluble phosphine coligand, and stannous chloride. The radiolabeling yield for RP444 is usually 90% or greater using 5 μg XV120 and 100 mCi $^{99m}TcO_4^-$. The XV120-to-technetium ratio is about 7:1 under these conditions. Since these complexes require no postlabeling purification, the procedure is amenable for the development of a kit formulation [42]. The ternary ligand system (HYNIC–BM, tricine, and TPPTS) has been successfully used for the ^{99m}Tc-labeling of a variety of HYNIC-conjugated small BMs (HYNIC–BMs) for the development of new receptor-based target-specific radiopharmaceuticals [21, 22, 32–35, 43, 44].

Ternary ligand ^{99m}Tc complexes, [^{99m}Tc(XV120)(tricine)(*L*)] (*L* is TPPTS, TPPDS, and TPPMS), are extremely stable in solution, and are formed as equal mixtures of two detectable isomeric forms. Using a chirality experiment, it has been demonstrated that the presence of two isomeric forms is due to the resolution of diastereomers resulting from a combination of the chiral technetium chelate and chiral centers in the cyclic peptide. It was also found that the two diastereomers of RP444 possess equal biological activity [46]. The chemical composition of these ternary ligand ^{99m}Tc complexes has been determined to be 1:1:1:1 for Tc:HYNIC:L:tricine through a series of mixed-ligand experiments [42], and was confirmed by LC–MS at both the tracer level and the macroscopic level [46, 54].

Apparently, the use of the ternary ligand system offers several advantages. First, bonding of the phosphine coligand to the technetium dramatically reduces the number of isomeric forms of the [^{99m}Tc]HYNIC complexes. Secondly, the solution stability of [^{99m}Tc]HYNIC complexes is dramatically improved. Finally, the hydrophilicity of [^{99m}Tc]HYNIC complexes with the new ternary ligand system can be tuned by either altering the number of sulfonato groups or by using water-soluble phosphines with other functionalities. The tricine coligand can also be substituted by other polyaminocarboxylates, such as *N*-bis(hydroxymethyl)methylglycine (dicine) and *N,N*-bis(hydroxymethyl)glycine (bicine). However, the specific activity of [^{99m}Tc]HYNIC complexes using dicine and bicine as coligands is not as high as that of the corresponding tricine complexes. This is consistent with the literature results reported by Larson et al. [55], who found that the [^{99m}Tc]tricine precursor has the highest reactivity with aromatic hydrazines.

Phosphine-P is known as a soft donor, and binds strongly to Tc(III). When bonded to the technetium, it uses its 3*p* electron lone-pair to form a Tc–P σ bond. The empty 3*d* orbitals can be used to accept the electrons from 4*d* orbitals of the technetium and form $d\pi$–$d\pi$ back-bonding. This suggests that monodentate ligands with other soft donors such as pyridine-N may also be used as coligands. Like phosphines, pyridine analogs also form ^{99m}Tc complexes, [^{99m}Tc(HYNIC–BM)(tricine)(L)] [Fig. 2: L is nicotinic acid, NIC, isonicotinic acid, ISONIC, ISONIC–L-Asp–OMe$_2$, *N*-(2-hydroxylethyl)isonicotinylamide, ISONIC-HE, 4-pyridyl acetic acid, PA, 2-(4-pyridyl)ethylsulfonic acid, PES, and pyridine-3-sulfonic acid, PSA], as equal mixtures of two detectable isomers, if the BM has one or more chiral centers. The bonding of pyridine-N to the technetium reduces the number of isomers and increases the solution stability of [^{99m}Tc]HYNIC complexes. In addition, pyridine analogs are much smaller, so the impact of the technetium chelate on receptor binding might be minimized. The hydrophilicity of [^{99m}Tc]HYNIC complexes can be tuned by changing the substituents on the pyridine ring. The 1:1:1:1 composition for Tc:HYNIC:L:tricine (L is pyridine analogs) was determined through the mixed-ligand experiments, and has been confirmed by the LC–MS spectral data [54].

The coligand ISONIC–Asp–OMe$_2$ is of particular interest because it contains two ester groups, which can be enzymatically hydrolyzed in biological systems. Since ISONIC–Asp-OMe$_2$ carries no charge, it is expected to form neutral ternary ligand ^{99m}Tc complexes, which can across the cell membrane and potentially bind to intracellular receptors, if the HYNIC–BM conjugate carries no charge. Once inside the cell, enzymatic hydrolysis of ester groups will result in the formation of negatively charged ^{99m}Tc species, which cannot be easily washed out from the cell. In this way, the target uptake may be improved. If ester groups are hydrolyzed in the blood stream, the negatively charged ^{99m}Tc species is expected to have faster renal clearance than the neutral precursor. Therefore, the introduction of two enzymatically active ester groups offers two potential advantages: increasing the target uptake and decreasing the background.

2.3 Macrocyclic Technetium Complexes

In principle, the ternary ligand system (HYNIC–BM, tricine, and TPPTS) can be used for ^{99m}Tc-labeling of any small BMs. However, problem may arise when it is used for ^{99m}Tc-labeling of small BMs containing one or more disulfide linkages, which are often vital to keep the rigid cyclic conformation of the BM and to maintain the high receptor binding affinity. The use of a large amount of TPPTS in combination with high-temperature heating may destroy the S–S disulfide bonds and cause an adverse effect on the biological properties of the ^{99m}Tc-labeled BM. Replacing TPPTS with a pyridine analog will definitely avoid the use of the phosphine coligand. However, a large amount (more than 10 mg/mL) of pyridine coligand has to be used in order to achieve high radiochemical purity (RCP) for the ternary ligand ^{99m}Tc complex. Therefore, there is a continuing need for a better chelating system which does not require the use of a large amount of phosphine or pyridine coligand.

One approach to minimize the use of a large amount of phosphine coligand is to attach the phosphine coligand onto HYNIC via a linker to form the corresponding phosphine-containing HYNIC chelator [56]. Two phosphine-

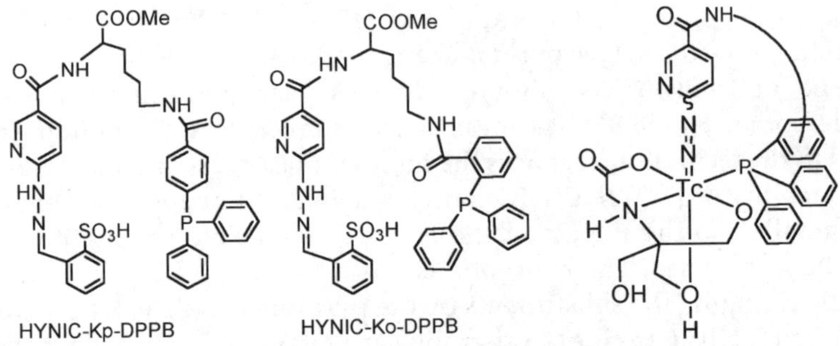

Fig. 3 Phosphine-containing HYNIC chelators and their macrocyclic technetium complexes. These chelators are designed in such a way that when HYNIC binds to the technetium center it becomes much easier for the phosphine-P to bond to the technetium and form a macrocyclic ^{99m}Tc chelate

containing HYNIC chelators which are designed in such a way that when HYNIC binds to the technetium center, the "effective concentration" of the phosphine-P donor in the vicinity of the technetium will be increased dramatically are shown in Fig 3. This makes it much easier for the phosphine-P to bond to the technetium and form a macrocyclic ^{99m}Tc chelate. The linker between HYNIC and the phosphine moiety may be a polymethylene chain or a small peptide sequence. The site for attachment of the linker may vary for the triphenylphosphine moiety. For example, the ω-amino group of lysine is attached to the triphenylphosphine moiety at either the para or the ortho position of the benzene ring relative to the P–C bond. Lysine was selected as a linker to connect HYNIC with the triphenylphosphine moiety, while the carboxylic group of the lysine linker can be used for conjugation of the targeting BM for the development of target-specific radiopharmaceuticals.

Macrocyclic ^{99m}Tc complexes, [^{99m}Tc(HYNIC–K*o*-DPPB)(tricine)] and [^{99m}Tc(HYNIC–K*p*-DPPB)(tricine)], where HYNIC–K*o*-DPPB is *N*-ε-(2-(diphenylphosphino)benzoyl)-*N*-*a*-(6-(2-(2-sulfonato-benzaldehyde)hydrazono)nicotinyl)lysine methyl ester and HYNIC–K*p*-DPPB is *N*-ε-(4-(diphenylphosphino)benzoyl)-*N*-*a*-(6-(2-(2-sulfonato-benzaldehyde)hydrazon o)nicotinyl)lysine methyl ester were prepared by reacting the phosphine-containing HYNIC chelator with $^{99m}TcO_4^-$ in the presence of excess tricine and stannous chloride [56]. It was found that both HYNIC–K*p*-DPPB and HYNIC–K*o*-DPPB are able to form highly stable ^{99m}Tc complexes, [^{99m}Tc(HYNIC–K*o*-DPPB)(tricine)] and [^{99m}Tc(HYNIC–K*p*-DPPB)(tricine)], when tricine is used as the coligand. The radio-high-performance liquid chromatography (HPLC) data suggest that [^{99m}Tc(HYNIC–K*p*-DPPB)(tricine)] exists as only one detectable isomer in solution, while the complex [^{99m}Tc(HYNIC–K*o*-DPPB)(tricine)] has three isomers (Fig. 4). It was also found that three isomers of the macrocyclic ^{99m}Tc complex [^{99m}Tc(HYNIC–K*o*-

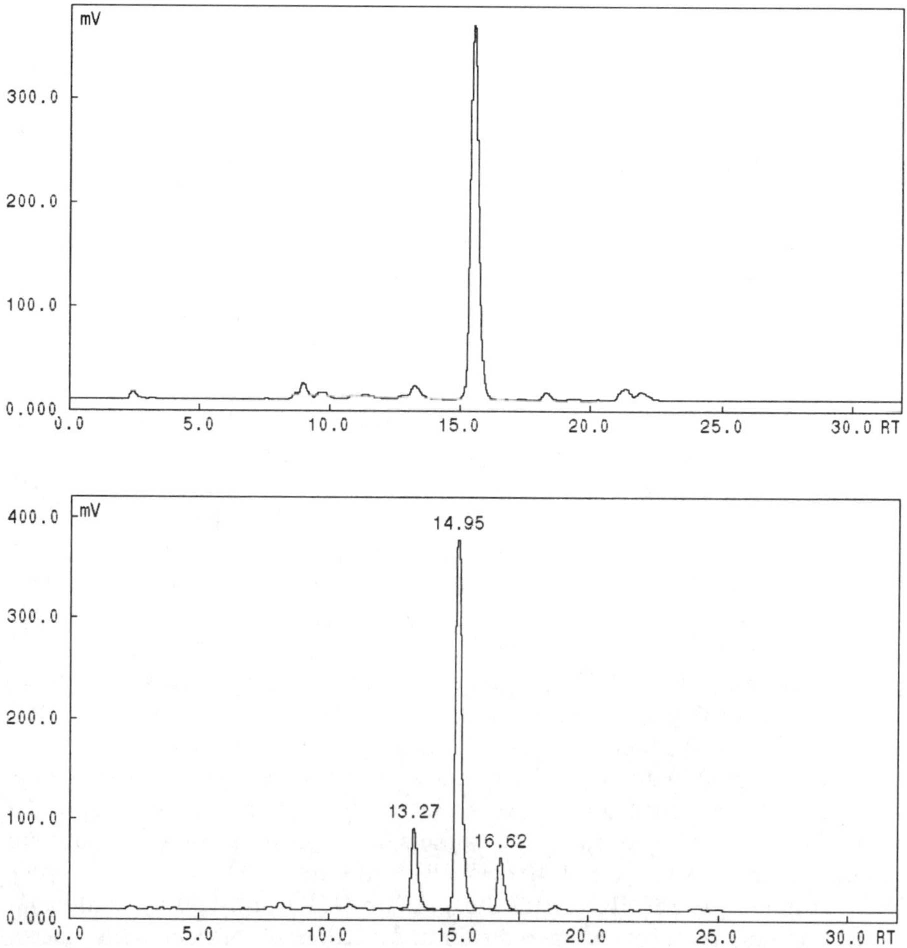

Fig. 4 Typical radio-high-performance liquid chromatography (*HPLC*) chromatograms of macrocyclic ^{99m}Tc complexes [^{99m}Tc(HYNIC–K*p*-DPPB)(tricine)] (*top*) and [^{99m}Tc(HYNIC–K*o*-DPPB)(tricine)] (*bottom*), where *HYNIC–Kp-DPPB* is *N*-ε-(2-(diphenylphosphino)benzoyl)-*N*-*a*-(6-(2-(2-sulfonato-benzaldehyde)hydrazono)nicotinyl)lysine methyl ester and *HYNIC–Ko-DPPB* is *N*-ε-(4-(diphenylphosphino)benzoyl)-*N*-*a*-(6-(2-(2-sulfonato-benzaldehyde)hydrazono)nicotinyl)lysine methyl ester

DPPB)(tricine)] interconvert at elevated temperatures, suggesting that the presence of these isomers might be due to conformational changes in the macrocyclic technetium chelate. The LC–MS data for these two macrocyclic ^{99m}Tc complexes are completely consistent with the proposed composition. These phosphine-containing HYNIC chelators may have the potential to act as bifunctional chelators for ^{99m}Tc labeling of small BMs [56].

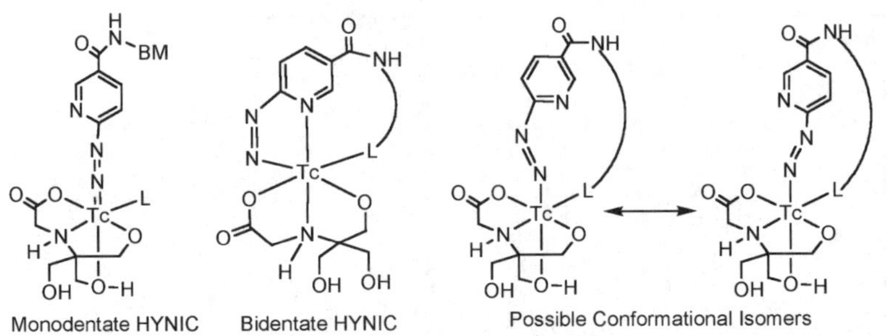

Fig. 5 Possible structures for macrocyclic ^{99m}Tc complexes [^{99m}Tc(HYNIC–P)(tricine)]

Several questions remain to be answered. These include (1) the bonding modality of the HYNIC chelator and the coordinated tricine, (2) the identity of the three isomers in the macrocyclic ^{99m}Tc complex [^{99m}Tc(HYNIC–K*o*-DPPB)(tricine)], (3) the reason for there being three isomers for [^{99m}Tc(HYNIC–K*o*-DPPB)(tricine)] while only one is detectable for the macrocyclic ^{99m}Tc complex [^{99m}Tc(HYNIC–K*p*-DPPB)(tricine)], and (4) the impact of linker length and attachment site on the coordination chemistry of phosphine-containing HYNIC chelators. In the ternary ligand ^{99m}Tc complex, [Tc(HYNIC–BM)(tricine)(TPPTS)], HYNIC is monodentate owing to the presence of the tetradentate tricine and monodentate phosphine coligands [46]. However, the HYNIC in macrocyclic ^{99m}Tc complexes, [Tc(HYNIC–*L*)(tricine)] (L is phosphines), may be forced to become bidentate (Fig. 5) owing to the bonding of the attached phosphine-P. The tridentate tricine may result in the formation of different coordination isomers owing to its asymmetric nature in bonding to the technetium [46]. If HYNIC is monodentate in the macrocyclic ^{99m}Tc complex, [Tc(HYNIC–*L*)(tricine)] (L is *o*-DPPB or *p*-DPPB), conformational isomers may form owing to the limited freedom of rotation in the macrocyclic technetium chelate. This may explain the presence of three different isomers, which interconvert in solution at elevated temperatures, for the macrocyclic ^{99m}Tc complex [^{99m}Tc(HYNIC–K*o*-DPPB)(tricine)]. However, in the absence of solid-state structure and variable temperature NMR studies, this explanation remains a speculation. Studies on structures of macrocyclic ^{99}Tc complexes will definitely help us to understand the fundamental coordination chemistry of HYNIC chelators in their macrocyclic ^{99m}Tc complexes.

3 Protection of the Hydrazine Moiety

The free hydrazine group in the HYNIC–BM conjugate is not stable to oxidation, particularly under basic conditions, and reacts readily with aldehydes or ketones, which are extracted from various plastic and rubber materials.

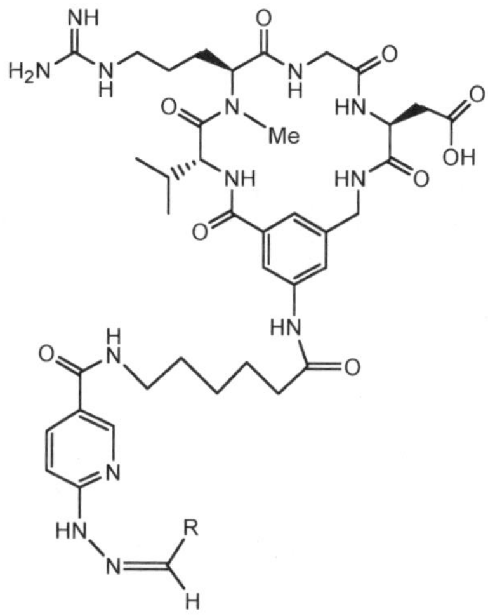

Hydrazone	Aldehyde	R
1	Benzaldehyde	C_6H_5
2	4-Dimethylaminobenzaldehyde	$4\text{-}NMe_2\text{-}C_6H_4$
3	4-Carboxybenzaldehyde	$4\text{-}HO_2C\text{-}C_6H_4$
4	2-Sulfonatobenzaldehyde	$2\text{-}NaO_3S\text{-}C_6H_4$
5	4-Pyridinecarboxaldehyde	$4\text{-}C_5H_4N$
6	Propyladehyde	C_2H_5
7	Crotonadehyde	$CH_3CH{=}CH\text{-}$
8	Glyoxylic Acid	COOH

Fig. 6 Stable hydrazones useful for synthesis of ternary ligand ^{99m}Tc complexes

This makes it very difficult to maintain the purity and the stability of the HYNIC–BM conjugate during the manufacturing process. In a US patent, Schwartz et al. [57] disclosed the use of propylaldehyde hydrazone to stabilize HYNIC-modified proteins. Although its use can prevent the cross-reaction of the hydrazine with other reactive groups on the protein, it can also be displaced by other aldehydes and ketones to form different hydrazones. This presents a significant problem in maintaining the purity of the final intermediate, and thus renders lower alkyl hydrazones unattractive as commercial reagents. To prevent the decomposition or degradation of the HYNIC–BM conjugate in the manufacturing process, several stable hydrazones (Fig. 6) have been synthesized [58] and studied for their solution stability in the presence of formaldehyde. It was found that for a hydrazone to be stable in aqueous solution there must be a conjugated π system [58].

Hydrazones are used only as protecting groups for the HYNIC–BM conjugate. During radiolabeling, the hydrazone moiety must hydrolyze (Scheme 2)

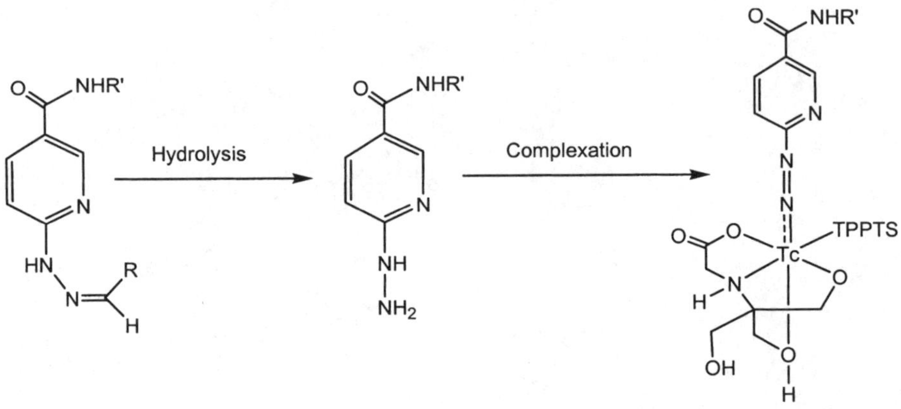

Scheme 2 Synthesis of ternary ligand ^{99m}TC complexes using stable hydrazones

to produce a sufficient quantity of free hydrazine to bind to the technetium via either a technetium-hydrazino or a technetium-diazenido bond [45]. The yield of the ternary ligand ^{99m}Tc complex depends largely on the kinetics of hydrolysis, as well as on the total concentration of the unprotected HYNIC–BM conjugate in the reaction mixture. The higher the degree of the hydrolysis of the hydrazone, the more free hydrazine groups are available for the technetium-binding, and the higher the radiolabeling yield of the ternary ligand ^{99m}Tc complex, within certain limits. However, the hydrolysis may not need to go to completion since there are typically 50–100 equivalents of the hydrazone to the total technetium in each reaction. The ternary ligand ^{99m}Tc complex obtained from the hydrazone should be the same as that prepared from the unprotected HYNIC–BM conjugate [45].

4 Formulation Development

The discovery of a new ^{99m}Tc radiopharmaceutical is just the beginning of a long process of radiopharmaceutical development. In addition to the biological efficacy (high target uptake, high target-to-background ratio, and favorable pharmacokinetics), the new ^{99m}Tc radiopharmaceutical must have very high radiochemical purity (90% or greater) and high solution stability (6 h or longer). A kit formulation is often required owing to the 6-h half-life of ^{99m}Tc. The kit formulation for rapid preparation of ^{99m}Tc radiopharmaceuticals has facilitated the practice of nuclear medicine, and is particularly important for the consistency and reproducibility in the RCP performance for the radiolabeling of the BFC–BM conjugate.

4.1 Kit Formulation and Kit Components

Kits are sterile, pyrogen-free, and nonradioactive mixtures, which are dried by lyophilization and stored under nitrogen in glass vials. For target-specific ^{99m}Tc radiopharmaceuticals, a kit contains a BFC–BM conjugate and a reducing agent, if necessary. Kit components are often dissolved in a buffer system, and then lyophilized to form a "cake". The buffer is often used for pH control, which is extremely important during manufacturing and radiolabeling processes. Sometimes a bulking agent is needed so that the kit components can crystallize on the crystal frameworks of the bulking agent during lyophilization. Other components (antioxidants, solubilizing agents, and weak transferring ligands) may be needed to improve the yield and solution stability of the ^{99m}Tc radiopharmaceutical. In many cases, the ^{99m}Tc-labeling can be accomplished simply by adding $^{99m}TcO_4^-$ to the kit.

In general, all the components must be nontoxic so that the kit matrix is suitable for intravenous injection. For receptor-based radiopharmaceuticals, the amount of the BFC–BM conjugate in the formulation has to be sufficiently high so that high RCP can be readily achieved for the desired ^{99m}Tc radiopharmaceutical. However, the use of a large amount of the BFC–BM conjugate may result in receptor-site saturation, blocking the docking of the ^{99m}Tc-labeled receptor ligand, as well as unwanted side effects. To avoid these problems, the BFC–BM concentration in each kit has to be very low (around 20 μg/mL, corresponding to 2×10^{-6} M for a BFC–BM conjugate with a molecular weight of 1,000 Da and IC_{50} values in the nanomolar range). Otherwise, postlabeling purification is needed to remove excess unlabeled BFC–BM conjugate, which is time-consuming and thus not amenable for the kit formulation. Compared with the total technetium concentration (around 5×10^{-7} M) in 100 mCi $^{99m}TcO_4^-$ eluant with 24-h prior elution time, 20 μg BFC–BM conjugate is not in overwhelming excess. In addition, the use of other kit components should not interfere with the ^{99m}Tc-labeling of the BFC–BM conjugate.

4.1.1 Reducing Agents

^{99m}Tc is obtained from the ^{99}Mo–^{99m}Tc generator as $^{99m}TcO_4^-$ in saline. Since there is no effective chemistry that can be used to attach $^{99m}TcO_4^-$ to a small BM, the Tc(VII) in $^{99m}TcO_4^-$ has to be reduced to a lower oxidation state in order to produce a stable ^{99m}Tc–BFC–BM complex. Various reducing agents, including stannous chloride, sodium borohydride, sodium dithionate, sodium dithionite, hydroxamine, formamidine sulfinic acid, dithiothreitol, and water-soluble phosphines, have been used for the reduction of $^{99m}TcO_4^-$. To date, stannous chloride remains the most commonly used reducing agent in commercial kits for rapid preparation of ^{99m}Tc radiopharmaceuticals. Other Sn(II) compounds, such as stannous citrate, stannous tartrate, stannous glucoheptonate, and stannous pyrophosphate, have also been used by different commercial manufacturers.

While most of these reducing agents can be used for ^{99m}Tc radiopharmaceuticals, only a few of them have been used for receptor-based ^{99m}Tc radiopharmaceuticals. This is mainly due to limitations imposed by the radiolabeling kinetics and the solution stability of the radiolabeled BFC–BM conjugate. For example, the use of borohydride may prove to be detrimental for a cyclic peptide containing S–S disulfide bonds because it is able to reduce the S–S disulfide bonds, particularly at elevated temperatures. Hydroxamine may not be able to reduce $^{99m}TcO_4^-$ to the required oxidation state in a short period of time (less than 30 min). Sodium dithionite and sodium dithionate are used at pH≥10 for effective reduction of $^{99m}TcO_4^-$. Harsh reaction conditions such as high pH should be avoided during the radiolabeling because the BFC–BM conjugate may decompose or undergo racemerization under these conditions.

4.1.2 Buffer Agent

The RCP of a ^{99m}Tc radiopharmaceutical is highly dependent on the pH of the kit mixture; thus, buffers can be important in the formulation. Sometimes the chelator itself acts as a buffer agent. The buffer agents most frequently used in commercial kits include sodium acetate, sodium benzoate, sodium bicarbonate, sodium citrate, sodium tartrate, and sodim suucinate. In general, the buffer agent should have a pK_a close to the pH at which the RCP of the expected ^{99m}Tc radiopharmaceutical is optimized. The use of a buffer agent should not interfere with radiolabeling of the BFC–BM conjugate and the RCP of the expected ^{99m}Tc radiopharmaceutical.

4.1.3 Transferring Agents

A weak transferring agent such as GH is often used to form an intermediate complex $[^{99m}TcO(GH)_2]^{n-}$, which undergoes ligand exchange with the BFC–BM conjugate under mild conditions to give the expected ^{99m}Tc radiopharmaceutical. The use of a transferring agent is particularly important for the ^{99m}Tc-labeling of BMs which are sensitive to harsh reaction conditions (e.g., high pH and heating at high temperatures).

4.1.4 Solublizing Agents

Solublizing agents are surfactants that help to increase dissolution of the lipophilic drug intermediate (the BFC–BM conjugate for receptor-based radiopharmaceuticals) so that a maximal amount of BFC–BM is available for the formation of the ^{99m}Tc radiopharmaceutical. Solublizing agents also help to reduce the glass surface absorption of the BFC–BM conjugate and the lipophilic ^{99m}Tc radiopharmaceutical. Reduction of the glass surface absorption of BFC–BM is particularly important in the manufacturing process.

4.1.5 Antioxidants

^{99m}Tc radiopharmaceuticals, when prepared in high quantity, may undergo autoradiolysis during preparation, release, and storage. Radiolytic degradation imposes serious safety concerns since the radioactivity that is not linked to the targeting BM will accumulate in nontargeted tissues. Decomposition of the radiopharmaceutical prior to or during administration will decrease the targeting capability of the BM. Therefore, it is very important to use a stabilizer to minimize autoradiolysis of the radiolabeled BMs. Radiolytic stabilizers are often antioxidants, which readily react with hydroxyl and superoxide radicals. Antioxidants, such as ascorbic acid, gentisic acid, and *p*-aminobenzoic acid, have been used as stabilizers to reduce the radiolytic degradation of ^{99m}Tc radiopharmaceuticals.

4.2 $SnCl_2$-Containing Formulation

Edwards et al. [45] recently reported a $SnCl_2$-containing formulation for the ^{99m}Tc-labeling of a GPIIb/IIIa receptor antagonist (XV066: 2-sulfonatobenzaldehyde hydrazone of XV120). Although it was first developed for routine preparation of RP444, a new ^{99m}Tc radiopharmaceutical useful for imaging thrombosis, the $SnCl_2$-containing formulation should be useful for the ^{99m}Tc labeling of any HYNIC–BM conjugate simply by replacing XV066 with the HYNIC–BM conjugate to be labeled.

There are many factors influencing the RCP of RP444 [45]. These include the hydrazone-protecting group, radiolabeling conditions (the temperature and heating time), and component levels. For example, if the amount of XV066 in the formulation is lower than 10 μg/vial for 50 mCi $^{99m}TcO_4^-$, the RCP of RP444 is dramatically reduced. Stannous chloride is used as the reducing agent. If the stannous chloride level is 5 μg/vial or lower, $^{99m}TcO_4^-$ is not completely reduced, and the RCP of RP444 is low. If a large amount (100 μg/vial or greater) of stannous chloride is used, the [^{99m}Tc]colloid formation becomes a major problem, and the corrected RCP of RP444 is also low. Tricine is a stabilizing agent for the [^{99m}Tc]tricine intermediate and a coligand in the ternary ligand system at the same time. At a lower tricine level (10 mg/vial or lower), the [^{99m}Tc]colloid formation becomes significant, and results in a lower RCP for RP444. TPPTS is also a coligand in the ternary ligand system. If the amount of TPPTS is less than 1.0 mg/vial, the RCP of RP444 is reduced. There is no significant difference in the RCP of RP444 if the TPPTS level is 1.0–7 mg/vial. The pH also has a significant impact on the RCP of RP444. At pH≤3.5, the tricine coligand may not able to stabilize the [^{99m}Tc]tricine intermediate, which decomposes to form a significant amount of $^{99m}TcO_4^-$. At pH≥5.0, the hydrazone group in XV066 remains highly stable in solution, and there is not enough hydrolyzed free hydrazine available for the ^{99m}Tc-labeling. Through a series of radiolabeling studies, it was found that the combination of 40 mg tricine, 1 mg TPPTS, 20 μg XV066 for 50 mCi $^{99m}TcO_4^-$, 20–50 μg stannous chloride, pH 4.5±0.5, and heating at 80 °C for 30 min gives the best yields for RP444. All kit components are water-soluble. Using the $SnCl_2$-containing formulation, RP444 can be readily prepared in high yield (85–92%), and excellent images in the canine deep vein thrombosis (DVT) and arteriovenous shunt models are produced. Injection of the whole or part of the formulation did not cause any unwanted side effects [59–62].

4.3 No-$SnCl_2$ Formulation

The $SnCl_2$-containing formulation suffers several shortcomings, including a large variation in RCP (85–92%) and a high failure rate (up to 50%) at high activity levels (100 mCi/vial). There is no firm pH control owing to the low buffer capacity of tricine. Stabilization of $SnCl_2$ always imposes a challenge in the manufacturing process. In addition, the $SnCl_2$-containing formulation typically results in the formation of a small amount of [^{99m}Tc]colloid. Therefore, there was a need to develop a new formulation, which gives better RCP performance (RCP≥90%), is more amenable to commercial manufacturing, and has better pH control. In developing the $SnCl_2$-containing formulation, it was found that $SnCl_2$ could be omitted if a large amount (more than 5 mg/mL) of TPPTS is used for the radiolabeling [45]. This is most likely due to the fact that TPPTS is an excellent reducing agent for $^{99m}TcO_4^-$ [63]. The [^{99m}Tc]colloid formation becomes minimal if TPPTS is used as the reducing agent. These observations eventually lead to the development of a "universal" no-$SnCl_2$ formulation for the ^{99m}Tc-labeling of HYNIC-conjugated small BMs.

Factors influencing the RCP of the ternary ligand ^{99m}Tc complex include heating temperature, heating time, pH, buffer agent, the bulking agent, and the component (HYNIC–BM, tricine, TPPTS, and $Na^{99m}TcO_4$) concentration [64]. For convenience, we will use RP444 as an example to illustrate the importance of the reaction conditions and component levels on the yield and RCP of the expected ternary ligand ^{99m}Tc complex. XV066 is the HYNIC–BM conjugate for RP444 [64]. A concentration of 5 μg/vial is too low to afford an RCP of more than 90%. There is no significant difference in the RCP performance between vials containing 20–40 μg XV066. The tricine concentration has no significant impact on the RCP of RP444. TPPTS serves as a reducing agent for $^{99m}TcO_4^-$ and as a coligand in the technetium chelate. The optimal concentration of TPPTS is 5–7 mg/vial. Pluoronic acid is a surfactant to reduce the glass absorption of XV066. The use of pluronic acid (0.01–1.0%) does not have a significantly impact on the RCP of RP444. Mannitol is a bulking agent to provide a "crystalline framework" for the other kit components to crystallize. Too much or too little mannitol in the formulation is detrimental for the RCP of RP444. Succinic acid is a buffer agent for pH control and has the buffer capacity at pH 4.5–5.5. The optimal pH is 5.0. There is no significant difference in the RCP at the buffer concentrations between 100 and 450 mM for succinic acid although there is an indication that the higher buffer concentration may give better RCP performance. Through a series of radiolabeling experiments, it was found that frozen vials containing 20 μg XV066, 5 mg TPPTS, 6.5 mg tricine, 40 mg mannitol, 38.5 mg disodium succinate hexahydrate, 12.7 mg succinic acid, and 0.1 mg pluronic acid (0.1%) give the best RCP (92–97%) for RP444. All the kit components are readily dissolved in 1 mL water. After lyophilization, all the components are crystallized to give a white solid firm "cake" [59]. The lyophilized vials can be stored at room temperature for more than 3 years without any significant degradation or increase in moisture. Using the no-$SnCl_2$ formulation, the radiolabeling is achieved by adding 1.5 mL of generator eluant (33–135 mCi/mL) and heating the reaction mixture at 100 °C for 10 min. RP444 is prepared in high yield with RCP≥90%. Formation of the [^{99m}Tc]colloid is minimal (less than 0.5%).

In principle, the no-$SnCl_2$ formulation can be used for the ^{99m}Tc-labeling of any HYNIC–BM conjugates, including small peptides and nonpeptide receptor ligands; however, there are also limitations using this no-$SnCl_2$ formulation. For example, SG380 is the HYNIC–BM conjugate for preparation of RP517, [^{99m}Tc(SG380)(tricine)(TPPTS)], a ^{99m}Tc-labeled LTB_4 receptor antagonist for imaging foci of infection and inflammation. Direct substitution of 20 μg SG380 for XV066 in the no-$SnCl_2$ formulation gave fairly low yields (60–70%) of RP517 [39]. SG380 is very lipophilic and has low solubility in the kit matrix. Even though a surfactant (Poloxamer-188, 0.1%) is used in the no-$SnCl_2$ formulation, both SG380 and RP517 showed significant adsorption on the glass surface of reaction vials. To solve the problem, a solubilizing agent or cosolvent such as ethanol has to be used to reduce glass-surface adsorption and to increase the solubility of SG380 in the kit matrix [39]. Therefore, the no-$SnCl_2$ formulation has to be modified according to the physical and chemical properties of the HYNIC–BM conjugate.

5
^{99m}Tc-Labeling Efficiency

The ^{99m}Tc-labeling efficiency is a term used to describe the ability of a BFC to achieve a high radiolabeling yield (more than 90%) of its ^{99m}Tc complex. Various chelators have been used for ^{99m}Tc-labeling of proteins and small BMs. These include N_3S triamidethiols [65–67], N_2S_2 diamidedithiols [67–70], N_2S_2 monoamidemonoaminedithiols [71], N_2S_2 diaminedithiols [72–77], boronic acid adducts of technetium dioximes [78, 79], tetraamines [80], and HYNIC

[1–48]. However, there is very little comparative data on these chelators with respect to their ^{99m}Tc-labeling efficiency.

Several factors may influence the labeling efficiency of a BFC. These include the identity of donor atoms, chelator concentration, and reaction conditions (e.g., temperature, time, and the pH). If the chelator concentration is fixed, the ^{99m}Tc-labeling conditions are largely dependent upon the nature of the donor atoms. A major advantage of HYNIC as the BFC is its high labeling efficiency (rapid and high-yield radiolabeling). For example, it has been reported that under optimized conditions, very a high specific activity (25,000 mCi/μmol and greater) can be achieved for ternary ligand ^{99m}Tc complexes, [^{99m}Tc(HYNIC–BM)(tricine)(TPPTS)] [42, 45, 47]. The HYNIC–BM-to-technetium ratio is about 6:1 under these conditions. The ^{99m}Tc-labeling efficiency of HYNIC is comparable to or better than that of N_2S_2 diaminedithiols [81].

For the HYNIC–BM conjugate, the ^{99m}Tc-labeling efficiency also depends on the hydrazone protecting group and the coligand used for preparation of the [^{99m}Tc]HYNIC complex. For example, it has been reported that tricine gives a much higher specific activity for the ^{99m}Tc-labeled HYNIC-IgG than GH [1, 2, 55]. Tricine also gives a higher specific activity for the ^{99m}Tc-labeled HYNIC–BM than EDDA analogues [42]. The use of the hydrazone protecting group usually slows down the rate of radiolabeling. That may explain why the elevated temperature (80–100 °C) is needed for successful ^{99m}Tc-labeling of the hydrazone-protected HYNIC–BM conjugate [39].

6 Effect of Coligand on Physical and Biological Properties of ^{99m}Tc Radiopharmaceuticals

In ternary ligand technetium complexes, [^{99m}Tc(HYNIC–BM)(tricine)(L)], the coligand may vary from phosphines to pyridine analogues (Fig. 2). The change of coligand will result in a dramatic change in the technetium chelate with respect to the size and charge. For small BMs, the technetium chelate contributes greatly to the overall size and molecular weight of the ^{99m}Tc-labeled HYNIC–BM conjugate. The change of coligand may have a significant impact on the physical (hydrophilicity or lipophilicity) and biological (biodistribution and the route of excretion) characteristics of the ternary ligand ^{99m}Tc complex [^{99m}Tc(HYNIC–BM)(tricine)(L)]. For example, the lipophilicity of RP444, RP445, and RP446 increases as the number of sulfonato groups on the phosphine coligand decreases [42]. As a result, RP444 and RP445 are preferentially excreted via the renal system, while RP446 is excreted via the hepatobiliary route [43]. The hydrophilicity of ternary ligand ^{99m}Tc complexes, [^{99m}Tc(XV120)(tricine)(L)] (L is NIC, ISONIC, ISONIC–L-Asp–OMe_2, ISONIC–HE, PA, and PSA) is readily altered simply by changing the functional groups on the pyridine ring [44].

Decristoforo and coworkers [20–23, 25] reported the ^{99m}Tc-labeling of several HYNIC-conjugated somatostatin analogs. It was found that the coligand has a significant impact on both the tumor accumulation and the biodistribution of ^{99m}Tc complexes. For example, the binary ligand ^{99m}Tc complex [^{99m}Tc(HYNIC–TOC)(tricine)$_2$] shows much higher protein binding than the complexes [^{99m}Tc(HYNIC–TOC)(EDDA)$_x$] and [^{99m}Tc(HYNIC–TOC)(tricine)(NIC)] [21–25]. The high serum protein binding was also observed for the ^{99m}Tc-labeled DNA oligonucleotides when tricine was used as the coligand [29–31]. Arano and his coworkers reported the ^{99m}Tc-labeling of HYNIC-derivatized small BMs and found that the combined use of tricine and nicotinic acid as coligands dramatically reduces the protein binding of the ^{99m}Tc-labeled HYNIC–BM conjugates.

The high serum protein binding of the ^{99m}Tc-labeled HYNIC–BM conjugates is most likely related to solution instability of the complex [^{99m}Tc(HYNIC–BM)(tricine)$_2$]. Liu et al. [44] have demonstrated that imine-N containing heterocycles, such as *N*-acetylhistidine (HIS-AC), can react with the complex [^{99m}Tc(XV120)(tricine)$_2$] and form the stabler ternary ligand ^{99m}Tc complex [^{99m}Tc(XV120)(tricine)(HIS-AC)]. Since proteins contain many histidine amino acid residues, it is quite possible that the binary ligand ^{99m}Tc complex [^{99m}Tc(HYNIC–BM)(tricine)$_2$] will react with the imidazole moiety of histidine residues, and form stable bonding between the [^{99m}Tc]HYNIC moiety and protein molecules. Protein binding usually leads to long blood retention time and lower target-to-background ratios.

Recently, Rennen et al. [82] reported the ^{99m}Tc-labeling of the HYNIC-conjugated interleukin-8 (HYNIC–IL-8) and the effect of coligands (tricine, EDDA, TPPTS/tricine, NIC/tricine and ISONIC/tricine) on in vivo characteristics of the ^{99m}Tc-labeled HYNIC-IL-8. The in vivo biodistribution of radiotracers in rabbits with *Escherichia coli* induced soft-tissue infection and was determined by γ-camera imaging as well as by tissue-counting at 6 h post injection. High-quality images with clear delineation of infectious foci could be obtained 2 h after injection of the ^{99m}Tc-labeled HYNIC–IL-8 conjugate. Ternary ligand ^{99m}Tc complexes [^{99m}Tc(HYNIC–IL-8)(tricine)(L)] (L is NIC and ISONIC) showed the highest abscess uptake (around 0.55% ID/g) and abscess-to-muscle ratios (above 200). The complex [^{99m}Tc(HYNIC–IL-8)(EDDA)$_x$] also showed highly favorable pharmacokinetic properties (rapid and high abscess uptake, rapid blood clearance and high abscess-to-background ratio) for infection imaging; however, the specific activity for the complex [^{99m}Tc(HYNIC–IL-8)(EDDA)$_x$] is much lower than that of ternary ligand ^{99m}Tc complexes [^{99m}Tc(HYNIC–IL-8)(tricine)(L)] (L is NIC and ISONIC). The use of the bulky TPPTS coligand resulted in a significant reduction of in vitro receptor binding affinity, much lower abscess uptake (0.075 ± 0.004% ID/g), and higher liver uptake (0.28 ± 0.01% ID/g) of the ^{99m}Tc-labeled HYNIC–IL-8 [82].

It is rather surprising that the in vivo and in vitro properties of a 72 amino acid protein such as IL-8 could be greatly affected by a small coligand such as nicotinic acid [82]. This study represents very few examples of in vivo studies to illustrate the importance of the ^{99m}Tc chelate on physical and biological properties of ^{99m}Tc complex radiopharmaceuticals. The results from this study also demonstrate the advantage of ternary ligand systems for the ^{99m}Tc-labeling of small BMs in developing target-specific radiopharmaceuticals because of the choice of different coligands.

7 Characterization of ^{99m}Tc Radiopharmaceuticals at the Tracer Level

7.1 Composition Determination of ^{99m}Tc Radiopharmaceuticals

One of the most important aspects of radiochemistry is to know the composition of the ^{99m}Tc radiopharmaceutical in the radiolabeled kit. A quick and accurate method would help radiochemists understand the fundamental coordination chemistry at the tracer (^{99m}Tc) level. Since the total technetium (^{99m}Tc and ^{99}Tc) concentration in the generator eluant is very low (10^{-8}–10^{-6} M), it is impossible to use spectroscopic (IR, UV/vis, and NMR) methods to characterize ^{99m}Tc radiopharmaceuticals. Therefore, there is a growing need for a quick and accurate method to determine the composition of the radiopharmaceutical at the tracer level.

7.1.1 Mixed-Ligand Experiment

A mixed ligand experiment has been used to determine the composition of ternary ligand ^{99m}Tc complexes, [^{99m}Tc(XV120)(tricine)(L)] (L is water-soluble phosphines and pyridine analogs) [42, 44]. In the first experiment, XV120 and *N*-(6-hydrazinonicotinyl)-D-phenylalanine methyl ester (HYNIC–D-Phe–OMe) were used in the same reaction mixture, while tricine and TPPTS were used as coligands [42]. After the radiolabeling, the reaction mixture was analyzed by radio-HPLC. If only one XV120 is bonded to the technetium, the chromatogram is expected to show two sets of radiometric peaks: one set from RP444 and the other from the complex [^{99m}Tc(HYNIC–D-Phe–OMe)(tricine)(TPPTS)]. If there are two HYNIC ligands in each complex, a third set of peaks from the mixed-ligand complex, [^{99m}Tc(XV120)(HYNIC–D-Phe–OMe)(tricine)(TPPTS)], is expected. The presence of only two sets of peaks in the radio-HPLC chromatogram (Fig. 7) demonstrates clearly that there is only one XV120 bonded to the technetium in [^{99m}Tc(XV120)(tricine)(L)]. The same experiment can be used to determine the number of tricine, phosphine, or pyridine coligands in these ternary ligand ^{99m}Tc complexes [42, 44]. This methodology should also apply to the composition determination of ^{99m}Tc complexes with other binary or ternary ligand systems. The mixed-ligand experiment is simple, easy to carry

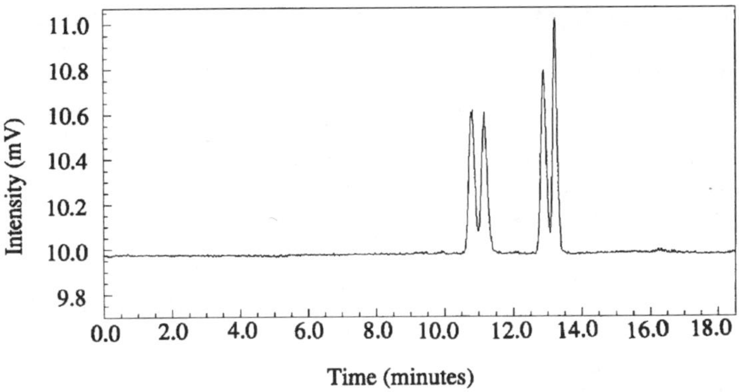

Fig. 7 Radio-HPLC chromatogram for RP444 and the complex [^{99m}Tc(HYNIC-D-Phe)(TPPTS)(tricine)]

out, and it only takes a few hours to know the composition of a new ^{99m}Tc radiopharmaceutical.

7.1.2
LC–MS Method

Traditionally, characterization of a new ^{99m}Tc radiopharmaceutical involves synthesis of the corresponding ^{99}Tc complex at the macroscopic level. The composition and the structure of ^{99}Tc complexes are determined by IR, NMR, fast atom bombardment MS, and X-ray crystallography. An HPLC concordance experiment is performed to demonstrate that the same technetium complex is prepared at both the tracer (^{99m}Tc) and macroscopic (^{99}Tc) levels. For peptide-based technetium radiopharmaceuticals, isolation of the ^{99}Tc-peptide complex usually involves multiple-step synthesis and tedious HPLC purification. It is also very difficult to grow single crystals for highly water soluble ^{99}Tc–peptide complexes. The chemistry of some chelating systems may differ at the tracer and macroscopic levels; thus, it is ideal to determine the composition of the ^{99m}Tc radiopharmaceutical at the tracer level.

MS has been used for a number of years as a powerful tool for the study of drug biodeposition and metabolism [83–87]. These types of studies have traditionally been carried out by gas chromatography–MS techniques. However, the use of LC–MS has been growing, especially since the introduction of the thermospray interface. With the advent of liquid-phase ionization techniques, such as electrospray, it has become possible to use LC–MS for the structural characterization of highly polar molecules at very low concentrations.

Liu et al. [54] first reported the use of LC–MS for characterization of ternary ligand ^{99m}Tc complexes [^{99m}Tc(XV120)(tricine)(L)] (L is TPPTS, TPPDS, TPPMS, ISONIC-HE, and ISONIC-Sorb) [54]. It was found that all the ^{99m}Tc complexes (Table 1) show the expected monoprotonated molecular ions,

Table 1 Liquid chromatography–mass spectrometry spectral data for ternary ligand^{99m}Tc complexes [^{99m}Tc(XV120)(tricine)(L)], where VX120 is cyclo(D-Val–NMeArg–Gly–Asp–Mamb(5-(6-(6-hydrazinonicotinamido)hexanamide))) and L is trisodium triphenylphosphine-3,3′,3″-trisulfonate (TPPTS), disodiumtriphenylphosphine-3,3′-disulfonate (TPPDS), sodium triphenylphosphine-3-monosulfonate (TPPMS), *N*-(2-hydroxylethyl)isonicotinylamide (ISONIC-HE), and ISONIC-Sorb. Data from Ref. [54]

Coligand (*L*)	Formula of the complex	Exact mass	Found $(M+1)^+$	Found $(M+2)^{2+}$
TPPS	$C_{62}H_{78}N_{14}O_{23}PS_3Tc$	1,613.4	1,613.8	807.5
TPPDS	$C_{62}H_{79}N_{14}O_{20}PS_2Tc$	1,533.4	1,534.5	767.7
TPPMS	$C_{62}H_{80}N_{14}O_{17}PSTc$	1,454.5	1,454.3	727.8
ISONIC-HE	$C_{52}H_{73}N_{16}O_{16}Tc$	1,277.1	1,277.3	639.3
ISONIC-Sorb	$C_{56}H_{81}N_{16}O_{20}Tc$	1,397.2	1,398.3	699.2

$(M+1)^+$, and diprotonated molecular ions, $(M+2)^{2+}$ (Fig. 8). The LC–MS data support the proposed structure, and are consistent with those obtained by fast atom bombardment MS for their ^{99}Tc analogs. The two isomers of RP444 show the same molecular ions with almost identical fragmentation patterns [54]. This is consistent with results from chirality experiments and provides direct evidence that the two radiometric peaks in the radio-HPLC chromatogram of ternary ligand ^{99m}Tc complexes [^{99m}Tc(HYNIC–BM)(tricine)(L)] (L is phosphines and pyridine analogs) are really due to resolution of two diastereomers [54].

The amount of injected activity necessary to obtain interpretable results is very much dependent on the charge and molecular weight of the ^{99m}Tc complex. For example, RP444 is highly charged, and usually requires a larger amount of activity to obtain the mass spectrum with a signal-to-noise ratio above 3. Thus, the decayed generator eluant is recommended to increase the technetium concentration. Using a fresh generator eluant (24-h prior elution), more than 10 mCi ^{99m}Tc (around 7×10^{-11} mol of technetium complex) is required to obtain an interpretable mass spectrum for RP444. Ternary ligand ^{99m}Tc complexes [^{99m}Tc(XV120)(tricine)(L)] (L is ISONIC-HE and ISONIC-Sorb) are neutral, and the molecular weights are also lower than that of RP444. Using a fresh generator eluant (24-h prior elution), only 1–2 mCi ^{99m}Tc (7×10^{-12}–1.5×10^{-11} mol of technetium complex) is required to obtain a reasonably clean mass spectrum [54]. LC-MS is particularly useful for ^{99m}Tc-labeled HYNIC conjugates, the ^{99}Tc analogs of which cannot be readily prepared. For example, [^{99}Tc(XV120)(tricine)(ISONIC-HE)] could not be readily prepared at the macroscopic level, owing to the competition between the ISONIC-HE coligand and XV120, which also contains a pyridine-N donor. At the tracer (^{99m}Tc) level, the molar ratio of ISONIC-HE to XV120 is around 2,000:1. At the macroscopic level, this ratio is very difficult to achieve. The synthesis of the ^{99}Tc complex [^{99}Tc(XV120)(tricine)(ISONIC-HE)] is extremely difficult. It is much

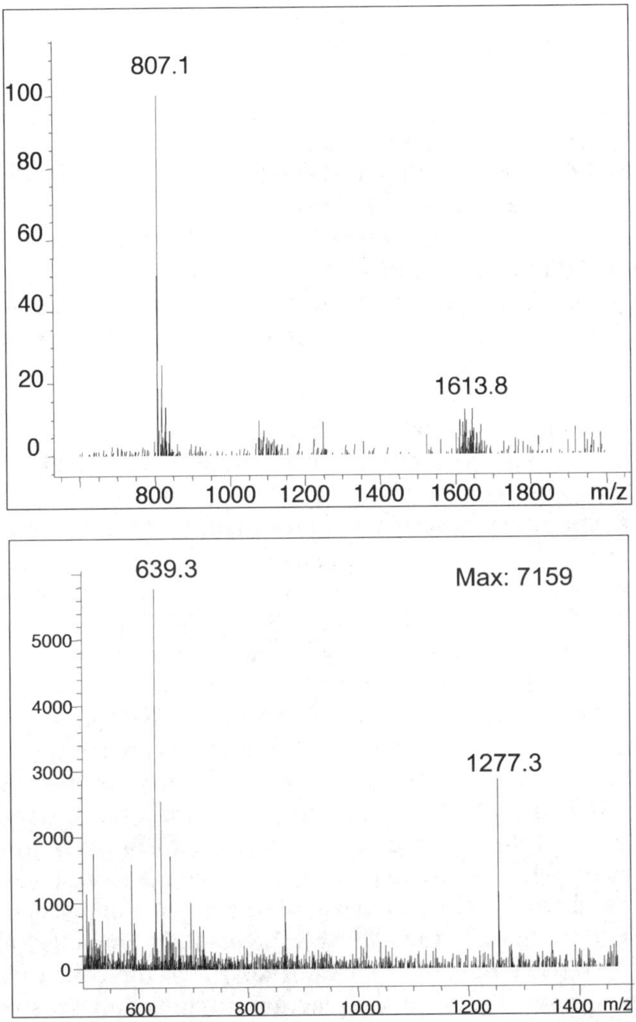

Fig. 8 Liquid chromatography–mass spectrometry spectrum of RP444 and [^{99m}Tc(HYNIC-tide)(tricine)(ISONIC-HE)], where ISONIC-HE is *N*-(2-hydroxylethyl)isonicotinylamide

easier to use LC–MS to determine the composition of [^{99m}Tc(XV120)(tricine)(ISONIC-HE)] at the tracer (^{99m}Tc) level.

LC–MS is a quick and accurate analytical tool for determining the composition of ^{99m}Tc radiopharmaceuticals. It also supplements structural information obtained from other analytical methods such as NMR. Therefore, LC–MS may also be a complimentary technique for structure determination of radiolabeled compounds by analyzing the fragmentation patterns of the mass spectra. LC–MS has the potential to become a routine analytical tool for radiopharmaceuticals labeled with ^{99m}Tc or other radioisotopes such as ^{186}Re.

7.2 Diastereomers and Chirality Experiments

For the last decade, HPLC has become a routine quality-control technique for ^{99m}Tc radiopharmaceuticals. The advantage of radio-HPLC is its capability to determine the RCP for the ^{99m}Tc-labeled BM, and to separate different ^{99m}Tc species in the kit matrix. It is also a very powerful tool for separation of different isomers, such as diastereomers. For example, HPLC chromatograms of ternary ligand ^{99m}Tc complexes [^{99m}Tc(HYNIC–BM)(tricine)(L)] (L is phosphines and pyridine analogs) always show two radiometric peaks with the area ratio being close to 1:1 if the HYNIC–BM conjugate contains a chiral center. These two radiometric peaks are most likely due to the resolution of either diastereomer resulting from the chiral centers on both the cyclic peptide backbone and the technetium chelate, or geometric isomers resulting from different geometric orientations of the diazenido ligand owing to restricted rotation around the Tc=N and N=N bonds.

A chirality experiment has been designed to understand the presence of two isomers in ternary ligand ^{99m}Tc complexes, [^{99m}Tc(HYNIC–BM)(tricine)(L)] [44]. In this experiment two closely related HYNIC derivatives, *N*-benzyl-6-(2-sulfobenzaldehydehydrazono)nicotinamide (HYNIC–BA) and *N*-((*R*)-(+)-α-methylbenzyl)-6-(2-sulfobenzaldehydehydrazono)nicotinamide (HYNIC–MBA), were used in the same reaction mixture; PSA (Fig. 2) was used as a coligand. HYNIC-MBA contains a chiral center and HYNIC–BA has no chiral center. Ternary ligand ^{99m}Tc complexes [^{99m}Tc(HYNIC–BA)(tricine)(PSA)] and [^{99m}Tc(HYNIC–MBA)(tricine)(PSA)] were synthesized by reacting $^{99m}TcO_4^-$ with HYNIC–BA and HYNIC–MBA, in the presence of excess tricine, PSA, and stannous chloride. If the existence of two isomeric forms were from different geometric orientations of the HYNIC moiety owing to restricted rotation around Tc=N and N=N bonds, there would be two pairs of peaks in the radio-HPLC chromatogram: one from [^{99m}Tc(HYNIC–BA)(tricine)(PSA)] and the other from [^{99m}Tc(HYNIC–MBA)(tricine)(PSA)]. On the contrary, the radio-HPLC chromatogram (Fig. 9) of the reaction mixture shows a single peak at 12.5 min from [^{99m}Tc(HYNIC–BA)(tricine)(PSA)], and a pair of peaks at 19.8 and 20.4 min from [^{99m}Tc(HYNIC–MBA)(tricine)(PSA)]. These results provides strong evidence that the isomerism seen in these ^{99m}Tc complexes is due to the resolution of two diastereomers [44].

In another chirality experiment, HYNIC–BA was used as the ligand while tricine and two pyridine analogs (ISONIC-HE without a chiral center and ISONIC–L-Asp–OMe$_2$ with a chiral center) were used as coligands. The radio-HPLC chromatogram (Fig. 10) of the reaction mixture shows a single peak at 9.0 min from [^{99m}Tc(HYNIC–BA)(tricine)(ISONIC-HE)], and a pair of peaks at 21.4 and 26.4 min from [^{99m}Tc(HYNIC–MBA)(tricine)(ISONIC–L-Asp–OMe$_2$)]. The area ratio of these two peaks is always close to 1:1, suggesting that they are indeed a result of resolution of the two diastereomers of [^{99m}Tc(HYNIC–MBA)-(tricine)(ISONIC–L-Asp–OMe$_2$)].

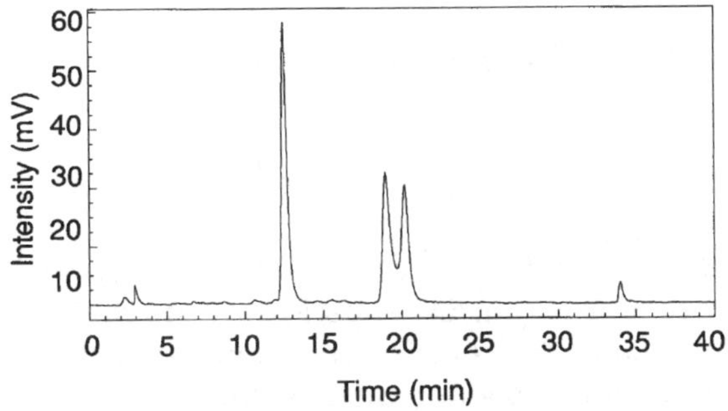

Fig. 9 Radio-HPLC chromatogram for the chirality experiment. The single peak at 12.5 min is due to [^{99m}Tc(HYNIC–BA)(tricine)(PSA)], where HYNIC–BA is *N*-benzyl-6-(2-sulfobenzaldehydehydrazono)nicotinamide and PSA is pyridine-3-sulfonic acid, and the peaks at 19.8 and 20.4 min are from [^{99m}Tc(HYNIC–MBA)(tricine)(PSA)], where HYNIC–MBA is *N*-((*R*)-(+)-α-methylbenzyl)-6-(2-sulfobenzaldehydehydrazono)nicotinamide

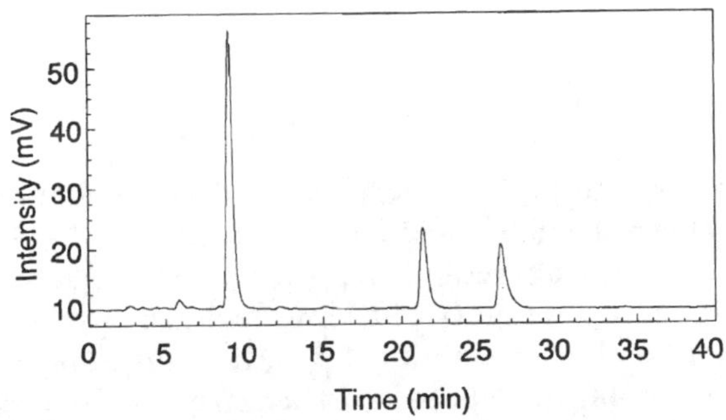

Fig. 10 Radio-HPLC chromatogram for the chirality experiment. The single peak at 11.9 min is due to [^{99m}Tc(HYNIC–BA)(tricine)(ISONIC-HE)], where ISONIC-HE is *N*-(2-hydroxylethyl)isonicotinylamide, and the doublet at 21.5 and 26.5 min is from [^{99m}Tc(HYNIC–BA)(tricine)(ISONIC–Asp–OMe_2)]

The technetium chelate is chiral owing to the tricine coligand being prochiral, and should be formed in an equal mixture of dextro and levo enantiomers. These enantiomers in combination with the chiral center(s) in the HYNIC–BM conjugate or coligand result in the formation of two diastereomers. If there is no chiral center in the HYNIC–BM conjugate or coligand, the ternary ligand ^{99m}Tc complex is expected to show only one radiometric peak unless a chiral chromatographic method is employed to separate the two enantiomers. The chirality experiment should also apply to other ^{99m}Tc complexes with other binary or ternary ligand systems. This experiment is particularly useful to distinguish diastereomers from geometric and conformational isomers.

8 Structural Characterization of Ternary Ligand Technetium Complexes

8.1 Technetium Oxidation State

While the composition of the ternary ligand ^{99m}Tc complexes has been determined via a series of mixed-ligand experiments and by LC–MS at the tracer (^{99m}Tc) level, the technetium oxidation state remains unclear. It has been proposed that the technetium oxidation state in the [^{99m}Tc]tricine complex is +5 [55, 88]. The oxidation state might change when HYNIC is bonded to the technetium center. Since organohydrazines are good reducing agents, the reaction of HYNIC with $^{99m}TcO_4^-$ or an intermediate [^{99m}TcO(tricine)$_2$] should be treated as a redox reaction instead of simple deprotonation of the hydrazine group. Therefore, the charge on the organohydrazine cannot be assigned simply as a function of the number of hydrogens lost when it is bonded to the technetium center.

8.2 Bonding Modality of HYNIC

Complexes containing technetium-hydrazido and technetium-diazenido bonds have previously been characterized by X-ray crystallography [89–102]. Nicholson and coworkers [95–102] recently reported a series of Tc(III)/Re(III) complexes of 2-hydrazinopyridine (HYPY) and hydrazinopyrimidine. Although there are several bonding modalities (Fig. 11) for organohydrazines such as HYPY, only four bonding modalities have been structurally characterized by X-ray crystallography, including neutral monodentate pyridiniumdiazenido (Fig. 11a), anionic monodentate pyridyldiazenido (Fig. 11b), neutral bidentate pyridyldiazene (Fig. 11c), and anionic bidentate pyridyldiazenido (Fig. 11d). There is no structural information available for the monodentate pyridylisodiazene (Fig. 11e) and anionic monodentate pyridyldiazenido (Fig. 11f).

The binding modality and the number of HYPY in the technetium complex also depend on the nature and availability of the coligand. For example, the reaction of NH_4[$^{99m}TcO_4^-$] with HYPY·2HCl produces a complex [$TcCl_3$(HN=NC_5H_4N)(N=NC_5H_4NH)], in which the technetium is bonded by three chlorides, a neutral bidentate pyridyldiazene and a neutral pyridiniumdiazenido. The reaction of HYPY·2HCl with [n-Bu_4N][$TcOCl_4$] in the presence of excess tricine and PPh_3 produces the complexes [Tc(HYPY)$(PPh_3)_2Cl_2$] (major product) and [Tc(HYPY)(tricine)(PPh_3)] (minor product).HYPY in the complex [Re(HYPY)$(PPh_3)_2Cl_2$] was found to be a neutral and bidentate pyridyldiazene [96].

[Tc(HYPY)(tricine)(PPh_3)] is a model compound for ternary ligand ^{99m}Tc complexes [^{99m}Tc(HYNIC-BM)(tricine)(L)] (L is TPPTS, TPPDS, and TPPMS). The NMR data suggest that the complex [Tc(HYPY)(tricine)(PPh_3)] has a dis-

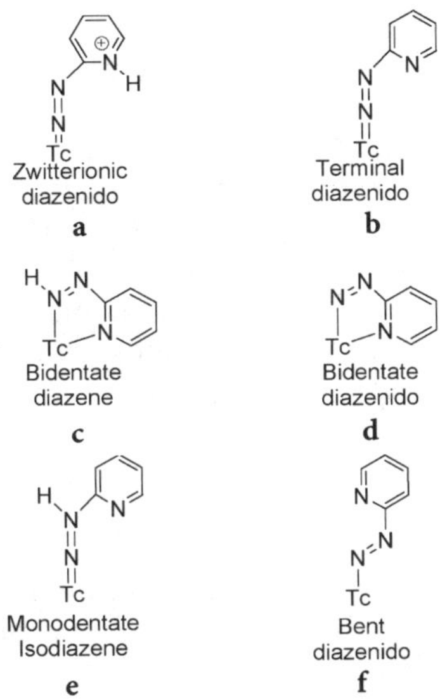

Fig. 11a–f Possible bonding modalities for 2-hydrazinopyridine (HYPY)

torted octahedral coordination sphere with a monodentate HYPY, a tetradentate tricine, and a monodentate PPh_3 coligand [46]; however, the exact boding modality of HYPY in [Tc(HYPY)(tricine)(PPh_3)] still remains unclear owing to the difficulty in determining which atom is bonded to the exchangeable hydrogen (Fig. 12). If the hydrogen is deprotonable in solution, these three different structures are indistinguishable.

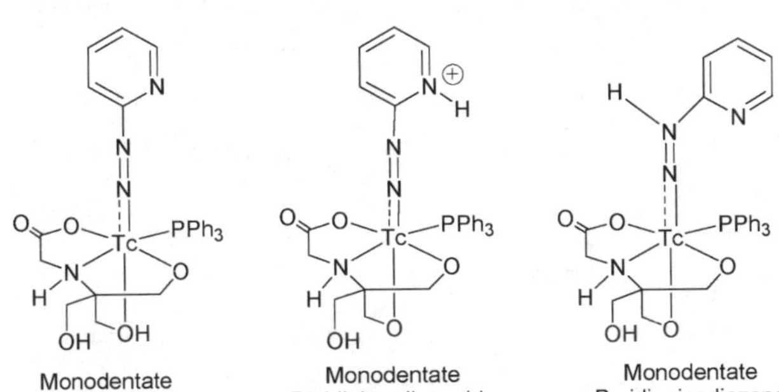

Fig. 12 Possible structures for the ternary ligand technetium complex [^{99m}Tc(HYPY)-(tricine)(PPh_3)]

9
^{99m}Tc-Labeled HYNIC–BM Conjugates as Radiopharmaceuticals

For the last several years, the ternary ligand system (HYNIC–BM, tricine, and TPPTS) has been used as a platform technology for the ^{99m}Tc-labeling of a variety of small BMs. These include a platelet GPIIb/IIIa receptor antagonist for imaging thrombosis, chemotactic peptides, LTB_4 receptor antagonists, and IL-8 for detection of infection and inflammation, integrin $\alpha_v\beta_3$ receptor antagonists, and somatostatin analogs for tumor imaging. Selected examples of targeting BMs are shown in Table 2.

9.1
Thrombus Imaging Agent

Venous and arterial thrombus formation is common and is potentially a life-threatening event. A venous thrombus is an intravascular deposit predomi-

Table 2 Ternary ligand ^{99m}Tc complexes [99mTc(6-hydrazinonicotinamide–biomolecule)-(tricene)(TPPTS)] as diagnostic radiopharmaceuticals

Compound	BM	Targeted Receptor	Targeted Disease
RP444	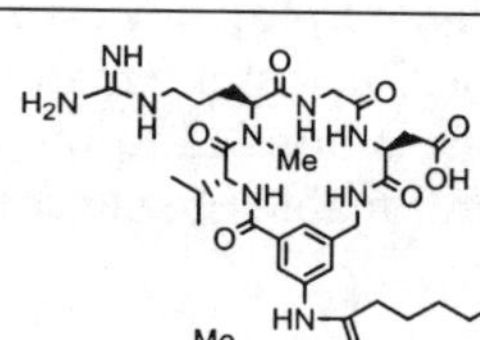	GPIIb/IIIa Platelets	Thrombosis
RP463	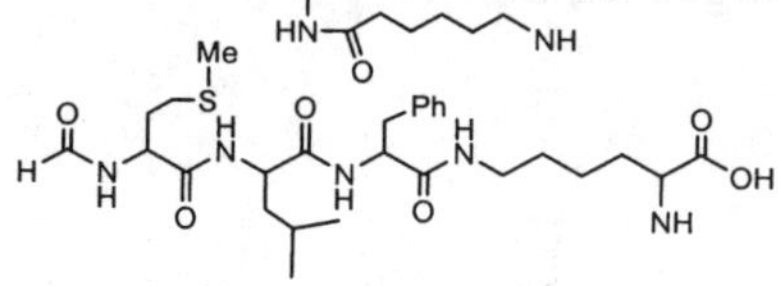	Chemotactic Peptide	Infection or Inflammation
RP517	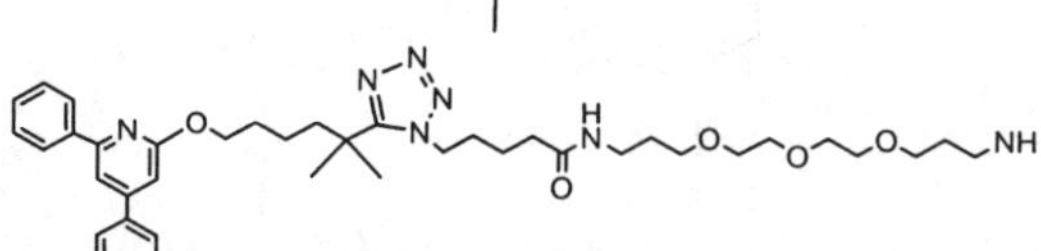	LTB4	Infection or Inflammation
RP593	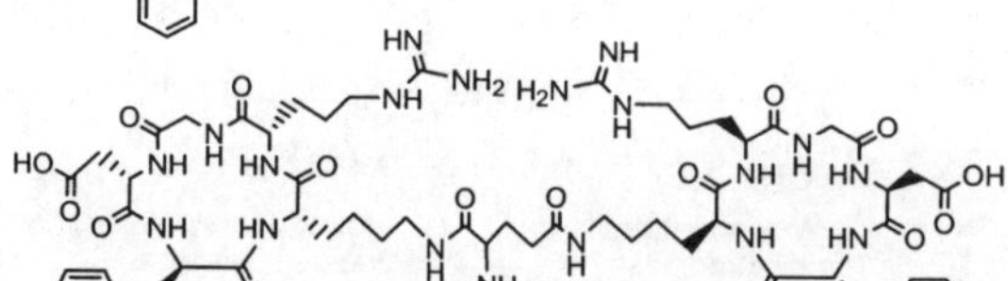	vitronectin αvβ3	Cancer Angiogenesis
RP678	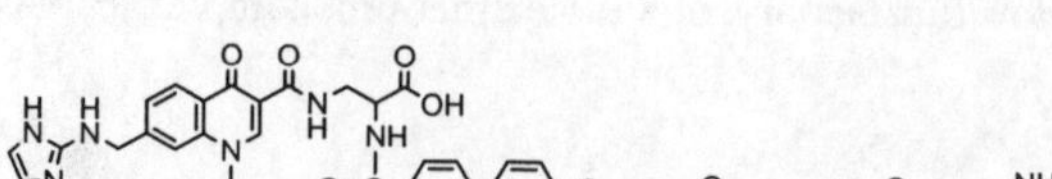	vitronectin αvβ3	Cancer Angiogenesis

nantly comprising fibrin and aggregates of platelets. Intraarterial platelet-fibrin thrombus formation is the final common pathway that leads to most complications of atherosclerosis, including acute myocardial infarction, stroke, and sudden death. In addition, intraarterial thrombus is a major cause of the complications associated with intravascular prosthetic materials such as arterial bypass grafts and heart valves. Thus, the development of an imaging agent that is able to detect thrombus with high accuracy would be a significant advance in diagnostic nuclear medicine.

RP444 (Table 2) is a ^{99m}Tc-labeled cyclic platelet GPIIb/IIIa receptor antagonist binding to the activated platelets. In the canine arteriovenous shunt model [43], RP444 was adequately incorporated into the arterial and venous portions of the rapidly growing thrombus (7.8–9.9 and 0.2–3.7% ID/g, respectively). In the canine DVT model, the thrombus uptake (2.86±0.4% ID/g) of RP444 was much higher than that of [^{99m}Tc]albumin. RP444 showed a moderate blood clearance with a $t_{1/2}$ of approximately 90 min. Visualization of DVT can be as early as 15 min post injection, and improves over time with a thrombus-to-muscle ratio of 9.7±1.9 at 120 min post injection. In a canine model of endothelial injury, RP444 is able to identify platelet-rich thrombus, and may have applications for detection of unstable coronary syndromes in humans [51]. Phase I clinical trials showed an excellent safety profile and acceptable dosimetry. Results from phase II trials clearly demonstrated the utility of RP444 as a new radiopharmaceutical for detection of DVT within 60 min with 93% sensitivity, 95% specificity, and 82% accuracy [52–54].

9.2 Infection/Inflammation Imaging Agents

White blood cells, particularly polymorphonuclear leukocytes and monocytes, accumulate in high concentrations at sites of infection. Therefore, research has been directed toward radiolabeling small molecules that bind to both circulating granulocytes and leukocytes. Recently, Edwards et al. [35] reported the radiolabeling of fMLFK–HYNIC using tricine and TPPTS as coligands. The ternary ligand ^{99m}Tc complex [^{99m}Tc(fMLFK–HYNIC)(tricine)(TPPTS)] (RP463, Table 2) was evaluated in two animal (guinea pig and rabbit) models of focal infection. It was also found that RP463 was incorporated rapidly into the infected tissues with a target-to-background ratio of 5–7:1 at 4 h post injection. In the rabbit infection model, a transient decrease in the white blood cell count of 35% was observed during the first 30 min after injection of the HPLC-purified RP463 [35]. Since RP463 was purified by HPLC, it should be free of unlabeled fMLFK–HYNIC. The neutropenic effect is probably due to the fact that RP463 is a very high affinity agonist for chemotactic peptide receptors on leukocytes. Therefore, future research should be focused on receptor antagonists as targeting BMs for imaging infection or inflammation.

LTB_4 is a product derived from the action of 5-lipoxygenase on arachidonic acid, and is known to stimulate degranulation, aggregation, chemotaxis, and chemokinesis of polymorphonuclear leukocytes, as well as to promote superoxide generation [36–39]. LTB_4 receptors are highly expressed on a number of inflammatory cells such as neutrophils and lymphocytes. Numerous LTB_4 receptor antagonists have been evaluated for their potential in the treatment of infectious and inflammatory diseases. Recently, we reported a series of ^{99m}Tc-labeled LTB_4 receptor antagonists [36–39]. The ternary ligand ^{99m}Tc complex RP517 (Table 2) was evaluated in two animal (guinea pig and rabbit) models of focal infection.

In the guinea pig peritonitis model, RP517 showed an uptake of 0.6± 0.05% ID/g at the site of infection 1 h after intravenous injection [36, 37]. The radioactivity was retained with an uptake of 0.5±0.08% ID/g at 4 h post injection. RP517 is excreted mainly via the hepatobiliary route. In the rabbit infection model, RP517 showed similar uptake at the infection site with a target-to-background (infected muscle-to-normal muscle) ratio of 16:1 at 1 h post injection. A primate dosimetry study indicated that the dose-limiting organ was the gall bladder. RP517 is under clinical investigation as a new imaging agent for acute infection and inflammation.

9.3 Tumor Imaging Agents

In spite of a number of imaging modalities currently available in clinics there is still an unmet need for the development of new diagnostic agents which can detect both primary and metastastic tumors with more specificity and accuracy. Radiolabeled receptor-binding BMs are of great interest because they have the potential to detect primary sites, identify occult metastatic lesions, guide surgical intervention, stage tumors, and predict the efficacy of therapeutic agents.

9.3.1 ^{99m}Tc-Labeled Integrin $\alpha_v\beta_3$ Receptor Antagonists for Tumor Imaging

The formation of new blood vessels (angiogenesis) is a requirement for malignant tumor growth and metastasis [103–106]. The angiogenic process depends on vascular endothelial cell migration and invasion, regulated by cell adhesion receptors.Integrins are a family of proteins that facilitate cellular adhesion to and migration on the extracellular matrix proteins found in intercellular spaces and basement membranes. They also regulate cellular entry and withdraw from the cell cycle. Integrin $\alpha_v\beta_3$ is a receptor for a wide variety of extracellular matrix proteins with the exposed RGD tripeptide sequence. These include vitronectin, fibronectin, fibrinogen, lamin, collagen, Von Willibrand's factor, osteoponin, and adenovirus particles. The expression of integrin $\alpha_v\beta_3$ enables a given cell to adhere to, migrate on, or respond to almost any matrix protein it may encounter. Despite its promiscuous behavior, integrin $\alpha_v\beta_3$ is generally expressed at low levels on epithelial cells and mature endothelial cells.

It has been reported that integrin $\alpha_v\beta_3$ is highly expressed on the neovasculature of tumors, including osteosarcomas, neuroblastomas, glioblastomas, melanomas, lung carcinomas, breast, prostate, and bladder cancers [107–112]. A recent study showed that integrin $\alpha_v\beta_3$ is overexpressed not only on endothelial cells but also on tumor cells in human breast cancer xenografts [113]. The expression of integrin $\alpha_v\beta_3$ correlates with tumor progression in melanoma, glioma, and ovarian and breast cancers [107–113]. The highly restricted expression of integrin $\alpha_v\beta_3$ during tumor invasion and metastasis presents an interesting molecular target for diagnosis of rapidly growing solid tumors [104, 114, 115].

RP593 (Table 2) is a ^{99m}Tc-labeled integrin $\alpha_v\beta_3$ receptor antagonist with two RGD-containing cyclic peptide receptor-binding motifs [33]. In a c-Neu oncomouse model, RP593 showed rapid blood clearance via the renal system, and demonstrated good tumor uptake and tumor retention (3.41±0.5% ID/g at 2-h and 1.5±0.3% ID/g at 24 h post injection). In dogs with mammary adenocarcinoma, the tumor uptake was 0.7±0.1% ID/g at 15-min and 0.43±0.05% ID/g at 24 h post injection. The tumor-to-muscle ratio was about 8:1 at 15 min post injection, and increased to 20:1 at 24 h post injection. RP593 has also been evaluated in athymic female BALB/c mice with a subcutaneously growing OVCAR-3 ovarian carcinoma xenograft [116]. The tumor uptake was 5.8±0.7% ID/g with the tumor-to-blood ratio being 63:1 at 24 h post injection. The kidney activity retention for RP593 was high, and was attributed to the presence of two arginine amino acid residues. RP593 is a promising tracer to study changes in integrin $\alpha_v\beta_3$ expression during metastasis and tumor-induced angiogenesis.

RP678 (Table 2) is a ^{99m}Tc-labeled nonpeptide integrin $\alpha_v\beta_3$ receptor antagonist with a quinolone pharmacophore [117, 118]. Compared with that of RP593, the tumor uptake of RP678 is about 1.5 times higher than that of RP593, while the kidney and liver uptake of RP678 is much lower than that of RP593 in the same animal model. In this respect, RP678 offers advantages over RP593; however, RP678 still shows a certain degree of hepatobiliary clearance. Therefore, the pharmacokinetics of RP678 needs to be further modified before it is used for human studies.

9.3.2
^{99m}Tc-Labeled Somatostatin Analogs for Tumor Imaging

Somatostatin receptors are overexpressed on a number of human tumors and their metastases [119–121], thereby serving well as the target for tumor imaging. Although the first report of in vivo imaging with a somatostatin analog appeared in 1976 [122], further development was hampered owing to rapid degradation of the native peptide by plasma and tissue proteases.For this reason, analogs of somatostatin have been synthesized using D-amino acids to prolong the in vivo half-life by inhibiting the action of amino and carboxypeptidases. Octreotide (Sandostatin, SMS 201–995) is a metabolically stable analog of so-

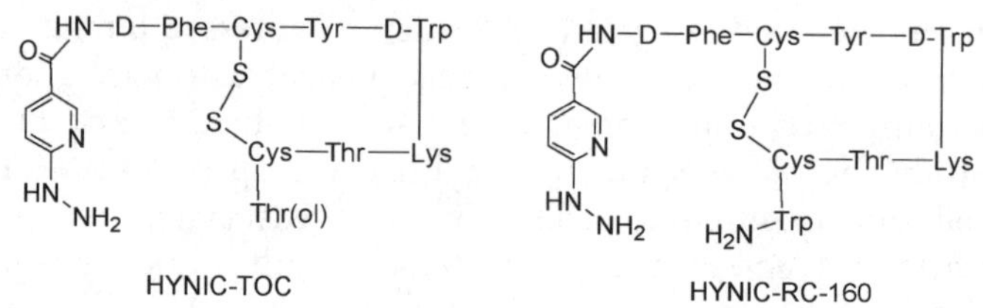

Fig. 13 Structures of two HYNIC-conjugated somatostatin analogs

matostatin, and has been successfully used for the treatment tumors. ^{111}In–DTPA–octreotide, where DTPA is diethylenetriaminepentaacetic acid, has become one of the most commonly used radiopharmaceuticals in clinics for tumor imaging; however, ^{111}In–DTPA–octreotide suffers several drawbacks such as a long half-life ($t_{1/2}$=67 h, g=173 keV, 89%, and 247 keV, 94%) and the high cost of ^{111}In. For diagnostic purposes, ^{99m}Tc is more desirable because of its low cost, easy availability, and ideal nuclear characteristics, which better match the rapid blood clearance and fast tumor accumulation of octreotide.

Decristoforo and coworkers [20–23, 25] have studied ^{99m}Tc-labeled somatostatin analogs to develop new diagnostic radiopharmaceuticals for imaging somatostatin-positive tumors. The structures of two HYNIC-conjugated somatostatin analogs (HYNIC–TOC and HYNIC–RC-160) are shown in Fig. 13; both analogs show extremely high receptor binding affinity with IC_{50} in the nanomolar range. In AR4-2J tumor-bearing nude mice, [^{99m}Tc(HYNIC–TOC)-(EDDA)$_n$] shows the most promising overall biodistribution profile with a rapid blood clearance and a high tumor uptake of 9.65±2.16% ID/g at 4 h post injection [22]. The tumor-to-liver and tumor-to-gastrointestinal tract ratios are similar to those of ^{111}In–DTPA–octreotide; but the tumor-to-kidney ratios are significantly higher in the same animal model [22]. Although the exact composition of the complex [^{99m}Tc(HYNIC–TOC)(EDDA)$_n$] is not known, recent LC–MS data seem to indicate that it contains two EDDA coligands [53].

In patients with somatostatin-positive tumors, [^{99m}Tc(HYNIC–TOC)-(EDDA)$_n$] has demonstrated rapid tumor uptake within the first hour after injection [23]. It also showed predominantly renal excretion. The accumulation patterns of [^{99m}Tc(HYNIC–TOC)(EDDA)$_n$] in tumors and normal organs are comparable to those of ^{111}In–DTPA–octreotide. A lesion-by-lesion comparison showed superior capabilities of [^{99m}Tc(HYNIC–TOC)(EDDA)$_n$] in respect of extrahepatic lesions. It allows much earlier diagnosis (10 min–4 h) compared with ^{111}In–DTPA–octreotide (4–24 h). [^{99m}Tc(HYNIC–TOC)(EDDA)$_n$] is under phase II clinical investigation as a new radiopharmaceutical for diagnosis of somatostatin receptor-positive tumors [53].

10 Conclusions

HYNIC is an extremely useful BFC for ^{99m}Tc-labeling of small biomolecules. The major advantage is its high ^{99m}Tc-labeling efficiency and the high in vivo stability of the resulting ^{99m}Tc complexes. Another advantage of HYNIC is the choice of different coligands, which allows easy modification of hydrophilicity and pharmacokinetics of ^{99m}Tc-labeled BMs. The binary ligand complex [^{99m}Tc(HYNIC–BM)(tricine)$_2$] can be prepared with a very high specific activity (20,000 mCi/µmol and greater); but it is unstable in solution and often exists in multiple isomeric forms in aqueous solution. The discovery of various ternary ligand systems represents a major achievement for the HYNIC technology by increasing the solution stability and reducing the number of isomeric forms in [^{99m}Tc–HYNIC] complexes. In principle, these ternary ligand systems can be used for the ^{99m}Tc-labeling of any HYNIC-conjugated small BMs. However, phosphine- or pyridine-containing HYNIC chelators may be the candidates of choice for ^{99m}Tc-labeling of small BMs containing one or more disulfide linkages, which are important to keep the rigid cyclic conformation of the BM and to maintain the high receptor binding affinity.

Acknowledgements The author would like to thank his former coworkers and supporting staff at Bristol–Myers Squibb Medical Imaging for their support and collaboration.

References

1. Abrams MJ, Juweid M, ten Kate CI, Schwartz DA, Hauser MM, Gaul FE, Fuccello, AJ, Rubin RH, Strauss HW, Fischman AJ (1990) J Nucl Med 31:2022
2. Schwartz DA, Abrams MJ, Hauser MM, Gaul FE, Larsen SK, Rauh D, Zubieta J (1991) Bioconjugate Chem 2:333
3. Ultee ME, Bridger GJ, Abrams MJ, Longley CB, Burton CA, Larsen S, Henson GW, Padmanabhan S, Gaul FE, Schwartz DA (1997) J Nucl Med 38:133
4. Bridger GJ, Abrams MJ, Padmanabhan S, Gaul FE, Larsen S, Henson GW, Schwartz DA, Burton CA, Ultee ME (1996) Bioconjugate Chem 7:255
5. Ono M, Arano Y, Uehara T, Fijioka Y, Ogawa K, Namba S, Mukai T, Nakayama M, Saji H (1999) Bioconjugate Chem 10:386
6. Lei K, Rusckowski M, Chang F, Qu T, Mardirossion G, Hnatowich DJ (1996) Nucl Med Biol 23:917
7. Fonti R, Cheung NKV, Bridger GJ, Guo HF, Abrams MJ, Larsen SM (1999) Nucl Med Biol 26:681
8. Zinn KR, Kelpke S, Chaudhuri TR, Sugg T, Mountz JM, Thompson JA (2000) Nucl Med Biol 27:407
9. Babich JW, Solomon H, Pike MC, Kroon D, Graham W, Abrams MJ, Tompkins RG, Rubin RH, Fischman AJ (1993) J Nucl Med 34:1964
10. Babich JW, Fischman AJ (1995) Nucl Med Biol 22:25
11. Fischman AJ, Rauh D, Solomon H, Babich JW, Tompkins RG; Kroon D, Strauss HW, Rubin HR (1993) J Nucl Med 34:2130
12. Babich JW, Graham WG, Barrow SA, Fischman AJ (1995) Nucl Med Biol 22:643

13. Babich JW, Graham W, Barrow SA, Dragotakes SC, Tompkins RH, Rubin RH, Fischman AJ (1993) J Nucl Med 34:2176
14. Babich JW, Tompkins RG, Graham W, Barrow SA, Fischman AJ (1997) J Nucl Med 38:1316
15. van der Laken CJ, Boerman OC, Oyen WJG, van de Ven MTP, Edwards DS, Barrett JA, van der Meer JW, Corstens FH (1997) J Nucl Med 38:1310
16. Claessens RAMJ, Boerman OC, Koenders EB, Oyen WJG, van der Meer JWM, Corstens FHM (1996) Eur J Nucl Med 23:414
17. van der Laken CJ, Boerman OC, Oyen WJG, van der Ven MPP, van der Meer JWM, Corstens FHM (1998) Eur J Nucl Med 25:347
18. Babich JW, Coco WG, Barrow SA, Fischman AJ, Femia FJ, Zibieta J (2000) Inorg Chim Acta 309:123
19. Krois D, Riedel C, Angelberger P, Kalchberger P, Virgolini I, Lehner H (1996) Liebigs Ann Chem 1463
20. Decristoforo C, Mather SJ (1999) Nucl Med Biol 26:389
21. Decristoforo C, Mather SJ (1999) Bioconjugate Chem 10:431
22. Decristoforo C, Mather SJ (1999) Eur J Nucl Med 26:869
23. Decristoforo C, Melendez L, Sosabowski JK, Mather SJ (2000) J Nucl Med 41:1114
24. Bangard M, Béhé M, Guhlke S, Otte R, Bender H, Maecke HR, Birsack HJ (2000) Eur J Nucl Med 27:628
25. Decristoforo C, Mather SJ, Cholewinski W, Donnemiller E, Riccabona G, Moncayo R (2000) Eur J Nucl Med 27:1318
26. Laverman P, Dams ETM, Oyen WJG, Storm G, Koenders EB, Prevost R, van der Meer JWM, Corstens FHM, Boerman OC (1998) Eur J Nucl Med 25:865
27. Laverman P, Dams ETM, Oyen WJG, Storm G, Koenders EB, Prevost R, van der Meer JWM, Corstens FHM, Boerman OC (1999) J Nucl Med 40:192
28. Zhang YM, Liu N, Zhu ZH, Rusckowski M, Hnatowich DJ (2000) Eur J Nucl Med 27:1707
29. Hnatowich DJ, Winnard P Jr, Virzi F, Santo T, Smith CL, Cantor CR, Rusckowski M (1995) J Nucl Med 36:2036
30. Hnatowich DJ (1996) Q J Nucl Med 40:202
31. Qu T, Wang Y, Zhu Z, Rusckowski M, Hnatowich DJ (2000) Nucl Med Commun 22:203
32. Guo W, Hinkle GH, Lee RJ (1999) J Nucl Med 40:1563
33. Ono M, Arano Y, Mukai T, Uehara T, Fujioka Y, Ogawa K, Namba S, Nakayama M, Saga T, Konishi J, Horiuchi K, Yokoyama A, Saji H (2001) Nucl Med Biol 28:155
34. Ono M, Arano Y, Mukai T, Fujioka Y, Ogawa K, Uehara T, Namba S, Saga T, Konishi J, Saji H (2001) Nucl Med Biol 28:215
35. Edwards DS, Liu S, Ziegler MC, Harris AR, Crocker AC, Heminway SJ, Barrett JA, Bridger GJ, Abrams MJ, Higgins JD III (1999) Bioconjugate Chem 10:884
36. Barrett JA, Harris TD, Heminway SJ, McKay L, Vining M, Crocker AC, Rajopadhye M, Liu S, Edwards DS (1998) J Nucl Med 39:215P
37. Barrett JA, Harris TD, Hemingway SJ, McKay L, Vining M, Crocker AC, Lazewasky J, Rajopadhye M, Liu S, Edwards DS (1998) Eur J Nucl Med 25:887
38. Brouwers AH, Laverman P, Boerman OC, Oyen WJG, Barrett JA, Harris TD, Edwards DS, Corstens FHM (2000) Nucl Med Commun 21:1043
39. Liu S, Harris AR, Williams NE, Edwards DS (2002) Bioconjugate Chem 13:881
40. Liu S, Edwards DS, Ziegler MC, Harris AR, Hemingway SJ, Barrett JA (2001) Bioconjugate Chem 12:624
41. Liu S, Edwards DS, Looby RJ, Harris AR, Poirier MJ, Barrett JA, Heminway SJ, Carroll TR (1996) Bioconjugate Chem 7:63
42. Edwards DS, Liu S, Barrett JA, Harris AR, Looby RJ, Ziegler MC, Heminway SJ, Carroll TR (1997) Bioconjugate Chem 8:146

43. Barrett JA, Crocker AC, Damphousse DJ, Heminway SJ, Liu S, Edwards DS, Lazewatsky JL, Kagan M, Mazaika TJ, Carroll TR (1997) Bioconjugate Chem 8:155
44. Liu S, Edwards DS, Harris AR (1998) Bioconjugate Chem 9:583
45. Edwards DS, Liu S, Harris AR, Poirier MJ, Ewels BA (1999) Bioconjugate Chem 10:803
46. Liu S, Edwards DS, Harris AR, Heminway SJ, Barrett JA (1999) Inorg Chem 38:1326
47. Edwards DS, Liu S, Harris AR, Ewels BA (1999) Bioconjugate Chem 10:803
48. Laverman P, Dams ETM, Oyen WJG, Storm G, Koenders EB, Prevost R, van der Meer JWM, Corstens FHM, Boerman OC (1999) J Nucl Med 40:192
49. Liu S, Edwards DS, Barrett JA (1997) Bioconjuate Chem 8:621
50. Edwards DS, Liu S (1997) Transition Met Chem 22:425
51. Liu S, Edwards DS (1999) Chem Rev 99:2235
52. Liu S, Edwards DS (2001) Drugs Future 26:375
53. von Guggenberg E, Sarg B, Lindner H, Alafort LM., Mather SJ, Moncayo R, Decristofo C (2003) J Labelled Compd Radiopharm 46:307
54. Liu S, Ziegler MC, Edwards DS (2000) Bioconjugate Chem 11:113
55. Larson SK, Solomon HF, Caldwell G, Abrams MJ (1995) Bioconjugate Chem 6:635
56. Purohit A, Liu S, Casebier D, Edwards DS (2003) Bioconjuate Chem 14:720
57. Schwartz DA, Abrams MJ, Gladomenico CM, Zubieta JA (1993) US Patent 5,206,370
58. Harris TD, Sworin M, Willianms N, Rajopadhye M, Damphousse PR, Glowacka D, Poirier MJ, Yu K (1998) Bioconjugate Chem 10:808
59. Line BR, Crane P, Lazewatsky J, Barrett JA, Cloutier D, Kagan M, Lukasiewicz R, Holmes RA (1996) J Nucl Med 37:117P
60. Line BR, Seabold JE, Edell SL, Weiland FL, Heit JA, Crane PD, Widner PL, Haber SJ. (1998) J Nucl Med 39:218P
61. Mitchel J, Waters D, Lai T, White M, Alberghini T, Salloum A, Knibbs D, Li A, Heller, GV (2000) Circulation 101:1643
62. Klem JA, Schaffer JV, Crane PD, Barrett JA, Henry GA, Canestri L, Ezekowitz MD (2000) J Nucl Cardiol 7:359
63. Pasqualini R, Duatti A, Bellande E, Comazzi V, Brucato V, Hoffschir D, Fagret D, Comet M (1994) J Nucl Med 35:334
64. Liu S, Edwards DS, Harris AR, Ziegler MC, Poirier MJ, Ewels BA, DiLuzio WR, Hui P (2001) J Pharm Sci 90:114
65. Breitz HB, Weiden PL, Vanderheyden JL, Appelbaum JW, Bjorn MJ, Fer MF, Wolf S B, Ratliff BA, Seiler CA, Foisie DC, Fisher DR, Schroff RW, Fritzberg AR, Abrams PG (1992) J Nucl Med 33:1099
66. Schroff RW, Weiden PL, Appelbaum JW, Fer MF, Breitz HB, Vanderheyden JL, Ratliff BA, Fisher DR, Foisie DC, Hanelin LG, Morgan AC Jr, Fritzberg AR, Abrams PG (1990) Antibody Immunoconjugates Radiopharm 3:99
67. Liu S, Edwards DS (1995) In: Nicolini M, Banoli G, Mazzi U (eds) Technetium and rhenium in chemistry and nuclear medicine, vol 4. SGEditorali, Padova, p 383
68. Fritzberg AR, Abrams PG, Beaumier PL, Kasina S, Morgan AC, Rao TN, Reno JM, Sanderson JA, Srinivasan A, Wilbur DS (1988) Proc Natl Acad Sci USA 85:402
69. Eary JF, Schroff RW, Abrams PG, Fritzberg AR, Morgan AC, Kasina S, Reno JM, Srinivasan A, Woodhouse CS, Wilbur DS, Natale RB, Collins C, Stehlin JS, Mitchell M, Nelp WB (1989) J Nucl Med 30:25
70. Kasina S, Rao TN, Srinivasan A, Sanderson JA, Fitzner JN, Reno JM, Beaumier PL, Fritzberg AR (1991) J Nucl Med 32:1445
71. Kung HF, Bradshaw JE, Chumpradit S, Zhuang ZP, Kung MP, Mu M, Frederik D (1995) In: Nicolini M, Banoli G, Mazzi U (eds) Technetium and rhenium in chemistry and nuclear medicine, vol 4. SGEditorali, Padova, pp 383–393

72. Eisenhut M, Mißfeldt M, Lehmann WD, Karas M (1991) J Labelled Compd Radiopharm 29:1283
73. Eisenhut M, Lehmann WD, Becker W, Behr T, Elser H, Strittmater W, Steinsträsser A, Baum RP, Valerius T, Repp R, Deo Y (1996) J Nucl Med 37:362
74. Baidoo KE, Lever SZ (1990) Bioconjugate Chem 1:132
75. DiZio JP, Fiashi R, Davison A, Jones AG, Katzenellenbogen JA (1991) Bioconjugate Chem 2:352
76. DiZio JP, Anderson CJ, Davison A, Ehrhardt GJ, Carlson KE, Welch MJ, Katzenellenbogen JA (1992) J Nucl Med 33:558
77. Baidoo KE, Lin KS, Zhan YG, Finley P, Scheffel U, Wagner HN Jr (1998) Bioconjugate Chem 9:218
78. Linder KE, Wen MD, Nowotnik DP, Malley MF, Gougoutas JZ, Nunn AD, Eckelman WC (1991) Bioconjugate Chem 2:160
79. Linder KE, Wen MD, Nowotnik DP, Ramalingam K, Sharkey RM, Yost F, Narra RK, Nunn AD, Eckelman WC (1991) Bioconjugate Chem 2:407
80. Maina T, Stolz B, Albert R, Nock B, Bruns C, Mäcke H (1995) In: Nicolini M, Banoli G, Mazzi U (eds) Technetium and rhenium in chemistry and nuclear medicine, vol 4. SGEditorali, Padova, pp 395–400
81. Liu S, Edwards SD, Harris AR, Singh PR (1997) Appl Radiat Isot 48:1103
82. Rennen HJJM, van Eerd JE, Oyen WJG, Corstens FHM, Edwards DS, Boerman OC (2002) Bioconjugate Chem 13:370
83. Lee MS, Kerns EH, Hail ME, Liu JP, Volk KJ (1997) LC-GC 15:542
84. Niessen WMA, Tinke AP (1995) J Chromatogr 703:37
85. Gelpé EJ (1995) J Chromatogr 703:59
86. Regnier F, Huang HJ (1996) J Chromatogr 750:3
87. Bruins APJ (1991) J Chromatogr 554:39
88. Abrams MJ, Larsen SK, Shaikh SN, Zubieta J (1991) Inorg Chim Acta 185:7
89. Archer CM, Dilworth JR, Jobanputra P, Thompson RM, McPartlin M, Povey DC, Smith GW, Kelly JD (1990) Polyhedron 9:1497
90. Dilworth JR, Jobanputra P, Thompson RM, Archer CM, Kelly JD, Hiller WZ (1991) Z Naturforsch 46b:449
91. Archer CM, Dilworth JR, Jobanputra P, Thompson RM, McPartin M, Hiller W (1993) J Chem Soc Dalton Trans 897
92. Dilworth JR, Jobanputra P, Thompson RM, Povey DC, Archer CM, Kelly JD (1994) J Chem Soc Dalton Trans 1251
93. Dilworth JR, Jobanputra P, Miller JR, Parrott SJ, Chen Q, Zubieta J (1993) Polyhedron 12:513
94. Coutinho B, Dilworth JR, Jobanputra P, Thompson RM, Schmid S, Strähle J, Archer CM (1995) J Chem Soc Dalton Trans1663
95. Nicholson T, Hirsch-Kuchma M, Freiberg E, Davison A, Jones AG (1998) Inorg Chim Acta 279:206
96. Rose DJ, Maresca KP, Nicholson T, Davison A, Jones AG, Babich J, Fischman A, Graham W, DeBord JRD, Zunieta JA (1998) Inorg Chem 37:2701
97. Nicholson T, Cook J, Davison A, Rose DJ, Maresca KP, Zubieta JA, Jones AG (1996) Inorg Chim Acta 252:421
98. Nicholson T, Cook J, Davison A, Rose DJ, Maresca KP, Zubieta JA, Jones AG (1996) Inorg Chim Acta 252:427
99. Hirsch-Kuchma M, Nicholson T, Davison A, Davis WM, Jones AG (1997) Inorg Chem 36:3237
100. Nicholson T, Hirsch-Kuchma M, Davison A, Davis WM, Jones AG (1998) Inorg Chim Acta 271:191

101. Hirsch-Kuchma M, Nicholson T, Davison A, Jones AG (1997) J Chem Soc Dalton Trans 3189
102. Hirsch-Kuchma M, Nicholson T, Davison A, Davis WM, Jones AG (1997) J Chem Soc Dalton Trans 3185
103. Folkman J (1995) Nat Med 1:27
104. Mousa SA (1998) Drugs Future 23:51
105. Mousa SA (2000) Emerging Ther Targets 4:143
106. Carmeliet P (2000) Nat Med 6:389
107. Meitar D, Crawford SE, Rademaker AW, Cohn SL (1996) J Clinical Oncol 14:405
108. Gasparini G, Brooks PC, Biganzoli E, Vermeulen PB, Bonoldi E, Dirix LY, Ranieri G, Miceli R, Cheresh DA (1998) Clinical Can Res 4:2625
109. Albelda SM, Mette SA, Elder DE, Stewart R, Damjanovich L, Herlyn M, Buck CA (1990) Cancer Res 50:6757
110. Falcioni R, Cimino L, Gentileschi MP, D'Agnano I, Zupi G, Kennel SJ, Sacchi A (1994) Exp Cell Res 210:113
111. Felding-Haermann B, Mueller BM, Romerdahl CA, Cheresh DA (1992) J Clin Invest 89:2018
112. Sengupla S, Chattopadhyay N, Mitra A, Ray S, Dasgupta S, Chatterjee A (2001) J Exp Clin Cancer Res 20:585
113. Zitzmann S, Ethemann V, Schwab M (2002) Cancer Res 62:5139
114. Weber WA, Haubner R, Vabuliene E, Kuhnast B, Webster HJ, Schwaiger M (2001) Q J Nucl Med 45:179
115. Van de Wiele C, Oltenfreiter R, De Winter O, Signore A, Slegers G, Dieckx RA (2002) Eur J Nucl Med 29:699
116. Janssen M, Oyen WJG, Massuger LFAG, Frielink C, Dijkgraaf I, Edwards DS, Rajopadyhe M, Corsten FHM, Boerman OC (2002) Cancer Biother Radiopharm 17:641
117. Harris TD, Kalogeropoulos S, Bartis J, Edwards DS, Liu S, Othank DC, Barrett JA (2001) J Labelled Compd Radiopharm 44(Suppl 1):S60
118. Harris TD, Kalogeropoulos S, Nguyen T, Liu S, Bartis J, Ellars CE, Edwards DS, Onthanks D, Yalamanchili P, Robinson SP, Lazewatsky J, Barrett JA (2003) Cancer Biother Radiopharm 18:627
119. Reubi JC (1993) Endocrenol Metab Clin North Am 22:917
120. Maini CL, Tofani A, Scuito R, Carapella C, Cioffi R, Crecco M (1993) Nucl Med Commun 14:550
121. Reubi JC (1995) J Nucl Med 36:1825
122. Bardfeld PA, Chervu LR, Myrty DRK (1976) Br J Radiol 49:381

Top Curr Chem (2005) 252: 155–178
DOI 10.1007/b101227

Monitoring Multidrug Resistance P-Glycoprotein Drug Transport Activity with Single-Photon Emission Computed Tomography and Positron Emission Tomography Radiopharmaceuticals

Vijay Sharma[1] · David Piwnica-Worms[1,2] (✉)

[1] Molecular Imaging Center, Mallinckrodt Institute of Radiology, Washington University Medical School, 510 S. Kingshighway Blvd., Box 8225, St. Louis, MO 63110, USA

[2] Department of Molecular Biology and Pharmacology, Washington University Medical School, St. Louis, MO 63110, USA

piwnica-wormsd@mir.wustl.edu

Abstract Multidrug resistance (MDR) mediated by overexpression of *MDR1* P-glycoprotein (Pgp) is one of the best characterized transporter-mediated barriers to successful chemotherapy in cancer patients and there is an expanding requirement to noninvasively interrogate Pgp transport activity in vivo. Both small organic medicinals as well as coordination compounds characterized as transport substrates for MDR1 Pgp are amenable to incorporation of positron emission tomography or single-photon emission computed tomography radionuclides and may enable noninvasive diagnostic imaging of MDR in cancer patients. In particular, several clinically approved [^{99m}Tc] complexes already show promise for the functional evaluation of *MDR1* Pgp-mediated transport activity in human tumors in vivo.

Keywords Multidrug resistance · ATP-binding cassette transporters · P-glycoprotein · Single-photon emission computed tomography · Positron emission tomography · Metal complexes · Chemotherapeutic agents

Abbreviations

EC_{50}	Half-maximal effective concentration
Ga-ENBPI	(Bis(*R*-2-hydroxybenzylidene)-*N*,*N'*-bis(3-aminopropyl)ethylenediamine)^{68}Ga complex
LRP	Lung resistance protein
MDR	Multidrug resistance
MIBI	2-Methoxyisobutyl isonitrile
MRP1	Multidrug resistance-associated protein-1
PET	Positron emission tomography
Pgp	P-glycoprotein
SPECT	Single-photon emission computed tomography
Tc-CO-MIBI	Tricarbonyltris(2-methoxyisobutyl isonitrile)^{99m}Tc complex
Tc-EIBI	Hexakis(2-ethoxy-isobutylisonitrile)^{99m}Tc complex
Tc-Furifosmin	*trans*[(1,2-Bis(dihydro-2,2,5,5-tetramethyl-3(2*H*)furanone-4-methyleneimino)ethane) bis(tris(3-methoxy-1-propyl)phosphine)]^{99m}Tc complex
Tc-Sestamibi	Hexakis(2-methoxyisobutyl isonitrile)^{99m}Tc complex
Tc-Tetrofosmin	[1,2-Bis{bis(2-ethoxyethyl)phosphino}ethane]$_2$O$_2$^{99m}Tc complex

1 Introduction

Resistance to chemotherapy represents a major obstacle in the treatment of cancer. Many tumors are intrinsically resistant to chemotherapy, whereas others initially respond to treatment, but acquire resistance to selected cytotoxic drugs during chemotherapy. Multidrug resistance (MDR) is the phenomenon by which cultured cells in vitro and tumor cells in vivo show resistance simultaneously to a variety of structurally and functionally dissimilar cytotoxic and xenobiotic compounds [1–6]. While several different genes have been shown to be associated with an MDR phenotype [7–9], MDR mediated by overexpression of the *MDR1* gene product, P-glycoprotein (Pgp), represents one of the best characterized barriers to chemotherapeutic treatment in cancer [6, 10]. Pgp, a 170-kD plasma membrane protein, is predicted by sequence analysis to comprise two symmetrical halves that share both homology with a family of ATP-binding cassette membrane transport proteins and a common ancestral origin with bacterial transport systems [5, 11]. Characterized by 12 transmembrane domains and two nucleotide-binding folds [3, 5], the protein is thought to hydrolyze ATP to affect outward transport of substrates from the cell surface membrane [5, 12]. Although the specific protein domains and amino acids involved in substrate recognition have not been formally identified, genetic and biochemical evidence has conventionally been interpreted to show putative membrane-associated domains interacting directly with selected cytotoxic agents to affect transport [10, 13–15]. The net effect is that Pgp decreases the intracellular concentration of substrate compounds in Pgp-expressing multidrug resistant cells compared with non-Pgp-expressing drug sensitive cells.

In addition to the standard paradigm of *MDR1* Pgp as an ATP-dependent efflux transporter of xenobiotics and chemotherapeutic drugs, several studies provide evidence supporting alternative mechanisms for the diminished Pgp-mediated drug accumulation in MDR cells. For example, a flippase model has been proposed for Pgp [16]. This model suggests that Pgp flips hydrophobic cytotoxic compounds from the inner to outer leaflet of the lipid bilayer wherein the agents can diffuse away or acts as a phospholipid flippase altering membrane permeability, thereby accounting for the observed decrease in intracellular concentration of drug and providing an explanation for the broad specificity of Pgp. Hoffman and coworkers [17, 18] have shown a more alkaline internal pH and decreased membrane potential in selected cells transfected with *MDR1* Pgp compared with normal cells. These models suggest that *MDR1* Pgp indirectly alters partitioning of substrates within cells through effects on intracellular pH and/or the membrane potential and does not directly transport drugs. In addition, cells expressing *MDR1* may have intracellular compartments that are more acidic than non-Pgp-expressing cells [19, 20], altering intracellular distribution of the drug [21], or altered membrane permeability, resulting in decreased drug influx [22]. By comparison, using a bicistronic vector for selection of cells, a line of stable transfectants, MCF-7/*MDR1*, retained expression of Pgp without the use of chemotherapeutic drugs [23]. Electrical current and drug transport experiments showed no changes in membrane potential or membrane conductance between parental MCF-7 and MCF-7/*MDR1* cells, but reduced unidirectional influx and steady-state cellular content of Pgp substrates. There was no change in the unidirectional efflux of substrates from these cells. Thus, the dominant effect of *MDR1* Pgp in this system was the reduction of drug influx, possibly through an increase in the intramembranous dipole potential [23]. Furthermore, *MDR1* Pgp may interfere with or alter pathways of apoptosis (programmed cell death) [24, 25], therefore offering protection from cytotoxic compounds. Thus, although the exact mechanism remains to be biochemically elucidated, the observed combined net effect is a decreased intracellular concentration of cytotoxic drugs that results from overexpression of *MDR1* Pgp, thereby rendering chemotherapeutic treatment ineffective in cancer.

In addition to its overexpression in tumors, *MDR1* Pgp is normally located in several tissues involved in excretory functions, including the brush border of proximal tubule cells in the kidney, the biliary surface of hepatocytes, and the apical surface of mucosal cells in the small intestine and colon [26, 27]. *MDR1* Pgp is also located on the luminal surface of endothelial cells lining capillaries in the brain and in the testis [28–30] as well as on the apical surface of choroid plexus epithelial cells [29]. However, despite its widely disseminated expression, the function of *MDR1* Pgp in normal physiology has not been clearly defined, although Pgp may have a role in protection from xenobiotics [31] and in intracellular cholesterol trafficking [32]. Furthermore, inhibition of Pgp with an MDR modulator could provide an effective means for increasing oral absorption of drugs and reducing drug excretion, resulting in decreased dosing requirements. For example, Pgp modulation is currently under evaluation as a means to improve oral absorption of chemotherapeutics and HIV-1 protease inhibitors such as indinavir, nelfinavir, saquinavir, and rotonavir [33, 34]. Similarly, the use of MDR modulators could allow drug penetration into Pgp protected sites in the body, such as the brain and selected hematopoietic cells, as has been shown for penetration of protease inhibitors into the central nervous system [34]. In addition, transgenic expression of the *MDR1* gene has been explored for hematopoietic cell protection in the context of cancer chemotherapy [35–37], wherein Pgp could protect hematopoietic progenitor cells from chemotherapy-induced myelotoxicity. Hematopoietic cells transduced via retroviral-mediated transfer of the *MDR1* gene have shown preferential survival in vivo after treatment with MDR drugs [37] and pilot clinical data support the approach [38]. Thus, a noninvasive method of determining Pgp-mediated drug transport activity could be important in the use of new modulators, applications of gene therapy to chemotherapeutic protocols, as well as predicting oral absorption, pharmacokinetics, and penetrance of MDR drugs into target tissues and the brain.

2 *MDR1* Pgp-Mediated Drug Transport in Chemotherapy

Compounds recognized by Pgp are typically characterized as modestly hydrophobic (octanol-to-water partitioning coefficient, *log*P>1), often contain titratable protons with a net cationic charge under physiological conditions, and are predominately “natural products” with an aromatic moiety [39]. Among an extensive list of conventional cytotoxic compounds, anthracyclines (doxorubicin **1**, daunorubicin **2**, taxanes (paclitaxel **3**, docetaxel **4**), *Vinca* alkaloids (vincristine **5**, vinblastine **6**, vindesine **7**), and etoposides (VP-16 **8**) are examples of clinically important chemotherapeutic drugs recognized by *MDR1* Pgp [6, 7, 39, 40]. The diversity of these agents emphasizes the key

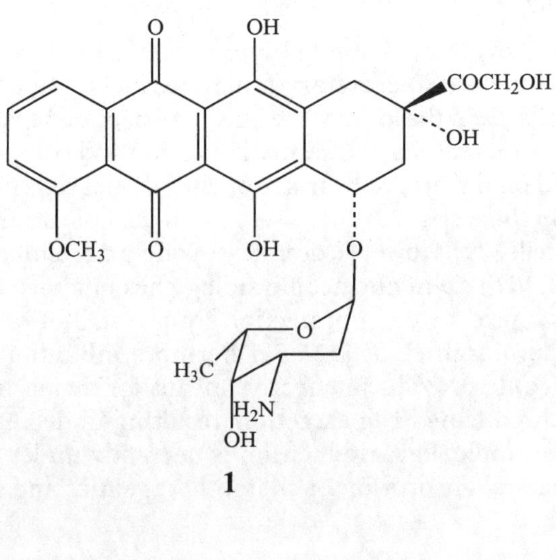

1

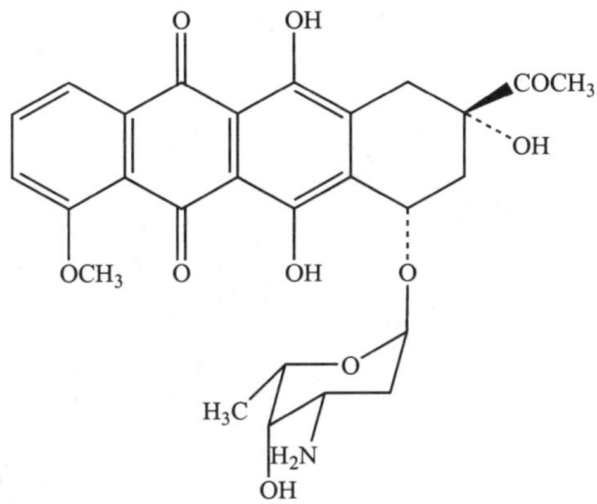

2

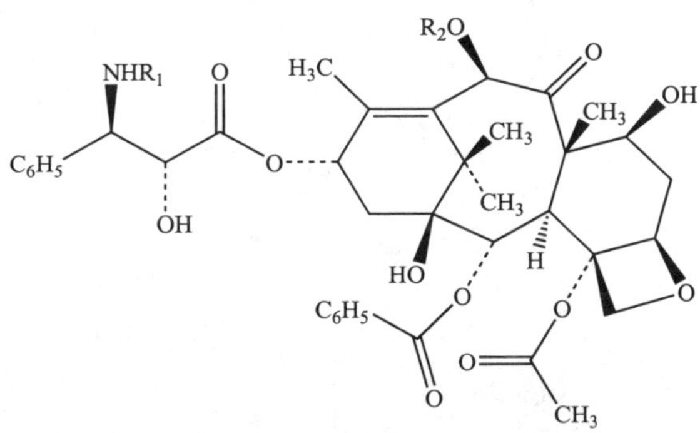

R_1 = COC_6H_5, R_2 = $COCH_3$ (Paclitaxel) **3**

R_1 = $COOC(CH_3)_3$, R_2 = H (Docetaxel) **4**

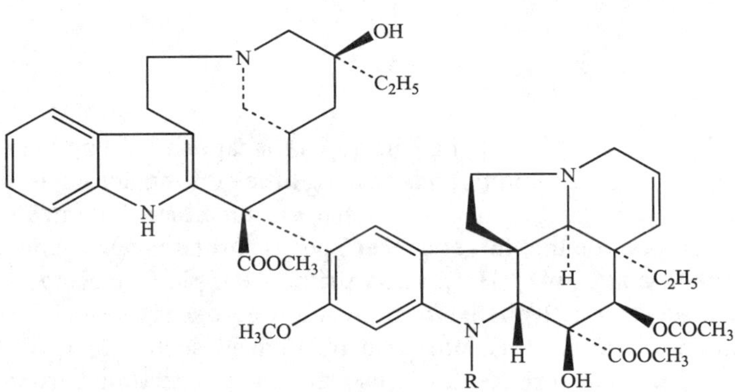

R = CHO (Vincristine) **5**

R = Me (Vinblastine) **6**

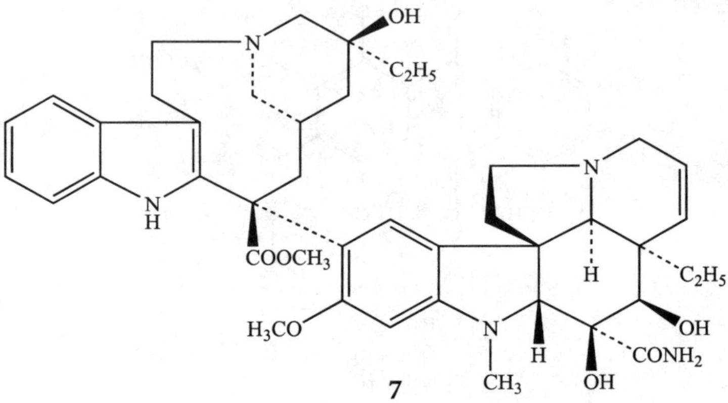

7

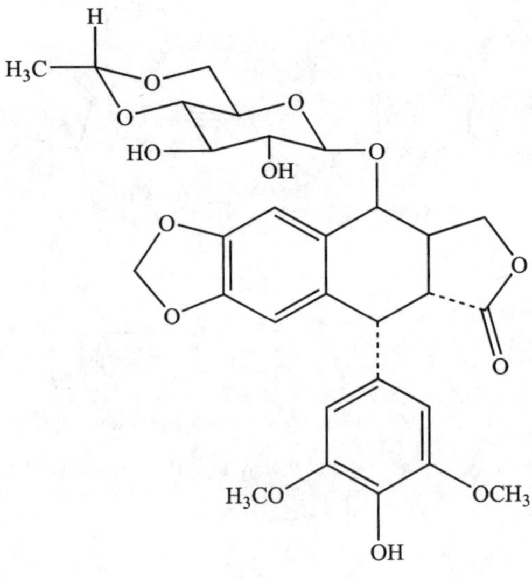

8

characteristic feature of MDR, i.e., the apparent capacity of Pgp to recognize a large group of cytotoxic compounds sharing little or no structural or functional similarities. Furthermore, in the modern era of molecular oncology, novel designer compounds that target cancer-specific proteins have generated wide excitement in clinics [41]. However, a prime example of targeted molecular therapy, Gleevec, **9**, a 2-phenylaminopyrimidine derivative [42, 43], while a highly potent inhibitor of receptor tyrosine kinases, such as BCR-Abl and platlet-derived growth factor receptor, is also recognized and transported by *MDR1* Pgp [6, 44]. Thus, it would appear that even novel targeted therapeutics will remain susceptible to broad-specificity transporter-mediated resistance mechanisms.

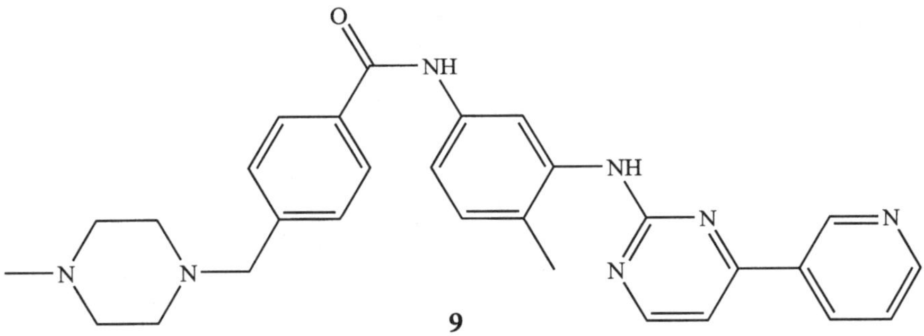

9

Because clinical studies have documented the poor outcomes associated with *MDR1* Pgp expression in tumors [40, 45], reversal of MDR by nontoxic agents that block the transport activity of *MDR1* Pgp has been an important target for pharmaceutical development. When coadministered with a cytotoxic agent, these nontoxic compounds, known as MDR modulators or reversal agents, enhance net accumulation of relevant cytotoxic drugs within the tumor cells. Many compounds known to have other pharmacological sites of action initially were used to reverse MDR in cancer cells grown in culture and several underwent pilot clinical trials [39]. These compounds included verapamil **10**, cyclosporin A **11**, quinidine **12**, trifluoperazine **13**, and their derivatives [39].

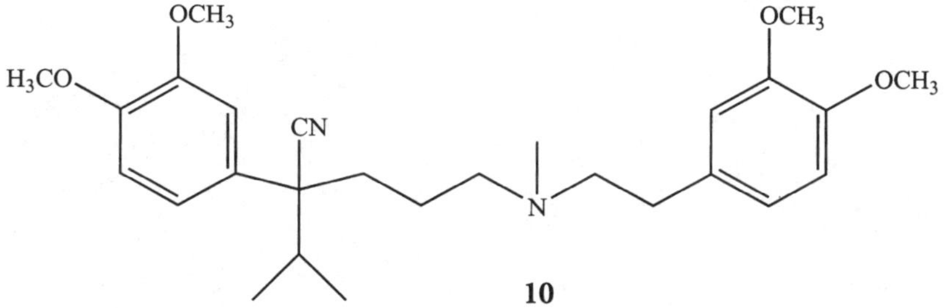

10

However, these agents had limited clinical utility because of unacceptable toxicities at serum levels of the drug needed to modulate *MDR1* Pgp [40]. Second-generation modulators, such as dexverapamil, an optically pure verapamil[46], and PSC 833 (Valspodar) **14**, a cyclic undecapeptide analogue of cyclosporin A [47]) were soon developed with some improvement in efficacy. These were followed by third-generation modulators, such as GF120918 **15**, a substituted isoquinolinyl acridonecarboxamide [48], LY335979 **16**, a difluorocyclopropyl dibenzosuberane [49], VX710 (biricodar) **17**, an amido-keto-pipecolinate [50], XR9576 (tariquidar) **18**, an analog of the anthranilamide pharmacophore [51–53], and MS209 **19**, a quinoline derivatized analog [54, 55], that have been developed more recently. Phase II/III clinical trials are currently in progress with these new, more specific, and more potent compounds. Thus, the MDR

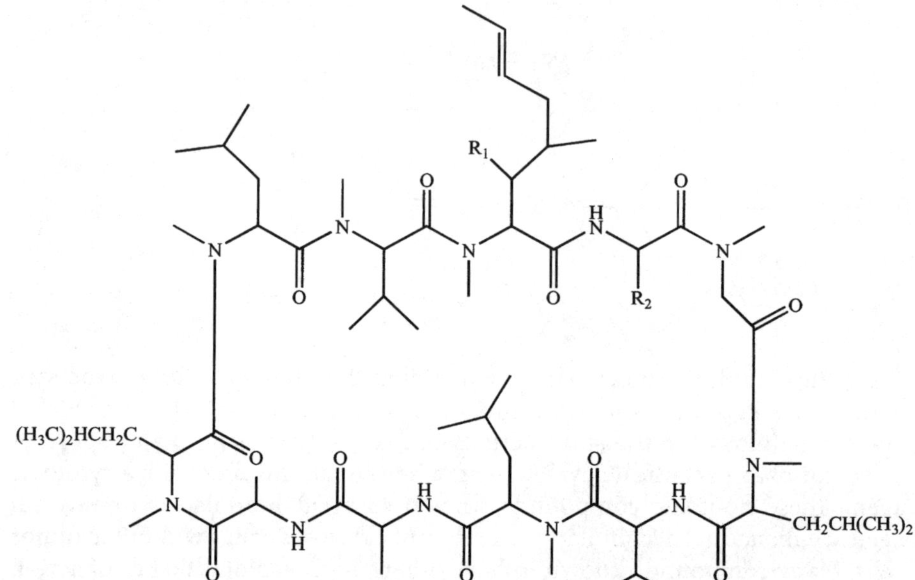

$R_1 = OH$, $R_2 = CH_2CH_3$ (Cyclosporin A) **11**

$R_1 = =O$, $R_2 = CH(CH_3)_2$ (PSC 833) **14**

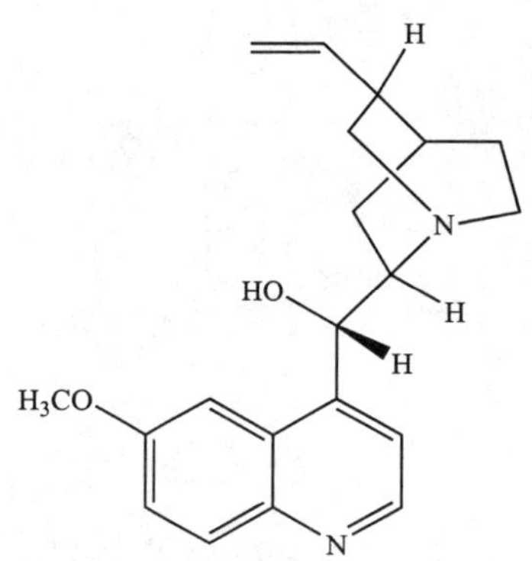

12

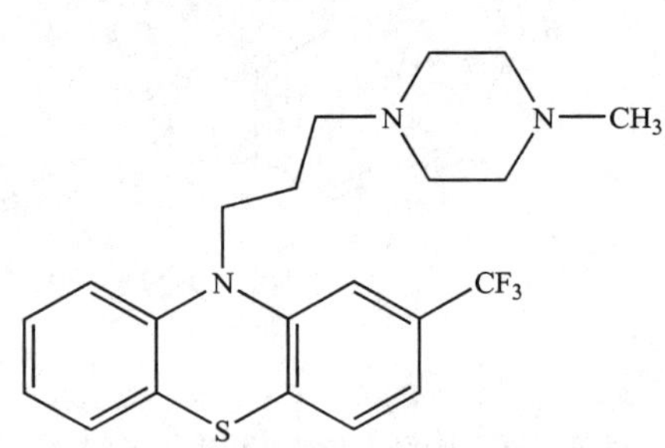

13

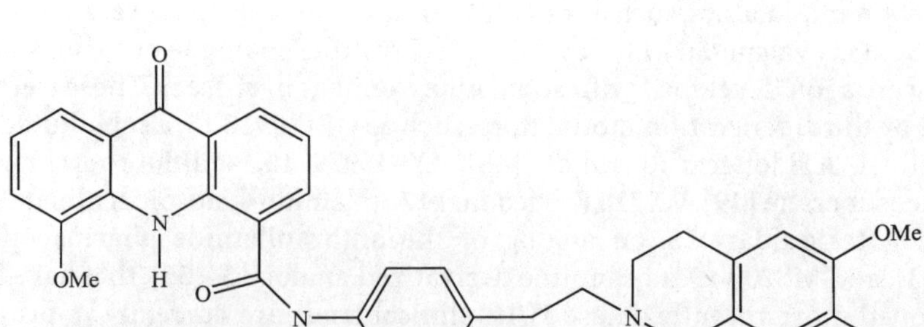

15

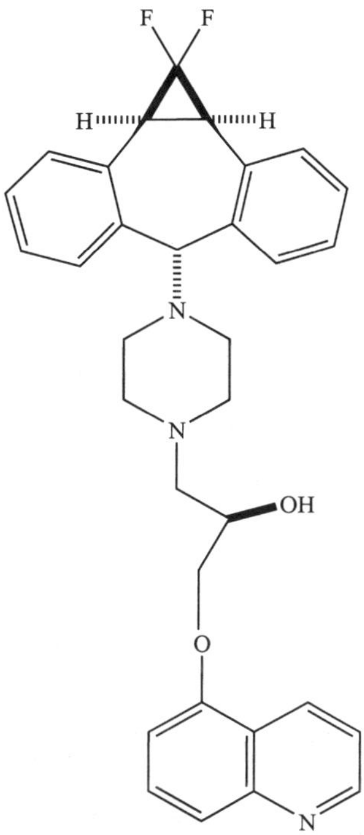

16

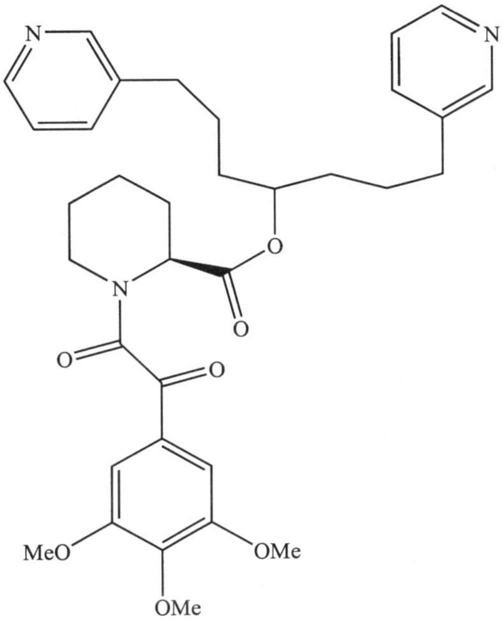

17

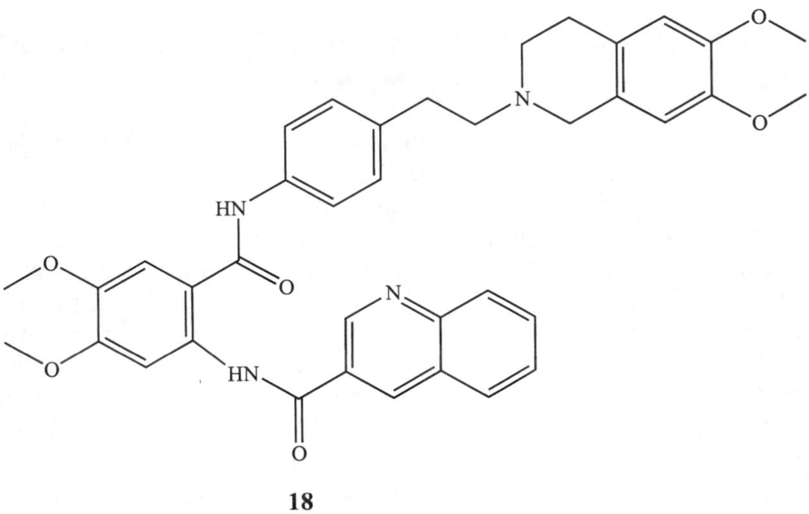

18

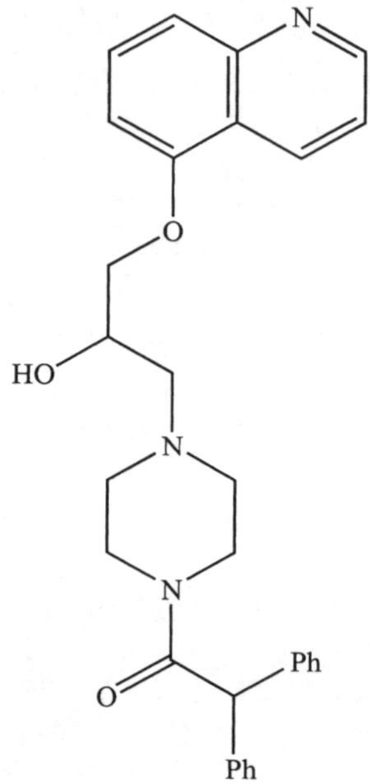

19

phenotype may be modulated more effectively with these targeted reversal agents to improve the efficacy of chemotherapeutic regimens.

3
Single-Photon Emission Computed Tomography Agents for Noninvasive Imaging of P-Glycoprotein Transport Activity

Increasingly, the choice of systemic therapy for cancer is based on a priori analysis of tumor markers to assess the presence or absence of a molecular pathway or target (such as a key receptor or enzyme activity) for a given therapeutic agent. Identification of tumor markers with diagnostic agents assists in the proper selection of patients most likely to benefit from targeted therapy. Measurement of MDR is one potentially important marker in planning systemic therapy. However, expression of *MDR1* Pgp, as detected at the level of mRNA or protein, does not always correlate with the functional assessment of Pgp-mediated transport activity. Because Pgp transport activity is affected by specific mutations as well as the phosphorylation state of the protein [5, 56, 57], altered or less-active forms of Pgp may be detected by polymerase chain reaction or immunohistochemistry which do not accurately reflect the status of tumor cell resistance. Thus, methods to functionally interrogate Pgp transport activity have been sought [58]. Imaging with a radiopharmaceutical that is transported by *MDR1* Pgp may identify noninvasively those tumors in which the transporter is not only expressed, but functional. Thus, significant effort has been directed toward the noninvasive detection of transporter-mediated resistance using planar scintigraphy or single-photon emission computed tomography (SPECT) employing radiolabeled metal complexes characterized as transport substrates for *MDR1* Pgp. These will now be reviewed in detail.

Hexakis(2-methoxyisobutylisonitrile)technetium-99m (commonly known as [^{99m}Tc]Sestamibi) (**20a**), although originally developed as a radiopharmaceutical for myocardial perfusion imaging [59, 60], was the first metal complex

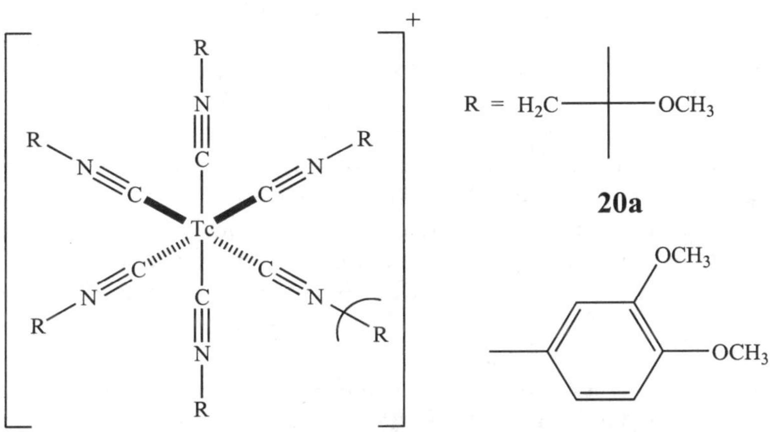

20 **20a** **20b**

shown to be a Pgp transport substrate [61]. Characterized by octahedral geometry around the central technetium(I) core [59, 62], the radiopharmaceutical possesses a cationic charge and modest hydrophobicity similar to many chemotherapeutic agents in the MDR phenotype. In the absence of Pgp expression, this [^{99m}Tc]isonitrile complex accumulates within the interior of cells in response to the physiologically negative mitochondrial and plasma membrane potentials maintained within cells [63, 64]. However, in Pgp-expressing multidrug resistant tumor cells, net cellular accumulation levels of [^{99m}Tc]Sestamibi are inversely proportional to the level of *MDR1* Pgp expression [61, 65–68]. Furthermore, complete reversal of the Pgp-mediated exclusion of [^{99m}Tc]Sestamibi has been affected by treatment with conventional *MDR1* Pgp inhibitors such as verapamil, cyclosporin A, and quinidine or recently discovered more-potent reversal agents such as GF120918, LY335979, PSC 833 or VX710 [32, 61, 65, 66, 69–75], and more recently, XR-9576 [6, 51]. Of interest, [^{99m}Tc]Sestamibi is also recognized as a transport substrate for the MDR-associated protein (MRP1) [76–78], a close homologue of Pgp [79], thereby providing a two-edged sword. On the one hand, cross-reactivity with MRP1 may reduce the specificity of the tracer for functional imaging of *MDR1* Pgp in tumor cells, but alternatively, this property may favorably enable [^{99m}Tc]Sestamibi to be a more general probe of transporter-mediated MDR in some cancer cell lines.

Furthermore, to optimize the transport and Pgp targeting characteristics of [^{99m}Tc]isonitrile complexes, several studies investigating structure–activity relationships have been performed. In one study, the alkyl chains in [^{99m}Tc]Sestamibi were replaced with longer-chain ether functionalities. The hexakis(2-ethoxy-isobutylisonitrile)[^{99m}Tc] complex ([^{99m}Tc]EIBI) was shown to be a transport substrate recognized by Pgp, but with slightly greater nonspecific cell binding than [^{99m}Tc]Sestamibi [80]. Substituted aromatic functionalities were also explored [81]. A series of substituted arylisonitrile analogs were synthesized from their corresponding amines through a reaction with dichlorocarbene under phase-transfer-catalyzed conditions, and noncarrier-added hexakis(arylisonitrile)[^{99m}Tc]complexes were produced by reaction with pertechnetate in the presence of sodium dithionite. The lead compound of the series, **20b**, demonstrated an overall encouraging transport profile in Pgp-expressing cells, but significant nonspecific adsorption to hydrophobic compartments was identified. The results also suggested that methoxy substituents, compared with other substituents, preferentially contributed to enhanced Pgp recognition for this class of compounds. However, none of these radiolabeled complexes exceeded [^{99m}Tc]Sestamibi in their Pgp-mediated transport properties.

In addition, several entirely different classes of technetium-complexes have been identified as Pgp transport substrates. Using a planar Schiff-base moiety and hydrophobic phosphines, nonreducible Tc(III) monocationic compounds known as "Q complexes" were developed a decade ago for applications in myocardial perfusion imaging [82, 83]. The lead complex for clinical development was trans[(1,2-bis(dihydro-2,2,5,5-tetramethyl-3(2*H*)furanone-4-methyleneimino)ethane) bis(tris(3-methoxy-1-propyl)phosphine)]Tc(III), known as [^{99m}Tc]Furifosmin **21** [84]. Because the hydrophobicity and Pgp targeting properties of these complexes could be readily adjusted by varying functionalities on the Schiff base or phosphine moieties independently, a variety of novel [^{99m}Tc]Q complexes with subtle structural variations were synthesized [69, 85]. This approach allowed the coordination environment of the Tc(III) metal core to be maintained while the overall electronic environment was altered, thereby enabling refined exploration and evaluation of features conferring Pgp-mediated transport properties. Ether functionalities can be incorporated into the equatorial Schiff base ligand by condensation of ethylenediamines with ether-containing β-dicarbonyl compounds [86]. The *gem*-dimethyl groups present sterically hinder the attack of diamine at the adjacent carbonyl and the strategy results in regioselective condensation at the less hindered carbonyl. Preparation of the tertiary phosphines was accomplished in a two-step, one-pot reaction involving treatment of 1-chloro-3-methoxy propane with magnesium in tetrahydrofuran and subsequent reaction of the reagent with dimethylchlorophosphines or dichloromethylphosphines to provide the necessary

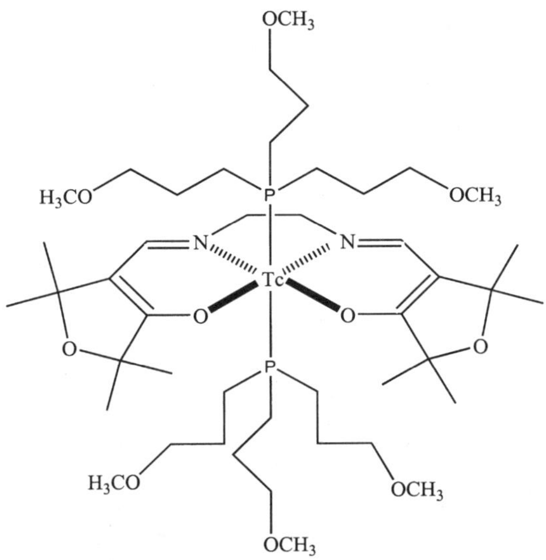

21

substituted phosphines with overall yield of 50–70% [85]. The desired [^{99m}Tc]Q complexes were then obtained by a two-step synthetic approach using the phosphines as both reductants and ligands [87]. From MDR transport assays in vitro, the trans[2,2′-(1,2-ethanediyldiimino) bis(1,5-methoxy-5-methyl-4-oxo-hexenyl)] bis[methyl-bis(3-methoxy-1-propyl)phosphine]Tc(III) complex **22** and the trans[5,5′-(1,2-ethanediyldiimino) bis(2-ethoxy-2-methyl-

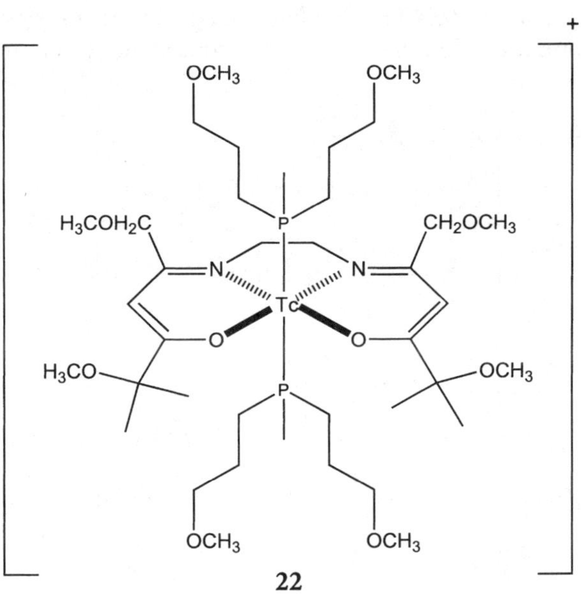

22

3-oxo-4-pentenyl)] bis[dimethyl(3-methoxy-1-propyl)phosphine]Tc(III) complex **23** were discovered. These complexes are nearly identical to [^{99m}Tc]Sestamibi in their Pgp recognition properties in vitro [69, 85]. In addition, a Tc(V)

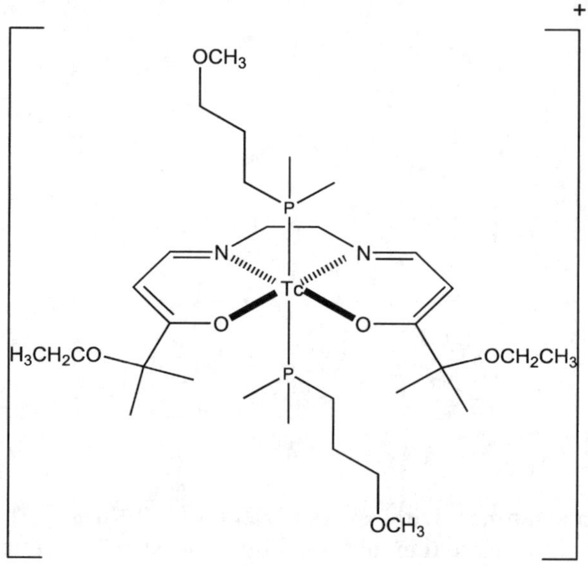

23

complex known as [^{99m}Tc]Tetrofosmin [88], bis[1,2-bis{bis(2-ethoxyethyl)-phosphino]ethane]Tc(V) **24**, has been identified as another [^{99m}Tc]complex with highly favorable *MDR1* Pgp-mediated transport properties [85, 89]. While these metal complexes do not share any obvious structural homology, they do

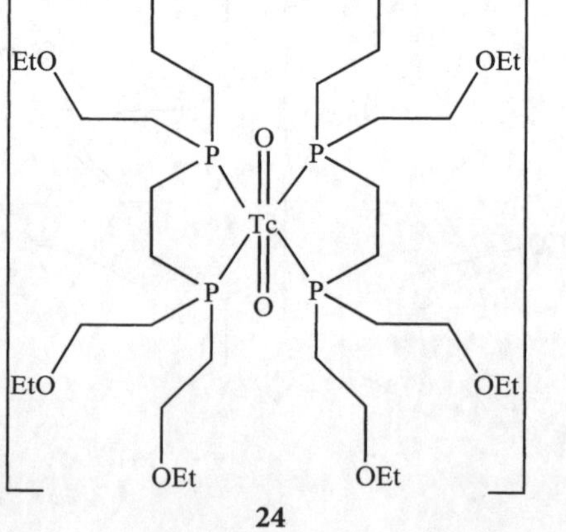

24

share the common features of a delocalized cationic charge and modest hydrophobicity. Overall, which of these selected [^{99m}Tc]complexes may be most clinically useful in evaluation of the Pgp status of tumors by SPECT imaging remains under intense investigation.

Another class of technetium-based radiopharmaceuticals has emerged on the basis of pioneering work done on the development of an air- and water-stable organometallic aqua complex [$^{99m}Tc(OH_2)_3(CO)_3$]$^+$ **25** obtained from re-

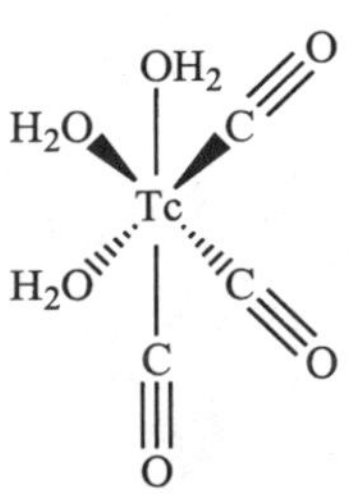

25

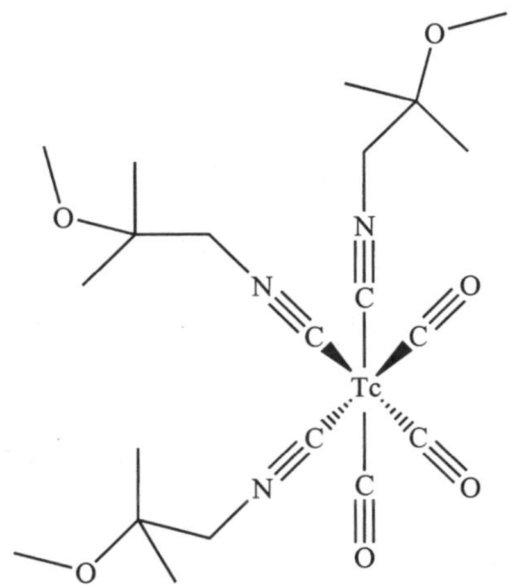

26

action of pertechnetate in saline under 1 atm CO [90]. It was shown that the coordinated water molecules were labile, thus exchangeable with other ligands, generating complexes with heterogeneous ligands. On the basis of these observations, these water molecules were substituted with methoxy-isobutylisonitrile ligands to generate another novel class of Pgp-targeted radiopharmaceuticals, tris(carbonyl)tris(2-methoxy-isobutylisonitrile)technetium(I); [$^{99m}Tc(CO)_3(MIBI)_3$]$^+$ (**26**) [91]. Recently, to assess the potential of this radiolabeled technetium complex, commonly known as Tc-CO-MIBI, as a noninvasive probe of *MDR1* Pgp activity in tumors, cellular accumulation in human epidermal carcinoma drug-sensitive KB 3-1 (–Pgp) and colchicine-derived drug-resistant KB 8-5 (+Pgp) cells was evaluated [92]. The radiotracer demonstrated 60-fold higher accumulation in drug-sensitive KB 3-1 cells compared with drug-resistant KB 8-5 cells. In KB 8-5 cells, tracer enhancement was observed with the potent *MDR1* Pgp modulator LY335979 (half-maximal effective concentration, EC_{50}=62 nM). Similar behavior was observed using drug-sensitive MCF-7 breast adenocarcinoma cells and MCF-7/*MDR1* stable transfectants, confirming that Tc-CO-MIBI is specifically excluded by overexpression of *MDR1* Pgp. By comparison, net accumulation in control H69 lung tumor cells was 9 times higher than in MDR-associated protein (*MRP1*)-expressing H69AR cells, indicating only modest transport by *MRP1*. Biodistribution analysis following tail vein injection of Tc-CO-MIBI showed delayed liver washout as well as enhanced brain uptake and retention in *mdr1a/1b*(–/–) gene-deleted mice versus wild-type mice, directly demonstrating that Tc-CO-MIBI is a functional probe of Pgp transport activity in vivo [92].

Organic scaffolds capable of coordinating other metals have also been explored. Multidentate ligands with an N_4O_2 donor core have the ability to form stable monomeric, monocationic, hydrophobic complexes with a variety of main-group [93, 94] and transition metals [95–97]. Schiff-base Ga(III) complexes were previously reported as potential positron-emitting radiopharmaceuticals with utility as myocardial perfusion imaging agents [98, 99]. These complexes, exemplified by the lead compound **27**, demonstrated pharma-

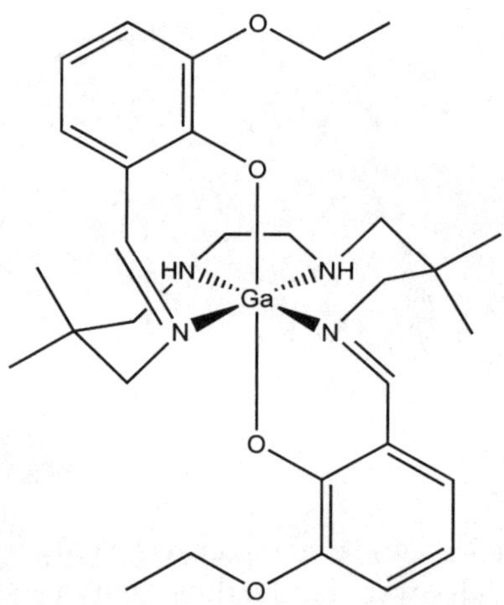

27

cological profiles consistent with their potential utility as positron emission tomography (PET) probes of *MDR1* Pgp activity in tumors [100–102]. The triaryl precursors containing a central imidazolidine ring were synthesized by condensation of an appropriate substituted linear tetramine with substituted salicylaldehydes. The desired metal(III) complexes were obtained by cleavage of the imidazolidine ring using transmetallation reactions using appropriate metal(III) acetylacetonates. Selected gallium(III) complexes were able to differentiate between human epidermal carcinoma KB 3-1 (–Pgp) and KB 8-5 (+Pgp) cells through differential cytotoxicity profiles indicating *MDR1* Pgp-mediated transport [101]. The active metal complexes containing 4,6-dimethoxy substituted aromatic rings were more potent than their corresponding 3-methoxy analogs. In addition, radiolabeled analogs of these compounds, commonly called Ga-ENBPI [(bis(*R*-2-hydroxybenzylidene)-*N*,*N′*-bis(3-aminopropyl)ethylenediamine)^{67}Ga] complexes, accumulated in cells in inverse proportion to expression of Pgp. Furthermore, uptake in MDR cells was completely reversed compared with that of control cells in the presence of potent MDR inhibitors, further demonstrating target specificity [100]. These results suggested that radiolabeled analogues of these Ga(III) complexes also could provide templates for ^{68}Ga PET radiopharmaceuticals to probe Pgp transport activity in tumors [102].

4 PET Agents for Noninvasive Imaging of P-Glycoprotein Transport Activity

PET radiopharmaceuticals offer enhanced spatial resolution and quantification capabilities compared with SPECT agents. To probe Pgp transport activity, PET-based radiopharmaceuticals have been actively investigated on three fronts: (1) use of SPECT scaffolds capable of accommodating PET radionuclides, (2) bioinorganic radiochemistry, and (3) conventional PET organic radiochemistry.

Regarding the use of scaffolds that coordinate both SPECT and PET radionuclides, two validated examples make use of the PET radionuclides ^{94m}Tc and ^{68}Ga. The radiosynthesis and biochemical validation of ^{94m}Tc-Sestamibi and ^{68}Ga-ENBPI complexes have been reported [100, 103]. As expected chemically, the highly favorable Pgp targeting properties of these complexes were retained upon transformation from SPECT to PET imaging agents.

On another front, organic scaffolds capable of accommodating PET radionuclides that generate novel metallopharmaceuticals through short synthetic routes have been reported. Based upon rigorous prior contributions [104, 105], a stable, monocationic radiolabeled complex of copper(II) **28** was obtained as a potential ^{64}Cu PET radiopharmaceutical targeting Pgp [106]. The desired diiminedioxime ligand was synthesized from 2,3-dimethyl-propane-1,2-diamine and heptane-2,3-dione-3-oxime. Cellular accumulation studies demonstrated significantly more accumulation of the radiolabeled compound

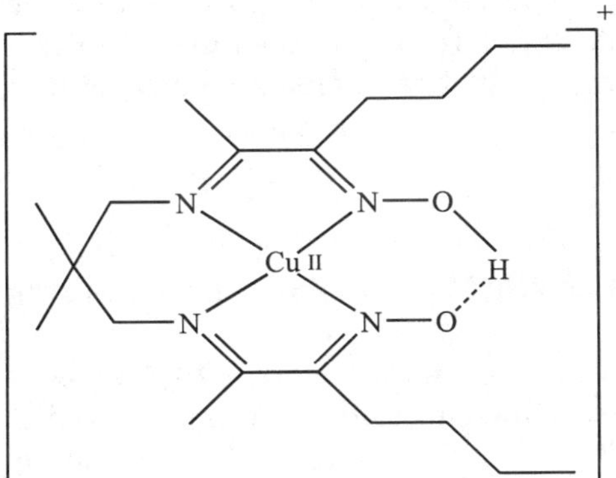

28

in MES-SA (–Pgp) cells compared with MES-SA (+Pgp) cells in vitro. Addition of the MDR reversal agent cyclosporin A completely reversed the accumulation profile in MES-SA cells, rendering uptake comparable to control (non-Pgp) cells.

Bidentate tertiary phosphine ligands have the ability to generate stable copper(I) complexes through a one step synthesis in quantitative yields [107, 108] and represent another class of potential ^{64}Cu PET radiopharmaceuticals targeting Pgp. These complexes previously demonstrated potent antitumor properties compared with their free ligands alone [109, 110]. As with [^{99m}Tc]Q complexes, phosphines were exploited as both ligands and reducing agents to generate cationic, hydrophobic, tetrahedral copper(I) complexes **29** with 1,2-bis(diphenylphosphino)ethane. These potential PET radiopharmaceuticals show evidence of Pgp targeting properties [111, 112]. However, although several leads exist for a ^{64}Cu-based radiopharmaceutical for interrogation of Pgp by PET, these radiopharmaceuticals show only modest Pgp targeting properties compared with ^{94m}Tc-Sestamibi or ^{68}Ga-ENBPI.

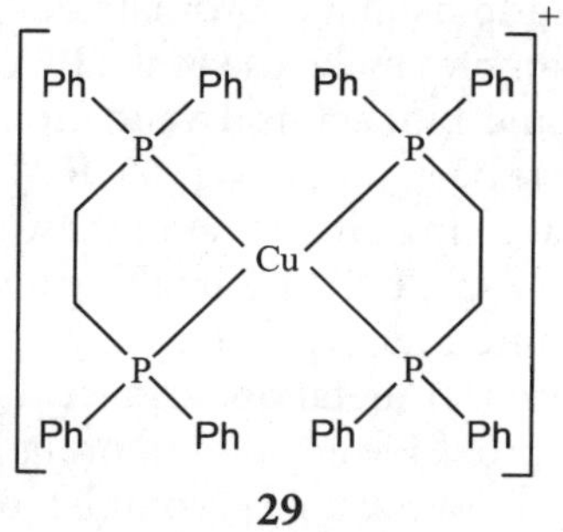

29

A third focus has been directed toward incorporation of conventional PET radionuclides ^{11}C or ^{18}F into existing substrates or inhibitors known to interact with Pgp [113–115]. This strategy has been employed to produce various PET agents, including ^{11}C-colchicine, ^{11}C-verapamil, ^{11}C-daunomycin, and ^{11}C/^{18}F-paclitaxel[115–123]. While promising data have been generated, some of these PET agents suffer from modest radiochemical yields and others from complex pharmacokinetics in vivo mediated, at least in part, by rapid metabolism of the radiolabeled compounds.

5 Molecular Imaging of *MDR1* Pgp in Clinical Drug Resistance

Several clinical studies have shown the feasibility of interrogating Pgp transport activity in vivo with imaging cameras commonly available in nuclear medicine facilities [71–73, 124–129]. In general, ^{99m}Tc-Sestamibi pharmacokinetic data are extracted from tumor images over time and correlated with immunohistochemical assessment of Pgp expression in the tumor specimen. For example, in a pioneering clinical study, Del Vecchio et al. [124] determined rates of efflux of ^{99m}Tc-Sestamibi in 30 patients with untreated breast cancer. Dynamic imaging of the primary tumor was performed for 4 h following injection of ^{99m}Tc-Sestamibi and tumor specimens were obtained for quantitative autoradiography of Pgp. Rates of efflux of ^{99m}Tc-Sestamibi were 2.7 times greater in tumors overexpressing *MDR1* Pgp compared with tumors that expressed Pgp at a level comparable to benign breast lesions. Estimates of sensitivity and specificity for in vivo detection of *MDR1* Pgp using ^{99m}Tc-Sestamibi were 80% and 95%, respectively. From these data, the authors concluded that efflux rate constants of ^{99m}Tc-Sestamibi may be used for noninvasive identification of *MDR1* Pgp in breast cancer.

Several additional clinical studies have examined the relationship between tumor retention of ^{99m}Tc-complexes and MDR [51, 128–136]. Analysis of washout rates by comparing early (less than 30 min) and delayed (180–240 min) SPECT and planar images have shown a strong inverse correlation between tumor-to-background ratios and levels of Pgp expression. The relation between therapeutic response and tumor retention of tracer also was analyzed in some studies and showed that most tumors with high tracer retention exhibited a favorable response to chemotherapy, whereas most tumors with low tracer retention did not respond to chemotherapy. These investigators have concluded that ^{99m}Tc-Sestamibi and ^{99m}Tc-Tetrofosmin are useful tools in vivo for prediction of MDR in a variety of cancers.

Early uptake, delayed uptake, and washout rates of ^{99m}Tc-Sestamibi have also been obtained for correlation with *MDR1* Pgp, MRP1, and lung resistance protein (LRP) expression determined by immunohistochemistry and *MDR1* Pgp, MRP1, and LRP mRNA levels determined by real-time reverse-transcription polymerase chain reaction [137]. These investigators found that increased levels of Pgp expression correlated with a low accumulation of ^{99m}Tc-Sestamibi on delayed scans and a high washout rate of ^{99m}Tc-Sestamibi. Interestingly, neither MRP1 nor LRP expression on the level of either protein or mRNA correlated significantly with tumor accumulation or efflux of ^{99m}Tc-Sestamibi in lung cancer. Thus, although, ^{99m}Tc-Sestamibi has been shown to modestly cross-react with MRP1 using cells in culture under ideal laboratory conditions [77, 138], unlike the case with *MDR1* Pgp, this lower level of MRP1 transport activity cannot be detected in patients in vivo. As a practical consequence, when used in patients, ^{99m}Tc-Sestamibi may be specific for detection of *MDR1* Pgp transport activity in vivo.

Furthermore, in addition to the in vitro and animal laboratory studies cited earlier, clinical blockade of Pgp-mediated extrusion of ^{99m}Tc-Sestamibi from tissues and tumors of patients has been imaged following treatment with highly selective and potent Pgp modulators such as PSC 833 (valspodar), VX-710 (biricodar), and XR-9576 (tariquidar) [6, 51, 71, 72, 74]. A two-step protocol for imaging MDR reversal in patients is under evaluation and this may provide a noninvasive method to determine the effectiveness of MDR modulation. After baseline imaging, administration of a potent modulating agent and reinjection of the tracer is performed. Those tissues or tumors showing higher accumulation of the tracer and/or reduced washout rates would indicate specific blockade of *MDR1* Pgp. Thus, these radiopharmaceuticals enable noninvasive mapping of specific transport events in vivo and provide a basis for further exploration of targeted molecular imaging of Pgp in cancer and gene therapy.

6 Conclusion

In summary, *MDR1* Pgp recognizes and outwardly transports a wide variety of bioinorganic complexes with a broad diversity of chelation scaffolds as well as conventional anticancer drugs. Both small organic medicinals as well as coordination compounds are amenable to incorporation of PET or SPECT radionuclides that may enable diagnostic imaging technologies to be exploited for functional interrogation of the MDR phenotype in cancer patients. In particular, several clinically approved [^{99m}Tc] complexes have already shown promise for use in the functional evaluation of *MDR1* Pgp-mediated transport activity in human tumors in vivo. Therefore, contributions of bioinorganic radiochemistry and conventional organic radiochemistry are rapidly growing to enhance the existing armamentarium of targeted diagnostic agents that would be capable of functionally analyzing the presence of Pgp in human subjects.

Acknowledgements The authors thank all colleagues and coworkers (both past and present) for their contributions that led to successful execution of this chapter. Financial assistance for this work was provided by grants from the US Department of Energy (DE-FG02-94ER61885) and National Institutes of Health (P50 CA94056).

References

1. Ling V, Thompson LH (1974) J Cell Physiol 83:103
2. Gerlach JH, Endicott JA, Juranka PF, Henderson G, Sarangi F, Deuchars KL, Ling V (1986) Nature 324:485
3. Gros P, Ben Neriah Y, Croop JM, Housman DE (1986) Nature 323:728
4. Endicott JA, Ling V (1989) Annu Rev Biochem 58:137
5. Gottesman MM, Pastan I (1993) Annu Rev Biochem 62:385
6. Gottesman M, Fojo T, Bates S (2002) Nat Rev Cancer 2:48
7. Bosch I, Croop J (1996) Biochim Biophys Acta 1288:F37
8. Borst P, Evers R, Kool M, Wijnholds J (2000) J Natl Cancer Inst 92:1295
9. Kruh G, Belinsky M (2003) Oncogene 22:7537
10. Ambudkar S, Dey S, Hrycyna C, Ramachandra M, Pastan I, Gottesman M (1999) Annu Rev Pharmacol Toxicol 31:361
11. Ames G (1986) Annu Rev Biochem 55:397
12. Gros P, Dhir R, Croop J, Talbot F (1991) Proc Natl Acad Sci USA 88:7289

13. Bruggemann EP, Germann UA, Gottesman MM, Pastan I (1989) J Biol Chem 264:15483
14. Yoshimura A, Kuwazuru Y, Sumizawa T, Ichikawa M, Ikeda S, Uda T, Akiyama S (1990) J Biol Chem 264:16282
15. Raviv Y, Pollard H, Bruggemann E, Pastan I, Gottesman M (1990) J Biol Chem 265:3975
16. Higgins CF, Gottesman MM (1992) Trends Biochem Sci 17:18
17. Hoffman M, Wei L, Roepe P (1996) J Gen Physiol 108:295
18. Hoffman M, Roepe P (1997) Biochemistry 36:11153
19. Altan N, Chen Y, Schindler M, Simon S (1998) J Exp Med 187:1583
20. Hurwitz S, Terashima M, Mizunuma N, Slapak C (1997) Blood 89:3745
21. Lautier D, Canitrot Y, Deeley R, Cole S (1996) Biochem Pharmacol 52:967
22. Stein WD, Cardarelli C, Pastan I, Gottesman MM (1994) Mol Pharmacol 45:763
23. Luker GD, Flagg TP, Sha Q, Luker KE, Pica CM, Nichols CG, Piwnica-Worms D (2001) J Biol Chem 276:49053
24. Robinson L, Roberts W, Ling T, Lamming D, Sternberg S, Roepe P (1997) Biochemistry 36:11169
25. Smyth MJ, Krasovskis E, Sutton VR, Johnstone RW (1998) Proc Natl Acad Sci USA 95:7024
26. Thiebaut F, Tsuruo T, Hamada H, Gottesman MM, Pastan I, Willingham MC (1987) Proc Natl Acad Sci USA 84:7735
27. Hitchins RN, Harman DH, Davey RA, Bell DR (1988) Eur J Cancer Clin Oncol 24:449
28. Cordon-Cardo C, OBrien JP, Casals D, Rittman GL, Biedler JL, Melamed MR, Bertino JR (1989) Proc Natl Acad Sci USA 86:695
29. Rao V, Dahlheimer J, Bardgett M, Snyder A, Finch R, Sartorelli A, Piwnica-Worms D (1999) Proc Natl Acad Sci USA 96:3900
30. Croop JM, Raymond M, Haber D, Devault A, Arceci RJ, Gros P, Housman DE (1989) Mol Cell Biol 9:1346
31. Schinkel A, Mayer U, Wagenaar E, Mol C, van Deemer L, Smit J, van der Valk M, Voordouw A, Spits H, van Tellingen O, Zijlmans J, Fibbe W, Borst P (1997) Proc Natl Acad Sci USA 94:4028
32. Luker G, Nilsson K, Covey D, Piwnica-Worms D (1999) J Biol Chem 274:6979
33. van Asperen J, van Tellingen O, van der Valk M, Rozenhart M, Beijnen J (1998) Clin Cancer Res 4:2293
34. Kim RB, Fromm MF, Wandel C, Leake B, Wood AJJ, Roden DM, Wilkinson GR (1998) J Clin Invest 101:289
35. Sorrentino B, Brandt S, Bodine D, Gottesman M, Pastan I, Cline A, Nienhuis A (1992) Science 257:99
36. Podda S, Ward M, Himelstein A, Richardson C, de la Flor-Weiss E, Smith L, Gottesman M, Pastan I, Bank A (1992) Proc Natl Acad Sci USA 89:9676
37. Hanania E, Fu S, Roninson I, Zu Z, Gottesman M, Deisseroth A (1995) Gene Ther 2:279
38. Moscow J, Huang H, Carter C, Hines K, Zujewsi J, Cusack G, Chow C, Venzon D, Sorrentino B, Chiang Y, Goldspiel B, Leitman S, Read E, Abati A, Gottesman M, Pastan I, Sellers S, Dunbar C, Cowan K (1999) Blood 94:52
39. Ford JM, Hait WN (1990) Pharmacol Rev 42:155
40. Sandor V, Fojo T, Bates S (1998) Drug Resistance Updates 1:190
41. Luo J, Manning B, Cantley L (2003) Cancer Cell 4:257
42. Buchdunger E, Zimmermann J, Mett H, Meyer T, Muller M, Druker BJ, Lydon NB (1996) Cancer Res 56:100
43. Schindler T, Bornmann W, Pellicena P, Miller WT, Clarkson B, Kuriyan J (2000) Science 289:1938
44. Mahon FX, Deininger MWN, Schultheis B, Chabrol J, Reiffers J, Goldman JM, Melo JV (2000) Blood 96:1070

45. Trock B, Leonessa F, Clarke R (1997) J Natl Cancer Inst 89:917
46. Wilson W, Jamis-Dow C, Bryant G, Balis F, Klecker R, Bates S, Chabner B, Steinberg S, Kohler D, Wittes R (1995) J Clin Oncol 13:1985
47. Gaveriaux C, Boesch D, Jachez B (1991) J Cell Pharmacol 2:225
48. Hyafil F, Vergely C, Du Vignaud P, Grand-Perret T (1993) Cancer Res 53:4595
49. Dantzig A, Shepard R, Cao J, Law K, Ehlhardt W, Baughman T, Bumol T, Starling J (1996) Cancer Res 56:4171
50. Germann U, Ford P, Schlakhter D, Mason V, Harding M (1997) Anticancer Drugs 8:141
51. Agrawal M, Abraham J, Balis F, Edgerly M, Stein W, Bates S, Fojo T, Chen C (2003) Clin Cancer Res 9:650
52. Mistry P, Stewart A, Dangerfield W, Okiji S, Liddle C, Bootle D, Plumb J, Templeton D, Charlton P (2001) Cancer Res 61:749
53. Martin C, Berridge G, Mistry P, Higgins C, Charlton P, Callaghan R (1999) Br J Pharmacol 128:403
54. Suzuki T, Fukazawa N, San-nohe K (1997) J Med Chem 40:2047
55. Suzuki T, Mabuchi K, Fukazawa N (1999) Biorg Med Chem Lett 9:659
56. Germann UA, Willingham MC, Pastan I, Gottesman MM (1990) Biochemistry 29: 2295
57. Ahmad S, Safa AR, Glazer RI (1994) Biochemistry 33:10313
58. Homolya L, Hollo Z, Germann UA, Pastan I, Gottesman MM, Sarkadi B (1993) J Biol Chem 268:21493
59. Abrams MA, Davison A, Jones AG, Costello CE, Pang H (1983) Inorg Chem 22:2798
60. Wackers FJ, Berman D, Maddahi J, Beller G, Straus H, Boucher C, Picard M, Holman B, Fridrich R (1989) J Nucl Med 30:301
61. Piwnica-Worms D, Chiu M, Budding M, Kronauge J, Kramer R, Croop J (1993) Cancer Res 53:977
62. Kronauge JF, Davison A, Roseberry AM, Costello CE, Maleknia S, Jones AG (1991) Inorg Chem 30:4265
63. Piwnica-Worms D, Kronauge J, Chiu M (1990) Circulation 82:1826
64. Backus M, Piwnica-Worms D, Hockett D, Kronauge J, Lieberman M, Ingram P, LeFurgey A (1993) Am J Physiol Cell 265:C178
65. Piwnica-Worms D, Rao V, Kronauge J, Croop J (1995) Biochemistry 34:12210
66. Ballinger J, Hua H, Berry B, Firby P, Boxen I (1995) Nucl Med Commun 16:253
67. Ballinger JR, Sheldon KM, Boxen I, Erlichman C, Ling V (1995) Q J Nucl Med 39:122
68. Cordobes M, Starzec A, Delmon-Moingeon L, Blanchot C, Kouyoumdjian J-C, Prevost G, Caglar M, Moretti J-L (1996) J Nucl Med 37:286
69. Luker G, Rao V, Crankshaw C, Dahlheimer J, Piwnica-Worms D (1997) Biochemistry 36:14218
70. Bosch I, Crankshaw C, Piwnica-Worms D, Croop J (1997) Leukemia 11:1131
71. Luker GD, Fracasso PM, Dobkin J, Piwnica-Worms D (1997) J Nucl Med 38:369
72. Chen C, Meadows B, Regis J, Kalafsky G, Fojo T, Carrasquillo J, Bates S (1997) Clin Cancer Res 3:545
73. Barbarics E, Kronauge J, Cohen D, Davison A, Jones A, Croop J (1998) Cancer Res 58:276
74. Peck R, Hewett J, Harding M, Wang Y, Chaturvedi P, Bhatnagar A, Ziessman H, Atkins F, Hawkins M (2001) J Clin Oncol 19:3130
75. Pichler A, Prior J, Piwnica-Worms D (2004) Proc Natl Acad Sci USA 101:1702
76. Crankshaw C, Piwnica-Worms D (1996) J Nucl Med 37:247P
77. Hendrikse N, Franssen E, van der Graaf W, Meijer C, Piers D, Vaalburg W, de Vries E (1998) Br J Cancer 77:353
78. Moretti J-L, Cordobes M, Starzec A, de Beco V, Vergote J, Benazzouz F, Boissier B, Cohen H, Safi N, Piperno-Neumann S, Kouyoumdjian J-C (1998) J Nucl Med 39:1214

79. Cole SPC, Bhardwaj G, Gerlach JH, Mackie JE, Grant CE, Almquist KC, Stewart AJ, Kurz EU, Duncan AMV, Deeley RG (1992) Science 258:1650
80. Piwnica-Worms D, Kronauge J (1994) J Nucl Med 35:230P
81. Herman LW, Sharma V, Kronauge JF, Barbarics E, Herman LA, Piwnica-Worms D (1995) J Med Chem 38:2955
82. Jurisson S, Berning D, Jia W, Ma D-S (1993) Chem Rev 93:1137
83. Deutsch E, Vanderheyden J-L, Gerundini P, Libson K, Hirth W, Colombo F, Savi A, Fazio F (1987) J Nucl Med 28:1870
84. Rossetti C, Vanoli G, Paganelli G, Kwiatkowski M, Zito F, Colombo F, Bonino C, Carpinelli A, Casati R, Deutsch K, Marmion M, Woulfe SR, Lunghi F, Deutsch E, Fazio F (1994) J Nucl Med 35:1571
85. Crankshaw C, Marmion M, Luker G, Rao V, Dahlheimer J, Burleigh B, Webb E, Deutsch K, Piwnica-Worms D (1998) J Nuc Med 39:77
86. Woulfe S, Dunn T, Marmion M, MacDonaold J, Rogic M, Deutsch E (1991) J Nucl Med 32:1101
87. Jurisson SS, Dancey K, McPartlin M, Tasker PA, Deutsch E (1984) Inorg Chem 23:4743
88. Kelly J, Forster A, Higley B, Archer C, Booker F, Canning L, Chiu K, Edwards B, Gill H, McPartlin M, Nagle K, Latham I, Pickett R, Storey A, Webbon P (1993) J Nucl Med 34:222
89. Ballinger JR, Bannerman J, Boxen I, Firby P, Hartman NG, Moore MJ (1996) J Nucl Med 37:1578
90. Alberto R, Schibli R, Egli A, Schubiger A (1998) J Am Chem Soc 120:7987
91. Marmion M, MacDonald J (2000) J Nucl Med 41:40P
92. Dyszlewski M, Blake H, Dalheimer J, Pica C, Piwnica-Worms D (2002) Mol Imaging 1:24
93. Wong E, Liu S, Lugger T, Hahn FE, Orvig C (1995) Inorg Chem 34:93
94. Yang LW, Liu S, Wong E, Rettig SJ, Orvig C (1995) Inorg Chem 34:2164
95. Sarma BD, Bailar JC (1955) J Am Chem Soc 77:5476
96. Sarma BD, Ray KR, Sievers RE, Bailar JC (1964) J Am Chem Soc 86:14
97. Tweedle MF, Wilson LJ (1976) J Am Chem Soc 98:4824
98. Tsang B, Mathias C, Green M (1993) J Nucl Med 34:1127
99. Tsang B, Mathias C, Fanwick P, Green M (1994) J Med Chem 37:4400
100. Sharma V, Beatty A, Wey S-P, Dahlheimer J, Pica C, Crankshaw C, Bass L, Green M, Welch M, Piwnica-Worms D (2000) Chem Biol 7:335
101. Sharma V, Crankshaw C, Piwnica-Worms D (1996) J Med Chem 39:3483
102. Sharma V, Wey SP, Bass L, Crankshaw CL, Green MA, Welch MJ, Piwnica-Worms D (1996) J Nucl Med 37:51P
103. Bigott H, McCarthy D, Wust F, Dahlheimer J, Piwnica-Worms D, Welch M (2001) J Labelled Compd Radiopharm 44:S119
104. Anderson O, Packard A (1980) Inorg Chem 19:2941
105. Anderson O, Packard A (1979) Inorg Chem 18:1940
106. Packard A, Kronauge J, Barbarics E, Kiani S, Treves S (2002) Nucl Med Biol 29:289
107. Lewis J, Zweit J, Dearling J, Rooney B, Blower P (1996) Chem Commun 1093
108. Blower P (1998) Transition Met Chem 23:109
109. Camus A, Marsich N, Nardin G (1976) Transition Met Chem 1:205
110. Berners-Price S, Johnson R, Mirabelli C, Faucette L, McCabe F, Sadler P (1987) Inorg Chem 26:3383
111. Zweit J, Lewis J, Dearling J, Rooney B, Coley H, Kelland L (1997) J Nucl Med 38:133P
112. Lewis J, Dearling J, Sosabowski J, Zweit J, Carnochan P, Kelland L, Coley H, Blower P (2000) Eur J Nucl Med 27:638
113. Mehta BM, Rosa E, Fissekis JD, Bading JR, Biedler JL, Larson SM (1992) J Nucl Med 33:1373
114. Mehta B, Rosa E, Biedler J, Larson S (1994) J Nucl Med 35:1179

115. Elsinga PH, Franssen JF, Hendrikse NH, Fluks L, Weemaes A-MA, van der Graaf WTA, de Vries EGE, Visser GM, Vaalburg W (1996) J Nucl Med 37:1571
116. Hendrikse N, Franssen E, van der Graaf W, Vaalburg W, de Vries E (1999) Eur J Nucl Med 26:283
117. Hendrikse N, EGE dV, Eriks-Fluks L, van der Graaf W, Hospers G, Willemsen A, Vaalburg W, Franssen E (1999) Cancer Res 59:2411
118. Levchenko A, Mehta B, Lee J-B, Humm J, Augensen F, Squire O, Kothari P, Finn R, Leonard E, Larson S (2000) J Nucl Med 41:493
119. Luurtsema G, Molthoff C, Windhorst A, Smit J, Keizer R, Boellaard R, Lammertsma A, Franssen E (2003) Nucl Med Biol 30:747
120. Hendrikse N, Vaalburg W (2002) Novartis Found Symp 243:137
121. Ravert H, Klecker R, Collins J, Mathews W, Pomper M, Wahl R, Dannals R (2002) J Labelled Compd Radiopharm 45:471
122. Kiesewetter D, Eckelman W (2001) J Labelled Compd Radiopharm 44: S903
123. Kiesewetter D, Jagoda E, Kao C, Ma Y, Ravasi L, Shimoji K, Szajek L, Eckelman W (2003) Nucl Med Biol 30:11
124. Del Vecchio S, Ciarmiello A, Potena MI, Carriero MV, Mainolfi C, Botti G, Thomas R, Cerra M, D'Aiuto G, Tsuruo T, Salvatore M (1997) Eur J Nucl Med 24:150
125. Del Vecchio S, Ciarmiello A, Pace L, Potena M, Carriero M, Mainolfi C, R T, D'Aiuto G, Tsuruo T, Salvatore M (1997) J Nucl Med 38:1348
126. Bom H, Kim Y, Lim S, Park K (1997) J Nucl Med 38:240P
127. Kostakoglu L, Elahi N, Kirarli P, Ruacan S, Sayek S, Baltali E, Sungur A, Hayran M, Bekdik C (1994) J Nucl Med 38:1003
128. Kostakoglu L, Kirath P, Ruacan S, Hayran M, Emri S, Ergun E, Bekdik C (1998) J Nucl Med 39:228
129. Ciarmiello A, Del Vecchio S, Silvestro P, Potena MI, Carriero MV, Thomas R, Botti G, D'Aiuto G, Salvatore M (1998) J Clin Oncol 16:1677
130. Fukumoto M, Yoshida D, Hayase N, Kurohara A, Akagi N, Yoshida S (1999) Cancer 86:1470
131. Nagamachi S, Jinnouchi S, Ohnishi T, Nakahara H, Flores L, Tamura S, Tokogami K, Kawano H, Wakisaka S (1999) Clin Nucl Med 24:765
132. Kao C, Hsieh J, Tsai S, Ho Y, Changlai S, Lee J (2001) J Nucl Med 42:17
133. Shiau Y, Tsai S, Wang J, Ho Y, Ho S, Kao C (2001) Lung Cancer 179:197
134. Sciuto R, Pasqualoni R, Bergomi S, Petrilli G, Vici P, Belli F, Botti C, Mottolese M, Maini C (2002) J Nucl Med 43:745
135. Shih C, Shiau Y, Wang J, Ho S, Kao A (2003) Lung 181:103
136. Ding H, Huang W, Tsai C, Change C, Kao A (2003) Nucl Med Biol 30:471
137. Zhou J, Higashi K, Ueda Y, Kodama Y, Guo D, Jisaki F, Sakurai A, Takegami T, Katsuda S, Tamamoto I (2001) J Nucl Med 42:1476
138. Chen W, Luker K, Dahlheimer J, Pica C, Luker G, Piwnica-Worms D (2000) Biochem Pharmacol 60:413

Top Curr Chem (2005) 252: 179–192
DOI 10.1007/b101228

Radiopharmaceuticals for Positron Emission Tomography Imaging of Somatostatin Receptor Positive Tumors

Wen Ping Li · Laura A. Meyer · Carolyn J. Anderson (✉)

Mallinckrodt Institute of Radiology, Washington University School of Medicine,
510 S. Kingshighway Blvd, Campus Box 8225, St. Louis, MO 63110, USA
andersoncj@wustl.edu

Abstract The targeting of somatostatin receptors in tumors has been a goal in cancer treatment and diagnosis since the 1980s. Over the past two decades, great strides have been made in the development of somatostatin analogs labeled with positron emitters for positron emission tomography (PET) imaging of somatostatin receptor positive tumors. In this review, the radionuclide production, radiochemistry, preclinical and clinical studies of somatostatin analogs labeled with several positron-emitting radionuclides are described. In the past 5 years, there have been clinical trials with ^{18}F-, ^{64}Cu-, ^{68}Ga-, ^{86}Y- and ^{110m}In-labeled somatostatin analogs. In addition, radiochemistry and preclinical studies have been performed with ^{76}Br and ^{66}Ga. The advantages of these positron-emitting somatostatin analogs include increased image resolution with PET compared with γ scintigraphy, improved quantification capabilities with PET, and the ability to use the positron emitters as companion isotopes to therapeutic radionuclides such as the ^{86}Y (PET)/^{90}Y (therapy) pair.

Keywords Somatostatin · Positron-emitting radionuclide · Cancer therapy · Diagnosis · Positron emission tomography

Abbreviations

^{76}BrNHS	*N*-Succinimidyl 4-[^{76}Br]bromobenzoate
CB-TE2A	4,11-Bis(carboxymethyl)-1,4,8,11-tetraazabicyclo[6.6.2]hexadecane
DFO	Desferrioxamine B
DOTA	1,4,7,10-Tetraazacyclododecane-*N*,*N*′,*N*″,*N*‴-tetraacetic acid
DTPA	Diethylenetriaminepentaacetic acid
^{111}In-DTPA-OC	^{111}In-DTPA-D-Phe^{1}-octreotide
[^{18}F]FBA	4-[^{18}F]fluorobenzoic acid
[^{18}F]FB-OC	4-[^{18}F]fluorobenzoyl-OC
[^{18}F]FP-Gluc-TOCA	N^{α}-(1-deoxy-D-fructosyl)-N^{ε}-(2-[^{18}F]fluoropropionyl)-Lys^{0}-Tyr^{3}-octreotate
OC	Octreotide
PET	Positron emission tomography
SPECT	Single-photon-emission computerized tomography
SSTr	Somatostatin receptor
SSTr2	Somatostatin receptor subtype 2
TATE	Octreotate
TETA	1,4,8,11-Tetraazacyclotetradecane-1,4,8,11-tetraacetic acid
TOC	Tyr^{3}-OC
Y3	Tyr^{3}

1 Introduction

Somatostatin is a 14-amino-acid peptide involved in the regulation and release of a number of hormones, including growth hormone, thyroid-stimulating hormone and prolactin. Somatostatin receptors (SSTr) are found on the cell surface and occur in a number of different normal organ systems, including the central nervous system, the gastrointestinal tract, and the exocrine and endocrine pancreas [1–3]. A large number of human tumors express SSTr [4], particularly SSTr subtype 2 (SSTr2). The targeting of SSTr in tumors has been a goal in cancer treatment and diagnosis since the 1980s. Somatostatin has a very short biological half-life, but analogs have been developed, such as octreotide (OC), that have much longer residence times [5].

OC has been labeled with ^{123}I and ^{111}In and used to image SSTr2-positive tumors in humans by conventional scintigraphy [6, 7]. ^{123}I-Tyr^{3}-OC has a high hepatobiliary excretion that hinders visualization of tumors in the abdomen, whereas ^{111}In-diethylenetriaminepentaacetic acid-D-Phe^{1}-OC (^{111}In-DTPA-OC) clears primarily through the kidneys [7]. In the USA and Europe, ^{111}In-DTPA-OC is currently a clinically approved agent for imaging neuroendocrine tumors. Because of the limited sensitivity and resolution of single-photon-emission computerized tomography (SPECT), several research groups have worked towards the development of a positron-emitting agent for positron emission tomography (PET) imaging of SSTr-positive tumors. Using somatostatin analogs labeled with a positron-emitting radionuclide and PET, a quantitative assessment of tracer accumulation within tissues can be achieved,

Table 1 Positron-emitting radionuclides

Isotope	$T_{1/2}$ (h)	Methods of production	Decay mode	$E_{\beta+}$ (keV)	Reference
^{18}F	1.83	Cyclotron $^{18}O(p,n)^{18}F$	β^+ (97%), EC (3%)	635	[27]
^{60}Cu	0.4	Cyclotron, $^{60}Ni(p,n)^{60}Cu$	β^+ (93%), EC (7%)	3,920, 3,000, 2,000	[16]
^{61}Cu	3.3	Cyclotron, $^{61}Ni(p,n)^{61}Cu$	β^+ (62%), EC (38%)	1,220, 1,150, 940, 560	[16]
^{64}Cu	12.7	Cyclotron, $^{64}Ni(p,n)^{64}Cu$	β^+ (19%), EC (41%), β^- (40%)	656	[15]
^{66}Ga	9.5	$^{66}Zn(p,n)^{66}Ga$	β^+ (57%), EC (43%)	4.15 (max)	[10, 12]
^{68}Ga	1.1	$^{68}Ge/^{68}Ga$ generator	β^+ (90%), EC (10%)	1,880, 770	[8]
^{76}Br	16.2	$^{76}Se(p,n)^{76}Br$	β^+ (54.7%), EC (45.3%)	3,941 (max)	
^{86}Y	14.7	Cyclotron, $^{86}Sr(p,n)^{86}Y$	β^+ (33%), EC (66%)	2,335, 2,019, 1,603, 1,248, 1,043	[20]
^{110m}In	1.15	Cyclotron, $^{110}Cd(p,n)^{110m}In$ or $^{110}Sn/^{110m}In$ generator	β^+ (62%), EC (38%)	1,010, 2,260 (max)	[22, 24]
^{124}I	100.2	Cyclotron	β^+ (25%), EC (75%)	2,134, 1,533	[64]

potentially offering improvements over SPECT where quantitative results are needed, and where tumor size is small or the tumor is deep within the body. This review will provide an overview of the progress thus far towards the development of a PET SSTr imaging agent, using positron-emitting radionuclides that include ^{18}F, ^{76}Br, ^{68}Ga, ^{66}Ga, ^{64}Cu, ^{110m}In and ^{86}Y (Table 1).

2 Radionuclides for PET Imaging

There are two gallium radionuclides with decay characteristics that are suitable for PET imaging. ^{68}Ga ($T_{1/2}$=68 min; 90% β^+) is produced from the $^{68}Ge/^{68}Ga$ generator. The long half-life of the parent nuclide ^{68}Ge ($T_{1/2}$=270·8 days) allows the generator to be used for 1–2 years, making ^{68}Ga radiopharmaceuticals relatively economical. A commercially available generator was based on the design of Loc'h et al. [8], where the generator was eluted in 3 mL 1 N HCl. More recently, another $^{68}Ge/^{68}Ga$ generator has recently been reported, where the generator is eluted in 1 mL 0.1 N HCl, making it more convenient for radiopharmaceutical preparation [9]. ^{66}Ga is a cyclotron-produced positron-emitting radionuclide that has been used in a limited number of studies requiring a medium half-life positron-emitting nuclide where a half-life longer than 1 h is needed. This nuclide can be produced in small biomedical cyclotrons, utilizing the ^{66}Zn(p,n)^{66}Ga reaction [10–13].

Positron-emitting copper radionuclides have a wide range of half-lives (10 min–12.7 h) and are cyclotron- or generator-produced. ^{64}Cu was initially produced using a reactor by the ^{64}Zn(n,p)^{64}Cu nuclear reaction [14] but more recently is produced by the ^{64}Ni(p,n)^{64}Cu nuclear reaction using a biomedical cyclotron [15]. A target has been specifically designed for the production of this nuclide [15], and by altering the enriched isotope of nickel used as the target, large quantities of ^{64}Cu, ^{60}Cu and ^{61}Cu have been produced [16]. ^{62}Cu is generator-produced from the decay of ^{62}Zn [17], and ^{60}Cu is produced with a biomedical cyclotron [16], but with 10- and 23-min half-lives, respectively, these radionuclides are not readily applicable to labeling somatostatin analogs or other peptide-based tumor imaging agents. ^{64}Cu is becoming widely-used for labeling peptide-based agents.

There are two radioisotopes of yttrium that are utilized in radiopharmaceuticals: ^{90}Y ($T_{1/2}$=64.06 h) and ^{86}Y ($T_{1/2}$=14.7 h). ^{90}Y is a pure β^- emitter produced following the decay of the parent nuclide ^{90}Sr ($T_{1/2}$=27.8 years) and has been widely used for targeted radiotherapy of cancer. ^{86}Y is produced by the ^{86}Sr(p,n)^{86}Y nuclear reaction and is a positron-emitting nuclide that has been used in PET imaging of patients that are undergoing therapy with ^{90}Y radiopharmaceuticals [18, 19]. No-carrier-added ^{86}Y can be produced with a biomedical cyclotron by the ^{86}Sr(p,n)^{86}Y reaction. As described by Rösch et al. [20], an enriched $^{86}SrCO_3$ target was irradiated yielding ^{86}Y, which was separated by a combined coprecipitation and ion-exchange purification process [20]. An improved procedure for separation and purification of ^{86}Y is an electrochemical separation that was recently reported [21], which produced ^{86}Y more rapidly and in high chemical purity, with the residual strontium content reduced to less than 0.1 ppm.

^{111}In-OC has been used extensively in the imaging with SPECT for SSTr-positive tumors. ^{110m}In ($T_{1/2}$=69 min) is a short-lived positron-emitting isotope of indium and can be produced either from a ^{110}Sn/^{110m}In generator or by the nuclear reaction ^{110}Cd(p,n)^{110m}In with a low-energy cyclotron (11.8 MeV protons with beam currents up to 15 μA) [22–24]. Using a low-energy cyclotron, 1 h irradiation of ^{110}Cd produced more than 20 GBq (540 mCi) [25]. With its short half-life, ^{110m}In is suitable for the labeling of small peptides such as OC that have rapid kinetics.

The positron-emitting halogens with applications for labeling somatostatin analogs include ^{18}F, ^{76}Br and ^{124}I. From an imaging standpoint, ^{18}F ($T_{1/2}$= 110 min) is an ideal isotope among all other halogen radionuclides for PET imaging, since it has 97% positron abundance with 0.64-MeV positron energy [26]. ^{18}F had historically been produced by a variety of cyclotron, linear-accelerator and reactor methods [27]. Currently, nearly all ^{18}F is produced using a single nuclear reaction, $^{18}O(p,n)^{18}F$, with the majority of the radionuclide isolated as the [^{18}F]fluoride ion in aqueous solution [28]. More importantly, ^{18}F can be produced in relatively large quantities (gigabecquerels or curies) in standard commercial PET cyclotrons supplied by several manufacturers [29–31]. ^{76}Br ($T_{1/2}$=16.2 h) is practical for PET imaging applications and can be prepared by the direct reaction of $Cu_2{}^{76}Se$ alloy with protons at 16 MeV, producing ^{76}Br in high yield and low ^{77}Br contamination with a small medical cyclotron [32]. ^{124}I can be produced via the nuclear reaction $^{124}Te(p,n)^{124}I$ with a cyclotron [33]. Despite the disadvantage of ^{124}I, a lower positron abundance and subsequent emission of high-energy γ-rays, successful PET imaging with ^{124}I-labeled tracers is possible [34, 35].

3 Somatostatin Analogs Labeled with Positron-Emitting Metal Radionuclides

3.1 Somatostatin Analogs Labeled with Gallium Radionuclides

The first positron-emitting metal radionuclide-labeled somatostatin analog was ^{68}Ga-[DFO]-OC [36], where DFO is desferrioxamine B. OC was modified with DFO, which is a chelator that complexes Fe^{3+} and $Ga^{3+}$with high stability. The binding affinity of [Ga]-DFO-OC was tested against [^{125}I-Tyr^3]-OC in rat cortex membranes and gave a pIC_{50} of 8.2. The biodistribution of [^{67}Ga]-DFO-OC in normal and in pancreatic islet cell-tumor-bearing rats showed renal clearance and tumor uptake similar to that of ^{111}In-DTPA-OC. PET scans with ^{68}Ga-DFO-OC revealed rapid tumor uptake that peaked at 0.9% ID/mL at 30 min post injection [36]. Since the mid-1990s there have been no other publications on ^{68}Ga-DFO-OC, and it is unknown whether this agent was evaluated in a successful clinical trial.

A promising new somatostatin analog for PET imaging with the gallium isotopes is DOTA-Tyr^3-OC (DOTATOC) (Fig. 1). OC was modified by replacing Phe in the 3-position of OC with tyrosine to increase the hydrophilicity for increased renal clearance and to allow dual labeling with ^{125}I for potential therapy applications [37]. Use of the bifunctional chelator 1,4,7,10-tetraazacyclododecane-*N*,*N*′,*N*″,*N*‴-tetraacetic acid (DOTA) allowed labeling with a broad spectrum of 2^+ and 3^+ radiometals. In a biodistribution study comparing ^{67}Ga-DOTATOC (^{67}Ga, $T_{1/2}$=78 h) with ^{111}In-DPTA-OC (^{111}In, $T_{1/2}$=2.8 days) in rat pancreatic AR42 J tumor-bearing nude mice, the gallium compound had a significantly higher tumor uptake and a lower kidney uptake. Examination of the crystal structure of a model peptide chelate, Ga-DOTA-D-PheNH$_2$, revealed that one carboxylate group is free and deprotonated at physiological pH; this increased hydrophilicity may contribute to the improved tumor uptake and re-

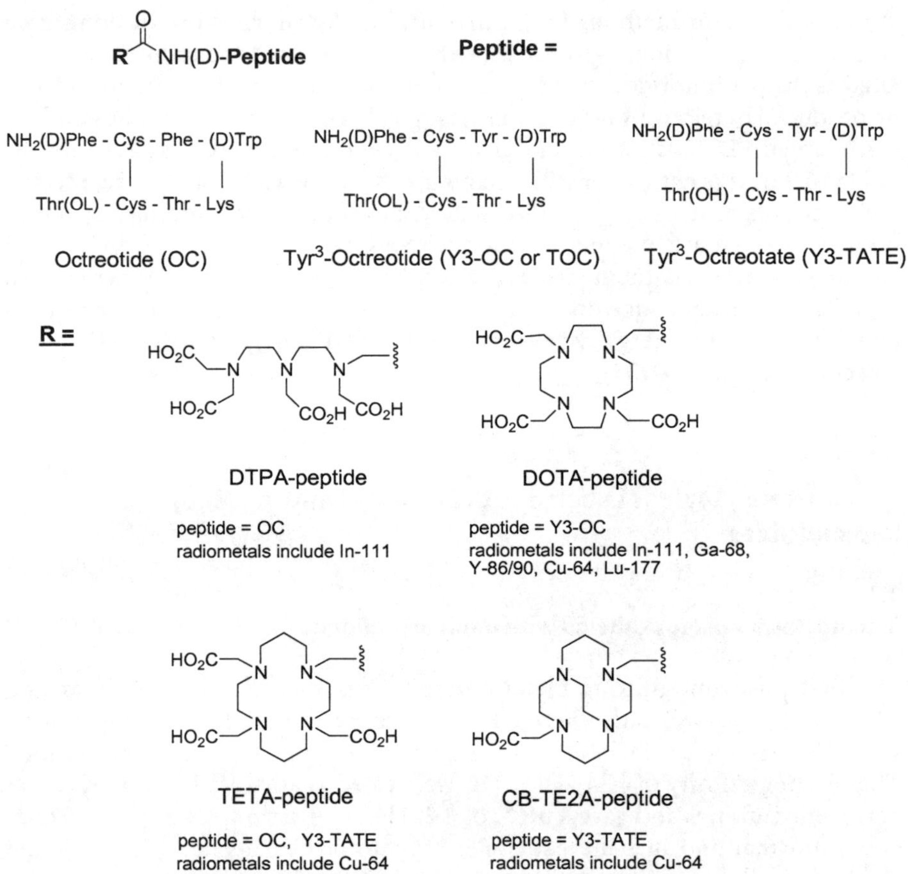

Fig. 1 Structures of somatostatin analogs and chelators for radiolabeling with metal positron-emitting radionuclides

nal clearance [37]. When affinities were compared in complete displacement experiments with ^{125}I-[Leu^8, D-Trp^{22}, Tyr^{25}]-somatostatin 28 using membranes from CCL39 cells stably transfected with human SSTr, Ga-DOTATOC was found to have a higher affinity (IC$_{50}$=2.5 nM) for SSTr2 than In-DTPA-OC (22 nM) [38].

Henze et al. [39] evaluated ^{68}Ga-DOTATOC as a potential agent for PET imaging of meningiomas. Dynamic PET scans were acquired over 120 min after intravenous injection of 175 MBq ^{68}Ga-DOTATOC in three patients with eight meningiomas. The radiolabeled compound showed rapid blood and renal clearance, with a half-life α of 3.5 min and a half-life β of 63 min. All meningiomas showed high uptake, with a mean standard uptake value of 10.6, and there was no tracer accumulation in the surrounding normal brain tissue. The ratio of uptake in tumor versus normal tissue reached 730 at 120 min post injection. The smallest legions detected were 7–8 mm, which is a vast improvement over the lower imaging threshold with ^{111}In-DTPA-OC by SPECT (2.7 cm) [39].

^{68}Ga-labeled DOTATOC was evaluated as a potential imaging agent for carcinoid tumors by Hofmann et al. [40]. Dynamic PET scans were acquired over 180 min after intravenous injection of 80–250 MBq in eight patients with carcinoid tumors. The group predefined 40 lesions by CT and/or magnetic resonance imaging, and of these lesions, the PET imaging of ^{68}Ga-DOTATOC identified 100%, whereas SPECT imaging of ^{111}In-DPTA-OC identified only 85%. The PET imaging of ^{68}Ga-DOTATOC also identified additional lesions not previously defined. Blood clearance was rapid and renal accumulation was low enough to allow delineation of adrenal glands [40].

Recently, ^{66}Ga has been evaluated for use in receptor-targeted PET imaging [12, 41]. Ugur et al. [12] evaluated ^{66}Ga-DOTATOC in animal biodistribution and micro PET studies and compared this with ^{67}Ga- and ^{68}Ga-labeled DOTATOC in AR42 J tumor-bearing nude mice. Tumor uptake was rapid and the maximum values of the percentage infective dose per gram were comparable for all three isotopes. Blood clearance was rapider and kidney accumulation was lower in the case of ^{66}Ga-DOTATOC, with a maximum percentage infective dose per gram in the kidney of 4.5 versus 9.18 for ^{67}Ga and 11.4 for ^{68}Ga.

3.2 Somatostatin Analogs Labeled with Copper Radionuclides

OC has been conjugated to the chelator 1,4,8,11-tetraazacyclotetradecane-1,4,8,11-tetraacetic acid (TETA) for labeling with ^{64}Cu [42]. ^{64}Cu-TETA-OC showed high affinity for SSTr both in vitro and in vivo and cleared primarily through the kidneys, with relatively low liver accumulation. ^{64}Cu-TETA-OC was evaluated as a PET imaging agent for neuroendocrine tumors in eight patients [43]. Preliminary results showed that in two patients, ^{64}Cu-TETA-OC and PET detected more SSTr-positive lesions than the currently used agent, ^{111}In-DTPA-OC, and γ scintigraphy/SPECT. In one patient, ^{111}In-DPTA-OC and γ scintigraphy showed low uptake in a lung lesion that was not observed with ^{64}Cu-TETA-OC and PET.

A series of ^{64}Cu-labeled somatostatin analogs was investigated to determine the optimal analog with respect to target tissue uptake and nontarget tissue clearance [44]. The chelator TETA was conjugated to OC, Tyr^3-OC (Y3-OC, also referred to as TOC), octreotate (TATE), and Y3-TATE and labeled with ^{64}Cu. Binding affinity and biodistribution studies were performed in rat pancreatic CA20948 tumor-bearing rats. Of these agents, ^{64}Cu-TETA-Y3-TATE showed the best targeting and clearance properties.

In addition to optimizing for the somatostatin analog, it is also necessary to optimize the chelator. It was reported that ^{64}Cu dissociated from TETA-OC in rat liver in vivo [45], and this therefore suggested that a stabler chelator for complexing Cu(II) was required. Additional support for this argument is that although ^{64}Cu-TETA-OC was reasonably successful in imaging SSTr-positive tumors in humans, there was a retention of activity in the blood and liver at longer times (from 4–24 h post injection) [43]. A new class of cross-bridged macrocyclic chelators for complexing copper radionuclides was reported by Sun et al. [46], and it was demonstrated in metabolism experiments in rats that ^{64}Cu-labeled 4,11-bis(carboxymethyl)-1,4,8,11-tetraazabicyclo[6.6.2]hexadecane (CB-TE2A) was dramatically stabler in vivo than ^{64}Cu-TETA [47]. A preliminary report described the synthesis and biological evaluation of ^{64}Cu-CB-TE2A-Y3-TATE, showing improved target tissue uptake and blood and liver clearance compared with ^{64}Cu-TETA-OC [48].

3.3 Somatostatin Analogs Labeled with Yttrium and Indium Radionuclides

^{90}Y ($T_{1/2}$=64.06 h) labeled DOTATOC has been suggested to be a promising targeted radiotherapy agent for SSTr-positive cancers (for a review, see Ref. [49]). Although ^{90}Y is highly advantageous for cancer therapy since it has 100% abundance of β^- emission, ^{90}Y does not emit γ-photons for imaging applications. Dosimetry calculations for clinical trials with ^{90}Y-DOTATOC have been made using ^{111}In-DOTATOC and γ scintigraphy [50]. The disadvantages of using ^{111}In are that it might behave somewhat differently than ^{90}Y, and gamma scintigraphy is not as accurate for quantification as PET. A more accurate surrogate for ^{90}Y is ^{86}Y ($T_{1/2}$=14.7 h), although the shorter half-life does not allow biodistribution at very long times (more than 60 h). Förster et al. [51] compared ^{86}Y-DOTATOC with ^{111}In-DTPA-OC in three patients with metastatic carcinoid tumors, and their data suggested that ^{111}In-DTPA-OC underestimated tumor uptake, and therefore ^{86}Y-DOTATOC was closer to ideal. The pharmacokinetics and dosimetry of ^{86}Y-DOTATOC were evaluated in a phase I PET study of 24 patients with SSTr-positive neuroendocrine tumors, along with the effect of amino acid coinfusion on renal and tumor uptake [52]. The use of ^{86}Y-DOTATOC allowed accurate dosimetry of tumor and kidneys, and was valuable in establishing the optimal amino acid regimen for decreasing renal absorbed dose.

^{110m}In, a positron-emitting radionuclide that is an alternative to ^{111}In, was used to label DTPA-OC in a clinical PET study with one patient [25]. ^{110m}In was prepared by irradiating an enriched ^{110}Cd target using the ^{110}Cd(p,n)^{110m}In nuclear reaction. In one patient, 175 MBq (4.7 mCi) ^{111}In-DTPA-OC was injected into a patient who had a small-intestine-carcinoma metastasis to the upper thorax, and the patient was imaged by SPECT. Into the same patient, 150 MBq (4.1 mCi) ^{110m}In-DTPA-OC was injected and the patient was imaged by PET. The 69-min half-life of ^{110m}In allowed the kinetics to be followed for 2 h. A comparison of the imaging studies clearly showed improvements in resolution and recovery with ^{110m}In PET versus ^{111}In SPECT.

4 Somatostatin Analogs Labeled with Positron-Emitting Halogen Radionuclides

Somatostatin analogs have been labeled with positron-emitting halogen radionuclides such as ^{18}F and ^{76}Br (Fig. 2). Another potential positron emitting halogen radionuclide for PET imaging applications is ^{124}I ($T_{1/2}$=4.2 days, 22.8% β^+) [35, 53], although at this time, ^{124}I-labeled somatostatin analogs have not yet been reported. A great deal of effort has been made to improve the in vivo behavior of radioiodine (^{123}I, ^{125}I, ^{131}I) labeled SSTr analogs by new strategies in SSTr ligand development [54]. With radioiodinated SSTr ligands, the major drawbacks include rapid clearance from the tumor, hepatic uptake, and high

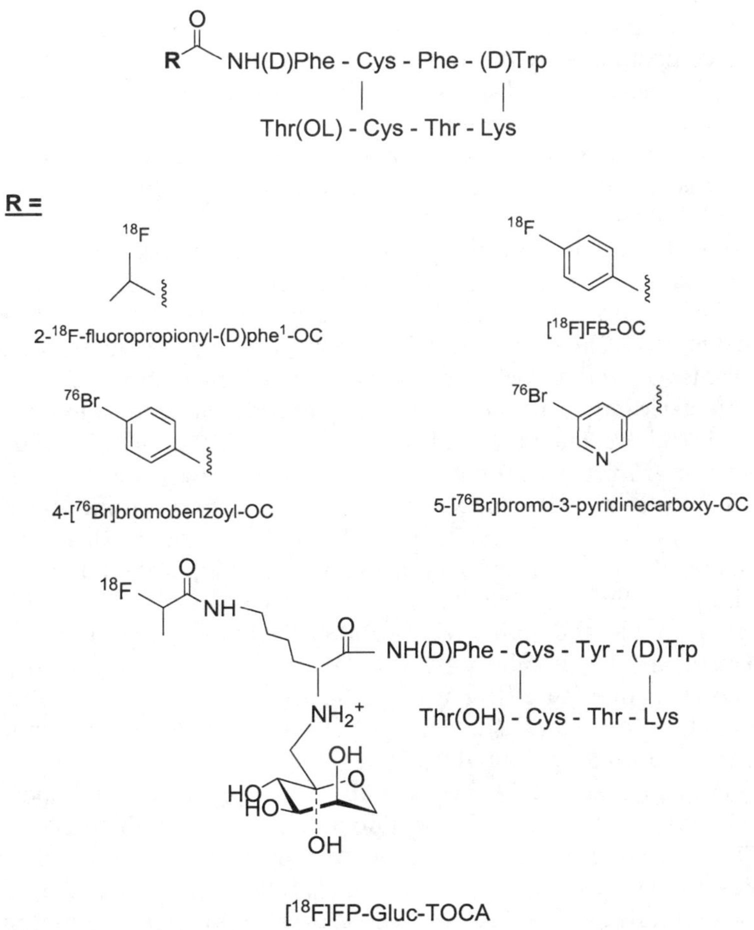

Fig. 2 Structures of somatostatin analogs and halogen-labeling synthons for radiolabeling with halogen positron-emitting radionuclides

thyroid and abdominal background activity [6, 55–57]. Radioiodinated carbohydrated N-terminal conjugates of SSTr analogs showed encouraging improvements in the biodistribution and tumor uptake kinetics [54]. With these improvements in developing iodinated SSTr analogs, it is anticipated that SSTr analogs labeled with ^{124}I will be reported in the near future.

4.1
^{18}F-labeled Somatostatin Analogs

Over the past 20 years a wide variety of ^{18}F-labeled peptide-based radiopharmaceuticals have been prepared and evaluated for their potential as diagnostic imaging agents. They have been used primarily for tumor imaging and, to a lesser extent, for infection/inflammation imaging. ^{18}F-labeled somatostatin analogs are examples of tumor-imaging agents.

Unlike radioiodination of somatostatin analogs, where the labeling conditions are compatible with the presence of an active ester moiety in the labeling agents, radiofluorination of somatostatin analogs requires strong basic conditions for the specific introduction of [^{18}F]fluoride and can only be achieved via a suitable prosthetic group [58]. Guhlke et al. [59] successfully radiolabeled OC with ^{18}F via 2-[^{18}F]fluoropropionic acid 4-nitrophenylester as a prosthetic group and evaluated its binding affinity and biological activity as well as its in vivo distribution. This radiolabeling method involved two steps including the acylation of ε-Boc-Lys^{5}-protected OC with the activated ester, and followed by an acidolytic deprotection leading to the desired product, 2-[^{18}F]fluoropropionyl-(D)Phe^{1}-OC. The use of the 2-[^{18}F]fluoropropionyl moiety provided certain advantages for the labeling of OC, such as high reactivity of the acylation agent, low steric impairment caused by the introduction of the fluoroacyl moiety and no side reactions. 2-[^{18}F]fluoropropionyl-(D)Phe^{1}-OC demonstrated high binding affinity to SSTr with slightly lower binding affinity than an iodinated somatostatin analog and a low nanomolar IC_{50} value (around 3 nM). Unfortunately, the PET pharmacokinetics of 2-[^{18}F]fluoropropionyl-(D)Phe^{1}-OC in SSTr-positive tumor-bearing rats was unfavorable [60]. Although 2-[^{18}F]fluoropropionyl-(D)Phe^{1}-OC exhibited rapid blood clearance (0.1% ID/g for 1 h, 0.04% ID/g for 2 h) and high in vivo stability, the lipophilicity of this ^{18}F compound resulted in hepatobiliary excretion followed by increasing activity in the intestines over time (40% ID/g at 1 h). The low tumor uptake along with the short retention (0.52±0.24% ID/g for 1 h, 0.17±0.04% ID/g for 2 h) further limited its application for potential PET imaging.

Another approach to labeling OC with ^{18}F was reported by Hostetler et al. [61]. OC was labeled via an in situ peptide coupling of 4-[^{18}F]fluorobenzoic acid ([^{18}F]FBA) with the N-terminus of OC to provide 4-[^{18}F]fluorobenzoyl-OC ([^{18}F]FB-OC). The process of synthesizing [^{18}F]FB-OC involved a new efficient, one-pot synthesis of [^{18}F]FBA using a microwave cavity and the combination of reagents,1,3-dicyclohexylcarbodiimide and 1-hydroxy-7-azabenzo-triazole. Unfortunately, [^{18}F]FB-OC also showed high liver uptake and low uptake in somatostatin-rich organs such as CA20948 tumor, pancreas, and adrenal glands in tumor-bearing male Lewis rats.

Improved and more efficient ^{18}F labeling methods for SSTr ligands via prosthetic groups have been developed in recent years, and successes in decreasing the lipophilicity of the ^{18}F-labeled SSTr analogs have also occurred, thereby improving target tissue uptake with decreased hepatobiliary clearance. Carbohydrated ^{18}F labeled octreotate, N^{α}-(1-deoxy-D-fructosyl)-N^{ε}-(2-[^{18}F]fluoropropionyl)-Lys^0-Tyr^3-octreotate ([^{18}F]FP-Gluc-TOCA) showed promising biokinetics as a potential PET imaging agent of SSTr [62]. The synthesis of [^{18}F]FP-Gluc-TOCA applied a trifunctional linker concept and it was completed in around 3 h with a specific activity of more than 37 GBq/μmol. [^{18}F]FP-Gluc-TOCA demonstrated higher selective binding affinity to human SSTr2 (IC_{50}=2.8±0.4 nM) over other SSTr subtypes such as human SSTr4 (IC_{50}=437±84 nM)and SSTr5 (IC_{50}=123±8.8 nM), and this agent had no affinity at all for human SSTr1 and SSTr3 ((IC_{50}>1,000 nM). The decreased lipophilicity of [^{18}F]FP-Gluc-TOCA with a log*P* of –1.70±0.02 led to rapid renal excretion (8.69±1.09% ID/g at 1 h) with lower liver (0.72±0.14% ID/g at 1 h) and intestinal uptake (1.88±0.52% ID/g at 1 h). Even more encouraging is that [^{18}F]FP-Gluc-TOCA showed impressive tumor uptake (13.54±1.47% ID/g at 1 h) in AR42 J tumor-bearing mice. Human PET imaging of [^{18}F]FP-Gluc-TOCA in a patient with a history of a histologically proven metastatic carcinoid in the liver revealed rapid clearance of the radioactivity from the blood pool within the first hour of injection by renal excretion [62]. Multiple liver metastases were clearly delineated with standard uptake values in the regions of interest, ranging from 21.4 to 38.0, which were only visible as regions of large, pronounced liver uptake by SPECT.

It is important to keep in mind that hydrophilic prosthetic groups are important for renal rather than hepatobiliary clearance. Wester et al. [62] have demonstrated this with their carbohydrated analogs. Since ^{18}F possesses inherent imaging advantages, the development of a simple and rapid synthesis which will enable ^{18}F-SSTr ligands to be clinically practical is also important.

4.2 ^{76}Br-labeled Octreotide

^{76}Br is an alternative to ^{18}F for labeling the SSTr2 ligand for PET imaging of SSTr-positive tumors. To the best of our knowledge, there is only one approach to ^{76}Br-labeling of OC using prosthetic groups to decrease the lipophilicity of the labeled peptide [63]. The OC was conjugated to *N*-succinimidyl 4-[^{76}Br]bromobenzoate (^{76}BrNHS) and *N*-succinimidyl 5-[^{76}Br]bromo-3-pyridinecarboxylate using microwave heating. The SSTr binding affinity of these two ^{76}Br conjugates was disappointing. The ^{76}Br-conjugate via ^{76}BrNHS, 4-[^{76}Br]bromobenzoyl-OC, showed lower binding to meningioma than the other ^{76}Br conjugate (5-[^{76}Br]bromo-3-pyridinecarboxy-OC). High nonspecific binding (more than 20%) to meningioma in heart tissue was observed for both ^{76}Br-

conjugates. Improved brominated analogs and radiochemistry methods will be needed for this area of research to grow.

5
Summary and Conclusions

Over the past 2 decades, great strides have been made in the development of somatostatin analogs labeled with positron emitters for PET imaging of SSTr-positive tumors. In this review, we described clinical trials with ^{18}F-, ^{64}Cu-, ^{68}Ga,- ^{86}Y- and ^{110m}In-labeled somatostatin analogs, all of which have been performed in the past 5 years, as well as discussed some radiochemistry and preclinical studies with ^{76}Br and ^{66}Ga. The advantages of these positron-emitting somatostatin analogs include increased image resolution with PET compared with SPECT, improved quantification capabilities with PET, and the ability to use the positron emitters as companion isotopes to therapeutic radionuclides such as the ^{86}Y/^{90}Y pair. The availability of a clinically approved PET somatostatin analog for diagnosis and/or as a companion to therapeutic studies with ^{90}Y will be a major benefit for cancer imaging and therapy.

Acknowledgements The authors acknowledge funding from the National Cancer Institute (CA64475), which supports their research in the development of somatostatin analogs for PET imaging and therapy.

References

1. Guillemin R (1978) Science 202:390
2. Reichlin S (1983) N Engl J Med 309:1495
3. Reichlin S (1983) N Engl J Med 309:1556
4. Reubi JC, Kvols LK, Krenning EP, Lamberts SWJ (1990) Metabolism (Suppl 2) 39:78
5. Bauer W, Briner U, Doepfner W, Haller R, Huguenin R, Marbach P, Petcher TJ, Pless J (1982) Life Sci 31:1133
6. Bakker WH, Krenning EP, Breeman WA, Kooij PPM, Reubi JC, Koper JW, de Jong M, Lameris JS, Visser TJ, Lamberts SW (1991) J Nucl Med 32:1184
7. Krenning EP, Bakker WH, Kooij PPM, Breeman WAP, Oei HY, de Jong M, Reubi JC, Visser TJ, Bruns C, Kwekkeboom DJ, Reijs AEM, van Hagen PM, Koper JW, Lamberts SWJ (1992) J Nucl Med 33:652
8. Loc'h C, Maziere B, Comar D (1980) J Nucl Med 21:171
9. Roivainen A, Tolvanen T, Salomaki S, Lendvai G, Velikyan I, Numminen P, Valila M, Sipila H, Bergstrom M, Harkonen P, Lonnberg H, Langstrom B (2004) J Nucl Med 45:347
10. Graham MC, Pentlow KS, Mawlawi O, Finn RD, Daghighian F, Larson SM (1997) Med Phys 24:317
11. Daube-Witherspoon ME, Green SL, Plascjak P, Wu C, Carson RE, Brechbiel M, Eckelman WC (1997) J Nucl Med 38:200P
12. Ugur O, Kothari PJ, Finn RD, Zanzonico P, Ruan S, Guenther I, Maecke HR, Larson SM (2002) Nucl Med Biol 29:147
13. Lewis MR, Reichert DE, Laforest R, Margenau WH, Shefer RE, Klinkowstein RE, Hughey BJ, Welch MJ (2002) Nucl Med Biol 29:701

14. Zinn KR, Chaudhuri TR, Cheng TP, Morris JS, Meyer WA (1994) Cancer 73:774
15. McCarthy DW, Shefer RE, Klinkowstein RE, Bass LA, Margenau WH, Cutler CS, Anderson CJ, Welch MJ (1997) Nucl Med Biol 24:35
16. McCarthy DW, Bass LA, Cutler PD, Shefer RE, Klinkowstein RE, Herrero P, Lewis JS, Cutler CS, Anderson CJ, Welch MJ (1999) Nucl Med Biol 26:351
17. Robinson J, G.D., Zielinski FW, Lee AW (1980) Int J Appl Radiat Isot 31:111
18. Herzog H, Rosch F, Stocklin G, Lueders C, Qaim SM, Feinendegen LE (1993) J Nucl Med 34:2222
19. Herzog H, Rosch F, Brockmann J, Muhlensiepen H, Kohle M, Stolz B, Marbach P, Muller-Gartner HW (1997) J Nucl Med 38:60P
20. Rösch F, Qaim SM, Stöcklin G (1993) Int J Appl Radiat Isot 44:677
21. Reischl G, Rosch F, Machulla H-J (2002) Radiochim Acta 90:225
22. Lundqvist H, Scott-Robson S, Einarsson L, Malmborg P (1991) Appl Radiat Isot 42:447
23. Lundqvist H, Tolmachev V, Bruskin A, Einarsson L, Malmborg P (1995) Appl Radiat Isot 46:859
24. Rösch F, Qaim SM, Sayed M, Novgorodov AF, Tsai Y-M (1996) Appl Radiat Isot 48:19
25. Lubberink M, Tolmachev V, Widstroem C, Bruskin A, Lundqvist H, Westlin J (2002) J Nucl Med 43:1391
26. Kilbourn MR, Dence CS, Welch MJ (1987) J Nucl Med 28:462
27. Wilbur DS (1992) Bioconjugate Chem 3:433
28. Kilbourn MR, Jerabek PA, Welch MJ (1985) Int J Appl Radiat Isot 36:327
29. Zeisler SK, Becker DW, Pavan RA, Moschel R, Ruhle H (2000) Appl Radiat Isot 53:449
30. Ruth TJ, Buckle KR, Chun KS, Hurtado ET, Jivan S, Zeisler SK (2001) Appl Radiat Isot 55:457
31. Helmeke HJ, Harms T, Knapp WH (2001) Appl Radiat Isot 54:753
32. Tolmachev V, Lovqvist A, Einarsson L, Schultz J, Lundqvist H (1998) Appl Radiat Isot 49:1537
33. Glaser M, Brown DJ, Law MP, Iozzo P, Waters SL, Poole K, Knickmeier M, Camici PG, Pike VW (2001) J Labelled Compd Radiopharm 44:465
34. Glaser M, Luthra SK, Brady F (2003) Int J Oncol 22:253
35. Blasberg RG, Roelcke U, Weinreich R, Beattie B, Von Ammon K, Yonekawa Y, Landolt H, Guenther I, Crompton NEA, Vontobel P, Missimer J, Maguire RP, Koziorowski J, Knust EJ, Finn RD, Leenders KL (2000) Cancer Res 60:624
36. Smith-Jones PM, Stolz B, Bruns C, Albert R, Reist HW, Fridrich R, Mäcke HR (1994) J Nucl Med 35:317
37. Heppeler A, Froidevaux S, Macke HR, Jermann E, Behe M, Powell P, Hennig M (1999) Chem Eur J 5:1974
38. Reubi JC, Schar J-C, Waser B, Wenger S, Heppeler A, Schmitt JS, Maecke HR (2000) Eur J Nucl Med 27:273
39. Henze M, Schuhmacher J, Hipp P, Kowalski J, Becker DW, Doll J, Macke H, Hofman M, Debus J, Haberkorn U (2001) J Nucl Med 42:1053
40. Hofman M, Maecke H, Borner AR, Weckesser E, Schoffski P, Oei ML, Schumacher J, Henze M, Heppeler A, Meyer GJ, Knapp WH (2001) Eur J Nucl Med 28:1751
41. Mathias C, Lewis MR, Reichert DE, Laforest R, Sharp TL, Lewis JS, Yang Z, Waters DJ, Snyder PW, Low PS, Welch MJ, Green MA (2003) Nucl Med Biol 30:725
42. Anderson CJ, Pajeau TS, Edwards WB, Sherman ELC, Rogers BE, Welch MJ (1995) J Nucl Med 36:2315
43. Anderson CJ, Dehdashti F, Cutler PD, Schwarz SW, Laforest R, Bass LA, Lewis JS, McCarthy DW (2001) J Nucl Med 42:213
44. Lewis JS, Lewis MR, Srinivasan A, Schmidt MA, Wang J, Anderson CJ (1999) J Med Chem 42:1341

45. Bass LA, Wang M, Welch MJ, Anderson CJ (2000) Bioconjugate Chem 11:527
46. Sun X, Wuest M, Weisman GR, Wong EH, Reed DP, Boswell CA, Motekaitis R, Martell AE, Welch MJ, Anderson CJ (2002) J Med Chem 45:469
47. Boswell CA, Sun X, Niu W, Wong EH, Weisman GR, Rheingold AL, Anderson CJ (2004) J Med Chem 47:1465
48. Sprague JE, Sun X, Chang CH, Weisman GR, Wong EH, Meyer LA, Achilefu S, Anderson CJ (2003) J Labelled Compd Radiopharm 46(suppl 1):S66
49. de Jong M, Kwekkeboom D, Valkema R, Krenning EP (2003) Eur J Nucl Med 30:463
50. Cremonesi M, Ferrari M, Zoboli S, Chinol M, Stabin MG, Orsi F, Maecke HR, Jermann E, Robertson C, Fiorenza M, Tosi G, Paganelli G (1999) Eur J Nucl Med 36:877
51. Förster GJ, Engelbach M, Brockmann J, Reber H, Buchholz H-G, Maecke H, Rosch F, Herzog H, Bartenstein P (2001) Eur J Nucl Med 28:1743
52. Jamar F, Barone R, Mathieu I, Walrand S, Labar M, Carlier P, De Camps J, Schran H, Chen T, Smith MC, Bouterfa H, Valkema R, Krenning EP, Kvols LK, Pauwels S (2003) Eur J Nucl Med 30:510
53. Glaser M, Collingridge DR, Aboagye EO, Bouchier-Hayes L, Hutchinson OC, Martin SJ, Price P, Brady F, Luthra SK (2003) Appl Radiat Isot 58:55
54. Wester H-J, Schottelius M, Scheidhauer K, Reubi J, Wolf AP, Schwaiger M (2002) Eur J Nucl Med 29:28
55. Krenning EP, Kwekkeboom DJ, Bakker WH, Breeman WAP, Kooij PPM, Oei HY, van Hagen M, Postema PTE, de Jong M, Reubi JC, Visser TJ, Reijs AEM, Hofland LJ, Koper JW, Lamberts SWJ (1993) Eur J Nucl Med 20:716
56. O'Connor MK, Kvols LK, Brown ML, Hung JC, Hayostek RJ, Cho DS, Vetter RJ (1992) J Nucl Med 33:1613
57. Bakker WH, Breemann WA, de Jong M, Visser TJ, Krenning EP (1996) Eur J Nucl Med 23:775
58. Coenen HH (1989) In: Baille TA, Jones JR (eds) Synthesis and application of isotopically labelled compounds. Elsevier, Amsterdam, p 433
59. Guhlke S, Wester H-J, Bruns C, Stocklin G (1994) Nucl Med Biol 21:819
60. Wester H-J, Brockmann J, Rosch F, Wutz W, Herzog H, Smith-Jones P, Stolz B, Bruns C, Stocklin G (1997) Nucl Med Biol 24:275
61. Hostetler ED, Edwards WB, Anderson CJ, Welch MJ (1999) J Lab Comp Radiopharm 42(Suppl 1):S720
62. Wester H, Schottelius M, Scheidhauer K, Meisetschläger G, Herz M, Rau F, Reubi J, Schwaiger M (2003) Eur J Nucl Med 30:117
63. Yngve U, Khan TS, Bergstrom M, Langstrom B (2001) J Labelled Compd Radiopharm 44:561
64. Larson SM, Pentlow KS, Volkow ND, Wolf AP, Finn RD, Lambrecht RM, Graham MC, Di RG, Bendriem B, Daghighian F et al (1992) J Nucl Med 33:2020

Top Curr Chem (2005) 252: 193–216
DOI 10.1007/b101229

Radiolabeled Integrin $\alpha_v\beta_3$ Antagonists as Radiopharmaceuticals for Tumor Radiotherapy

Shuang Liu[1] (✉) · Simon P. Robinson[2] · D. Scott Edwards[2]

[1] Department of Industrial and Physical Pharmacy, School of Pharmacy, Purdue University, 575 Stadium Drive Mall, West Lafayette, IN47907-2091, USA
lius@pharmacy.purdue.edu

[2] Bristol-Myers Squibb Medical Imaging, 331 Treble Cove Road N., Billerica, MA01862, USA

Abstract Integrins are a family of proteins that facilitate cellular adhesion to and migration on the extracellular matrix proteins found in intercellular spaces and basement membranes. integrin $\alpha_v\beta_3$ is a receptor for a variety of extracellular matrix proteins with the exposed RGD tripeptide sequence. These include vitronectin, fibronectin, fibrinogen, lamin, collagen, Von Willibrand's factor, osteoponin and adenovirus particles. Integrin $\alpha_v\beta_3$ is generally expressed at low levels on epithelial cells and mature endothelial cells, but it is highly expressed on the surface of both endothelial cells in tumor neovasculature and some tumor cells, including osteosarcomas, neuroblastomas, glioblastomas, melanomas, and carcinomas of the lung, the breast, the prostate, and the bladder. The expression of integrin $\alpha_v\beta_3$ correlates with the tumor progression in melanoma, glioma, and ovarian and breast cancers. The highly restricted expression of integrin $\alpha_v\beta_3$ during tumor invasion and metastasis presents an interesting molecular target for both early detection and treatment of rapidly growing solid tumors. This chapter will discuss some important aspects associated with the radiolabeling of diethylenetriaminepentaacetic acid conjugated and 1,4,7,10-tetraazacyclododecane-*N,N′,N″,N‴*-tetraacetic acid conjugated integrin $\alpha_v\beta_3$ antagonists, and the use of ^{90}Y-labeled integrin $\alpha_v\beta_3$ receptor antagonists as potential therapeutic radiopharmaceuticals for the treatment of solid tumors. It will focus on radiopharmaceutical design, selection of the radionuclide, the bifunctional chelator, and targeting biomolecules, as well as modification of the pharmacokinetics.

Keywords ^{90}Y-labeled integrin $\alpha_v\beta_3$ receptor antagonists · Radiotherapy · Target-specific radiopharmaceuticals

1 Introduction

Therapeutic radiopharmaceuticals are radiolabeled molecules designed to deliver therapeutic doses of ionizing radiation to specific disease sites. The systemic administration of therapeutic radiopharmaceuticals that are designed for specific localization at tumor sites provides the opportunity for the treatment of both localized and disseminated disease [1–12]. Most systemically administered therapeutic radiopharmaceuticals are small organic or inorganic compounds with definite composition. They can also be macromolecules, such as monoclonal antibodies or antibody fragments labeled with a therapeutic radionuclide. Ideally, the radiopharmaceutical should localize at the tumor sites in sufficient concentration to deliver a cytotoxic radiation dose to the tumor cells and clear rapidly from the blood and other normal organs to minimize radiation damage to normal tissues [2, 8].

Since the introduction of monoclonal antibodies, much effort has been made to exploit their specificity [13–20]. It has been the Holy Grail in cancer research using radiolabeled monoclonal antibodies for radioimmunodetection and radioimmunotherapy. In the past decade, antibody and genetic technology advances have resulted not only in novel molecules but also in a dramatic change in the methods for making antibodies on a large scale. In 2002, the FDA approved the first ^{90}Y-labeled anti-CD20 monoclonal antibody (Zevalin, IDEC Pharmaceuticals) for radioimmunotherapy of non-Hodgkin's lymphoma [18–20]. Other radiolabeled antibodies are under clinical investigations for treatment of cancers. These include ^{131}I-labeled anti-CD20 monoclonal antibody (tositumomab, Bexxar Corixa Corporation) for the treatment of non-Hodgkin's lymphoma [21, 22], and radiolabeled anti-prostate-specific membrane antigen antibody (J591) for the treatment of prostate cancer [23–25]. It has been reported that injection of nanocurie levels of ^{225}Ac-labeled J591 into mice bearing prostate carcinoma or disseminated human lymphoma induced significant tumor regression and prolonged survival [23].

The introduction of ^{111}In-DTPA-octreotide, where DTPA is diethylenetriaminepentaacetic acid, for the diagnosis of somatostatin receptor positive tumors has spurred the search for new receptor-based target-specific therapeutic radiopharmaceuticals. A number of radiolabeled small peptides have been studied for their potential use in tumor therapy [26–34]. Several ^{90}Y-labeled somatostatin analogs have been under clinical investigation for their efficacy in the treatment of somatostatin-positive cancers [35–40]. Several excellent reviews have appeared recently covering a broad range of topics related to radiolabeled peptides as diagnostic and therapeutic radiopharmaceuticals [1–12, 41–47]. This chapter will focus on some important issues related to the radiolabeling of integrin $\alpha_v\beta_3$ antagonists (peptides and nonpeptides) and their application as potential therapeutic radiopharmaceuticals for the treatment of rapidly growing solid tumors.

2 Why Integrin $\alpha_v\beta_3$ Targeted Radiopharmaceuticals?

Tumors produce diffusible angiogenic factors, which activate endothelial cells in nearby established capillaries or venules and induce endothelial proliferation, migration, and new vessel formation through a series of sequential but partially overlapping steps. Once vascularized, previous dormant tumors begin to grow rapidly and their volumes increase exponentially. The formation of new blood vessels (angiogenesis) is a requirement for malignant tumor growth and metastasis [48–51]. The angiogenic process depends on vascular endothelial cell migration and invasion, regulated by cell adhesion receptors.

Integrins are a family of proteins that facilitate cellular adhesion to and migration on the extracellular matrix proteins found in intercellular spaces and basement membranes. They also regulate cellular entry and withdraw from the cell cycle. Integrin $\alpha_v\beta_3$ is a receptor for a variety of extracellular matrix proteins with the exposed RGD tripeptide sequence [48–51]. These include vitronectin, fibronectin, fibrinogen, lamin, collagen, Von Willibrand's factor, osteoponin, and adenovirus particles. The expression of integrin $\alpha_v\beta_3$ enables a given cell to adhere to, migrate on or respond to almost any matrix protein it may encounter. Despite its promiscuous behavior, integrin $\alpha_v\beta_3$ is generally expressed at low levels on epithelial cells and mature endothelial cells. It has been reported that integrin $\alpha_v\beta_3$ is highly expressed on the neovasculature of tumors, including osteosarcomas, neuroblastomas, glioblastomas, melanomas, lung carcinomas, and breast, prostate, and bladder cancers [52–57]. A recent study has shown that integrin $\alpha_v\beta_3$ is overexpressed not only on endothelial cells but also on tumor cells in human breast cancer xenografts [58]. The expression of integrin $\alpha_v\beta_3$ correlates with tumor progression in melanoma, glioma, and ovarian and breast cancers [52–58]. The highly restricted expression of integrin $\alpha_v\beta_3$ during tumor invasion and metastasis presents an interesting molecular target for diagnosis and radiotherapy of rapidly growing solid tumors [59–71].

3 Radiopharmaceutical Design

The new radiopharmaceutical targeting integrin $\alpha_v\beta_3$ can be divided into four parts (Fig. 1): an integrin $\alpha_v\beta_3$ targeting moiety, a linker, a bifunctional chelator (BFC), and a radionuclide. The $\alpha_v\beta_3$ targeting moiety serves as the vehicle to carry the radionuclide to the receptor site at the tumor neovasculature. The radionuclide is the radiation source. Between the targeting molecule and the radionuclide is the BFC, which strongly coordinates to the metal ion and is covalently attached to the targeting molecule either directly or through a linker. Selection of a BFC is largely determined by the nature and oxidation state of the metallic radionuclide. The linker can be a simple hydrocarbon chain or a small peptide sequence, which is often used for modification of pharmacokinetics.

3.1 Choice of Radionuclide

Some selected metallic radionuclides useful for radiotherapy are shown in Table 1. In general, identifying the most appropriate isotope for radiotherapy is often a difficult task and requires balancing a variety of factors. These include tumor uptake and retention of the radiolabeled BFC–biomolecule (BM)

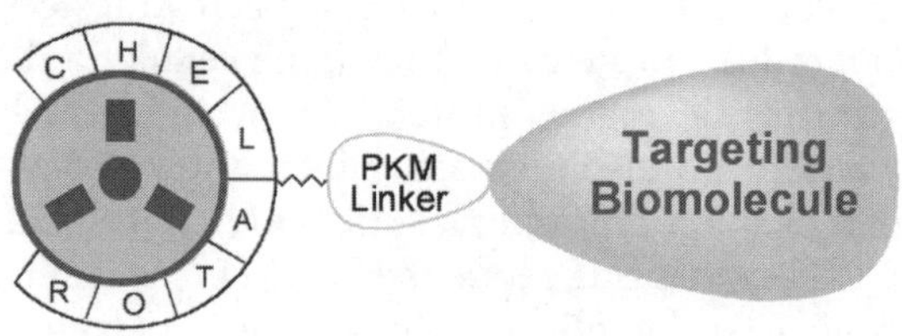

Therapeutic Radionuclide: ^{177}Lu and ^{90}Y

Bifunctional Chelators: DTPA and DOTA

Pharmacokinetic Linker: poly(amino acid)

Targeting Biomolecule: peptides and nonpeptides

Fig. 1 Schematic presentation of the radiopharmaceutical design. The targeting biomolecule serves as the vehicle to carry a radionuclide to the receptor site on tumor cells. The radionuclide is the radiation source. The bifunctional chelator (BFC) is used for radiometal chelation and attachment of the targeting molecule. The linker is often used for modification of pharmacokinetics. Diethylenetriaminepentaacetic acid (DTPA), 1,4,7,10-tetraazacyclododecane-*N*,*N′*,*N″*,*N‴*-tetraacetic acid (DOTA)

Table 1 Metallic radionuclides useful for radiotherapy

Radio-nuclide	Half-life (days)	Maximum energy (MeV)	Maximum range (mm)	γ radiation (keV)	Production source	Specific activity[a]
^{67}Cu	2.58	0.575	1.8	185 (40%)	Accelerator	Low
^{90}Y	2.66	2.27	12.0	–	Generator/reactor	High
^{109}Pd	0.56	1.03	5.0	88 (%)	Reactor	Low
^{111}Ag	7.47	1.05	4.8	–	Reactor	Low
^{149}Pm	2.21	1.07	5.0	286 (3%)	Reactor	Low
^{153}Sm	1.95	0.80	3.0	103 (28%)	Reactor	Low
^{166}Ho	1.1	1.60	8.0	81 (6.3%)	Reactor/generator	High
^{177}Lu	6.7	0.497	1.5	208 (28%)	Reactor	Medium
^{186}Re	3.7	1.02	5.0	137 (9%)	Reactor	Low
^{188}Re	0.71	2.12	11.0	155 (15%)	Reactor	Low
^{212}Bi	1.0 h	7.8 (α)	70 μm	2.25 (β^-, 64%)	Generator	Low

[a] The specific activity of a radionuclide depends on the source and method of production, as well as the technique of separation. In general, generator-produced radionuclides, such as ^{90}Y and ^{188}Re, have a higher specific activity than accelerator- or reactor-produced radioisotopes. High specific activity can also be achieved by chemical purification of the desired radionuclide from the parent element after direct (n,γ)-activation of the target.

conjugate, blood clearance, rate of radiation delivery, half-life and specific activity of the metallic radionuclide, and the feasibility of large-scale production of the radionuclide in an economical fashion. The key point for a target-specific therapeutic radiopharmaceutical is to deliver a tumoricidal dose of radiation to the tumor cells while not causing significant side effects to normal tissues.

Ideally, the physical half-life of the therapeutic radionuclide should match the biological half-life of the radiopharmaceutical at the tumor site. The radionuclide should have a half-life long enough so that a minimal dose rate (more than 0.4 Gy/h) is achieved and all cells are irradiated during the most radiation-sensitive phases of the cell cycle. At the same time, the half-life of the radionuclide has to be long enough to allow adequate time for manufacturing, release, and transportation. If the half-life of the radionuclide is too short, most of the radioactivity decays before it reaches the maximum tumor-to-background ratio. In this situation, attaining a sufficient dose for tumor ablation (more than 50 Gy) will be problematic. On the other hand, too long a half-life would cause unnecessary radiation burden to normal tissues. Other considerations include availability and quality of the radionuclide. The purity has to be sufficient and reproducible, as trace amounts of impurities can affect the radiochemical purity of the radiopharmaceutical. The chosen radionuclide should have high specific activity since the target receptor sites in tumors are often limited in number. The specific activity of a radionuclide depends primarily on the production method. Trace metal contaminants must be minimized as they often compete with the radionuclide for the BFC and their metal complexes compete for receptor binding with the radiolabeled BFC–BM conjugate.

β-Particle emitters have relatively long penetration ranges (2–12 mm in the tissue) depending on the energy level. The long-range penetration is particularly important for solid tumors with high heterogeneity. The β-particle emitters yield a more homogeneous dose distribution even when they are heterogeneously distributed within the target tissue. Among various metallic radionuclides, lanthanide radiometals are of particular interest. There are several lanthanide isotopes to choose, including low-energy β-emitter ^{177}Lu, medium-energy β-emitters ^{149}Pm and ^{153}Sm, and high-energy β-emitters ^{166}Ho and ^{90}Y. Yttrium and lanthanide metals share similar coordination chemistry. The chelator technology is well developed and well understood.

For systemic cancer radiotherapy, ^{90}Y is of particular interest because it is a high-energy pure β-particle emitter. The half-life of 2.7 days is short enough to achieve a critical dose rate and long enough to allow the radiopharmaceutical to be manufactured and delivered for clinical use. For quantitative imaging, the corresponding ^{111}In-BFC–BM complex is often utilized as a surrogate to determine the biodistribution and the dosimetry of ^{90}Y-BFC–BM. Depending on the tumor size and location, the choice of the β-emitter may be different. For example, low-energy β-emitters such as ^{177}Lu are better for smaller metastases, while high-energy β-emitters such as ^{90}Y are used for larger tumors. ^{177}Lu also has a small percentage of γ-emission, which can be used for dosimetry determination and staging of radiotherapy. The ^{90}Y-labeled anti-CD20 monoclonal antibody (Zevalin, IDEC Pharmaceuticals) has recently been approved by the FDA for the treatment of non-Hodgkin's lymphoma [18–20]. Several ^{90}Y-labeled somatostatin analogs are still in clinical trials to study their therapeutic efficacy in the treatment of somatostatin-positive cancers [35–40].

Rhenium has two isotopes (^{186}Re and ^{188}Re) useful for tumor radiotherapy. ^{186}Re has a half-life of 3.68 days with a β-emission (E_{max}=1.07 MeV, 91% abundance) and a γ-photon (E=137 keV, 9% abundance). ^{186}Re is produced by the irradiation of ^{185}Re with neutrons [^{185}Re(n,γ)^{186}Re]. The yield of ^{186}Re depends on the amount of ^{185}Re in the target, the energy of neutrons available, and the neutron reflux. The specific activity is from low to medium. ^{188}Re has a half-life of 16.98 h with a high-energy β-emission (E_{max}=2.12 MeV, 85% abundance) and 155-keV γ-photons (15% abundance). ^{188}Re can be prepared either from the nuclear reaction ^{187}Re(n,γ)^{188}Re or from the ^{188}W–^{186}Re generator. The generator-produced ^{188}Re is carrier-free and has very high specific activity. The ^{188}W–^{186}Re generator also has a very long useful shelf-life, and is inexpensive and readily available. The 15% 155-keV γ-photons also allow imaging for dosimetry determination.

^{212}Bi is a generator-produced α-emitter with a half-life of 1 h. The average α-particle energy is 7.8 MeV with a penetration range of 50–90 μm in tissue. The daughter isotope of ^{212}Bi is ^{208}Tl, which emits a 2.6-MeV γ-ray and along with other medium-to high-energy γ-emissions. This will result in a very high radiation dose to the patient. The half-life of 1 h for ^{212}Bi is probably too short for the receptor-based therapeutic radiopharmaceutical to reach maximum concentration at the tumor sites and deposit sufficient radiation dose to tumor cells.

3.2 Bifunctional Chelators

The use of metallic radionuclides offers many opportunities for designing new radiopharmaceuticals by modifying the coordination environment around the metal with a variety of chelators. The coordination chemistry of the metallic radionuclide will determine the geometry and solution stability of the metal chelate. Different metallic radionuclides have different coordination chemistries, and require BFCs with different donor atoms and ligand frameworks. The metal chelate can significantly impact the tumor uptake and biodistribution of radiolabeled small BMs owing to the fact that in many cases the metal chelate contributes greatly to the overall size of the radiopharmaceutical. Therefore, selection of the BFC is very important for the development of a clinically useful therapeutic radiopharmaceutical. In this chapter, we will focus only on DTPA- and 1,4,7,10-tetraazacyclododecane-*N*,*N*′,*N*″,*N*‴-tetraacetic acid (DOTA)-based BFCs useful for radiolabeling of BMs with yttrium and lanthanide isotopes.

Many acyclic and macrocyclic BFCs have been used for radiolabeling of antibodies and small BMs with yttrium and lanthanide isotopes [72–80]. BFCs for yttrium and lanthanide radiopharmaceuticals have recently been reviewed [8]. The structures of selected acyclic and macrocyclic BFCs, some of which are commercially available from Macrocyclics, Richardson, TX, are shown in Fig. 2. A major advantage of using DTPA analogs as BFCs is their extremely high radiolabeling efficiency (fast and high-yield radiolabeling) under mild conditions; but the kinetic lability of their metal chelates often results in dissociation of the radiometal from the metal chelate, and leads to radiation toxicity to non-target organs such as bone marrow. Radiolabeling kinetics of DOTA-based BFCs is usually slow, and largely depends on the radiolabeling conditions [8], including the DOTA-conjugate concentration, pH, reaction temperature and heating time, buffer agent, and presence of other contaminant metal ions. In the end, the choice of the BFC will be based on the BM to be labeled and the pharmacokinetics of the radiolabeled BM. There is always a balance between the high radiolabeling efficiency of DTPA derivatives and the high kinetic inertness of DOTA analogs.

DOTA is of particular interest as a BFC for radiolabeling of small BMs with yttrium and lanthanide isotopes. The macrocyclic framework is well organized so that it forms yttrium and indium complexes with high thermodynamic stability and kinetic inertness. The pK_a values of the carboxylic groups are in the range 2–5. Lower pK_a values result in less competition from protons, high stability of the metal complex, and minimum acid-assisted demetallation, even

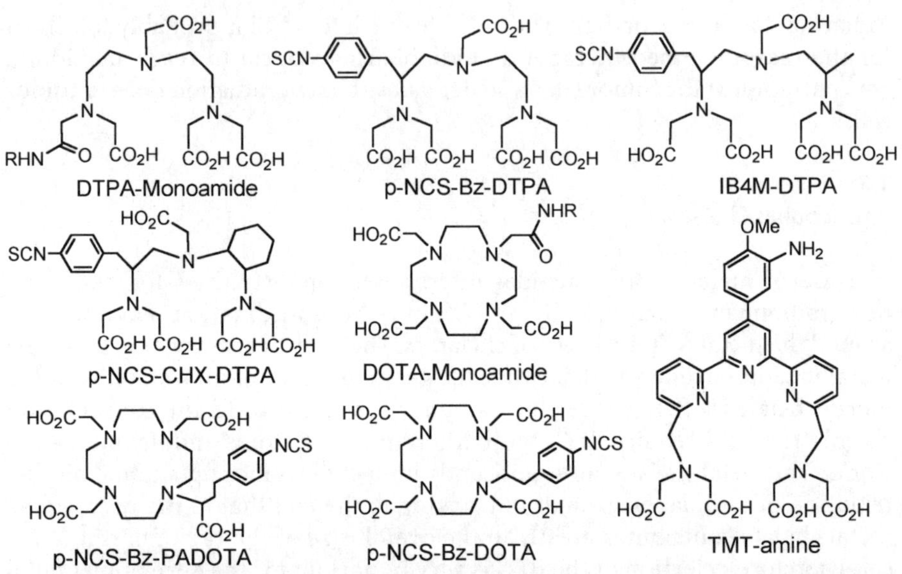

Fig. 2 Structures of selected acyclic and macrocyclic BFCs useful for the radiometal labeling of biomolecules, including antibodies, antibody fragments, small peptides, and nonpeptide receptor ligands

under acidic (pH≤2) conditions. Conjugation of BMs can be easily achieved using one of the carboxylic groups. The acetate groups attached to the nitrogen donor atoms have low molecular weight, and are very hydrophilic; therefore, the contribution of the BFC to the overall molecular weight of the DOTA–BM conjugate can be minimized. The high hydrophilicity improves the pharmacokinetics of the radiopharmaceutical. High radiolabeling efficiency can be achieved at elevated temperatures without causing any significant effects on the chemical and biological characteristics of the radiolabeled DOTA–BM conjugates.

3.3 Targeting Biomolecule

The targeting BMs can be small peptides or nonpeptide heterocycles. Many peptide and nonpeptide integrin $\alpha_v\beta_3$ receptor antagonists have been studied for their potential use as therapeutic drugs for the treatment of cancer [81–93], some of which have demonstrated very high binding affinity and selectivity for integrin $\alpha_v\beta_3$, and have been shown to inhibit neovascularization, tumor-induced angiogenesis, and tumor growth.

Selected examples of peptide and nonpeptide leads that have high affinity for integrin $\alpha_v\beta_3$ with IC_{50} values in the nanomolar or subnanomolar range, and are potential candidates as targeting BMs to carry the radionuclide to tumors,

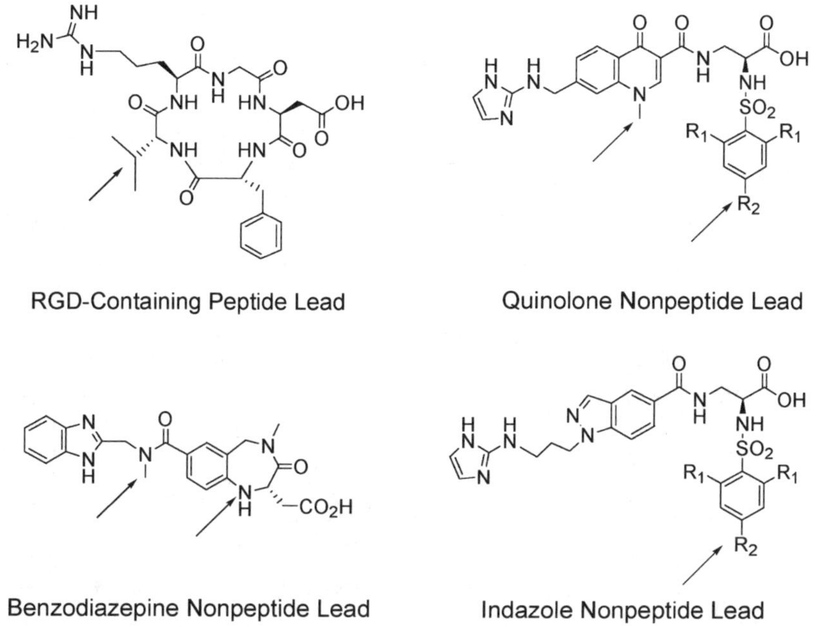

Fig. 3 Selected examples of peptide and nonpeptide leads. *Arrows* indicate possible sites for modification and attachment of BFCs

are shown in Fig. 3. Arrows indicate possible sites for conjugation of the radiometal chelate. Examples of DTPA- and DOTA-conjugated integrin $\alpha_v\beta_3$ receptor antagonists are shown in Figs. 4 and 5. In general, the selection of a specific targeting BM depends largely on the receptor-binding affinity and selectivity for integrin $\alpha_v\beta_3$ over GPIIb/IIIa. The selected integrin $\alpha_v\beta_3$ receptor antagonists should have high receptor binding affinity with an IC_{50} value in the nanomolar range, and high selectivity for integrin $\alpha_v\beta_3$ with a high IC_{50} $\alpha_v\beta_3$-to-GPIIb/IIIa ratio. Attachment of the DTPA or DOTA chelator should not interfere with the receptor binding of the selected BMs.

4
^{90}Y- and ^{111}In-Labeling of DTPA-Conjugated Integrin $\alpha_v\beta_3$ Receptor Antagonists

Recently, Liu et al. [94] reported the synthesis of DTPA–c(RGDfK) (Fig. 4: SQ169) and DTPA[c(RGDfK)]$_2$ (Fig. 4: SQ170), and their ^{90}Y complexes (^{90}Y-SQ169 and ^{90}Y-SQ170). ^{90}Y-SQ169 and ^{90}Y-SQ170 can be prepared in high yield (more than 95%) by reacting 2 μg SQ169 and SQ170, respectively, with 20 mCi $^{90}YCl_3$ (SQ169:Y ratio approximately 4:1) in a 0.5 M ammonium acetate buffer (pH 6.5–8.0) at room temperature. In both cases, the ^{90}Y chelation was instan-

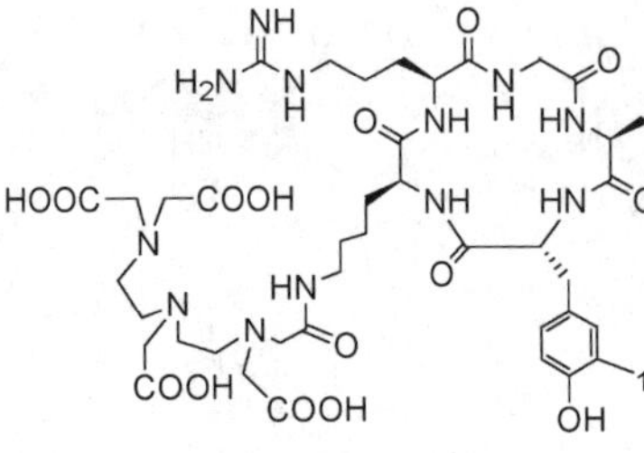

DTPA-D-[^{125}I]Tyr^4-c(RGDyK)

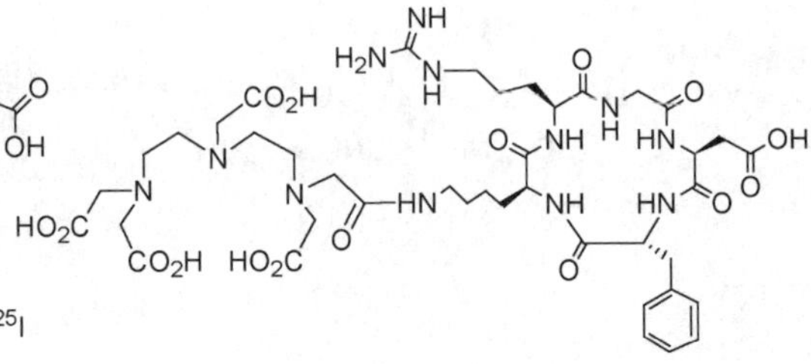

SQ169: DTPA-c(RGDfK)

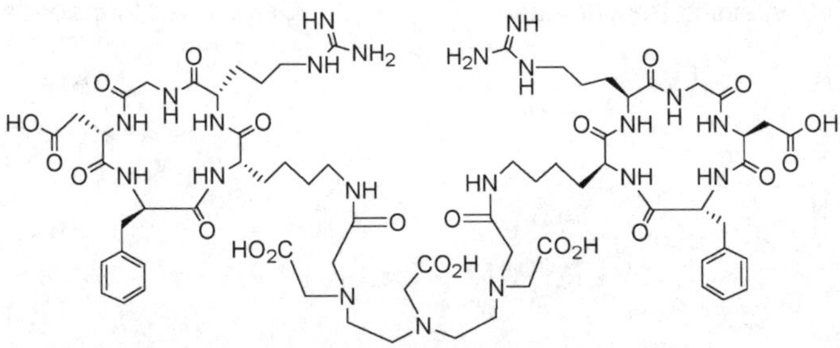

SQ170: DTPA-[c(RGDfK)]$_2$

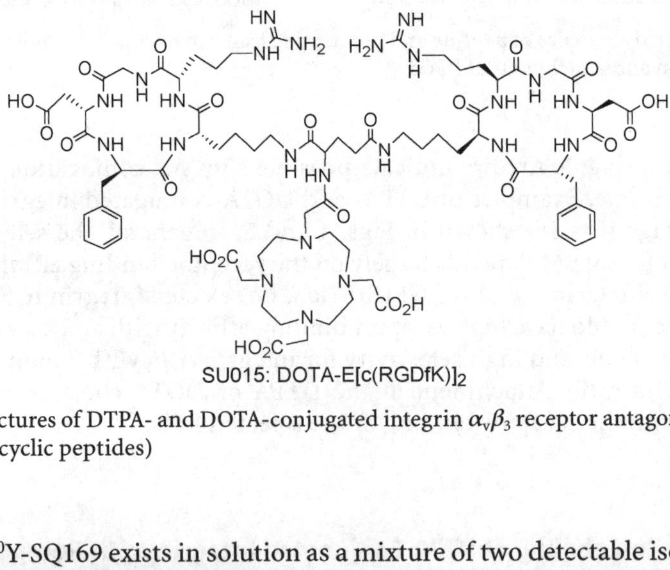

SU015: DOTA-E[c(RGDfK)]$_2$

Fig. 4 Structures of DTPA- and DOTA-conjugated integrin $\alpha_v\beta_3$ receptor antagonists (RGD-containing cyclic peptides)

taneous. ^{90}Y-SQ169 exists in solution as a mixture of two detectable isomers (cis and trans isomers), which slowly interconvert at room temperature. The interconversion of different isomers in ^{90}Y-SQ169 involves exchange of "wrapping isomers" via the "wagging" of the diethylenetriamine backbone, "shuffling" of the two NO_2 donor sets, and inversion at the terminal amine-nitrogen atom. For ^{90}Y-SQ170, the interconversion of different isomers is much faster owing to the weak bonding of two carbonyl-oxygen donors [94]; therefore, ^{90}Y-SQ170 shows only one broad radiometric peak in its high-performance liquid chromatography (HPLC) chromatogram.

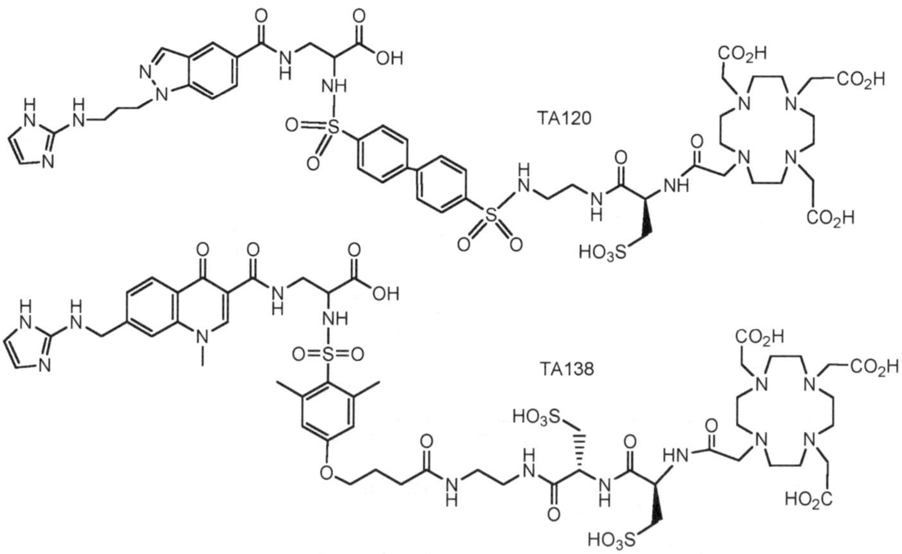

Fig. 5 Structures of DOTA-conjugated integrin $\alpha_v\beta_3$ receptor antagonists (nonpeptides)

^{111}In-SQ169 and ^{111}In-SQ170 can be prepared in high yield by reacting SQ169 and SQ170, respectively, with $^{111}InCl_3$ in the 0.5 M ammonium acetate buffer (pH 6.0–7.0) at room temperature [95]. The ^{111}In chelation was almost instantaneous for both SQ169 and SQ170. ^{111}In-SQ169 and ^{111}In-SQ170 are stable in solution for more than 24 h without significant decomposition. It is surprising to find that ^{111}In-SQ169 and ^{111}In-SQ170 are more hydrophilic than their corresponding ^{90}Y analogs (Fig. 6), suggesting a different coordination sphere in ^{111}In and ^{90}Y complexes of the same DTPA–BM conjugate. By a reversed-phase HPLC method, both ^{111}In-SQ169 and ^{111}In-SQ170 showed only one radiometric peak in their respective HPLC chromatogram owing to the rapid interconversion of different isomers [95].

5 ^{90}Y- and ^{111}In-Labeling of DOTA-Conjugated Integrin $\alpha_v\beta_3$ Receptor Antagonists

SU015 (Fig. 4) is a DOTA-conjugated cyclic peptide dimer [96, 97]. ^{90}Y-SU015 can be prepared by reacting SU015 with $^{90}YCl_3$ in ammonium acetate buffer (pH 5.0–7.5). There are many factors influencing the rate of ^{90}Y chelation and the radiolabeling efficiency of SU015. These include the pH, reaction temperature and heating time, as well as the presence of trace metal contaminants, such as Ca^{2+}, Fe^{3+}, and Zn^{2+}. The chelation of ^{90}Y by SU015 is slow, so heating at elevated temperatures (50–100 °C) is needed to complete the ^{90}Y-labeling.

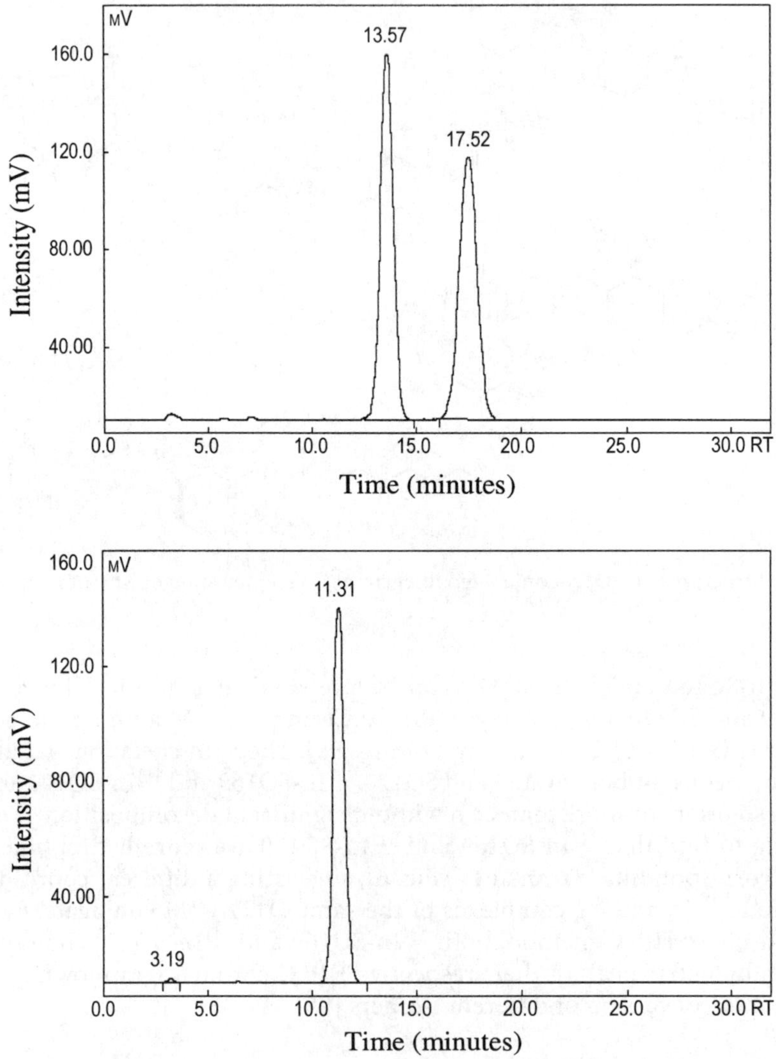

Fig. 6 Typical radio-high-performance liquid chromatography (*HPLC*) chromatograms for ^{90}Y-SQ169 (*top*) and ^{111}In-SQ169 (*bottom*)

Under optimized radiolabeling conditions (pH 7.2–7.8 and heating at 80–100 °C for 5–10 min), the minimum amount of SU015 required to achieve 95% radiochemical purity for ^{90}Y-SU015 is around 25 µg for 20 mCi $^{90}YCl_3$ corresponding to a SU015-to-^{90}Y ratio of around 30:1 [96].

Radiopharmaceuticals comprising β-emitting radionuclides may undergo autoradiolysis during preparation and storage. Radiolysis is caused by the formation of free radicals such as hydroxyl and superoxide radicals [98], and is dependent on the concentration of the radionuclide and oxygen dissolved in the reaction mixture. To prevent radiolysis, a radiolytic stabilizer is often used during or/and after the radiolabeling [99–101]. However, the use of the stabilizer cannot totally eliminate the oxygen dissolved in the reaction mixture. The combination of oxygen and the high-energy β-particles may still result in formation of a large number of superoxide radicals, which are very reactive toward organic molecules. In this situation, anaerobic conditions might be needed for successful radiolabeling [102].

Different BMs have different sensitivity toward radiolysis. For example, SU015 (Fig. 4) is a DOTA-conjugated cyclic peptide and its ^{90}Y complex (^{90}Y-SU015) can be prepared in high yield in the presence of oxygen [96]. TA138 (Fig. 5) is a DOTA-conjugated nonpeptide integrin $\alpha_v\beta_3$ receptor antagonist, and is very sensitive to radiolysis. ^{90}Y-TA138 has to be prepared under anaerobic conditions. Addition of a stabilizer (e.g., sodium gentisate) in the HPLC autosampler vial is also needed to maintain the solution stability of ^{90}Y-TA138 [102].

6
Differences Between ^{90}Y- and ^{111}In-Labeled DTPA and DOTA Bioconjugates

6.1
Structural Differences

Y^{3+} and In^{3+} are trivalent metal cations. The main difference is their size. As a result, Y^{3+} and In^{3+} often have different coordination chemistry, particularly with a polydentate chelator such as DTPA and DOTA derivatives. For example, Y^{3+} has an ionic radius of 0.87 Å [103], and fits perfectly into the coordination cavity of the DOTA-monoamide chelator. Most yttrium complexes of DOTA derivatives are able to maintain their rigid eight-coordinated structure in solution [104–106]. In^{3+} has an ionic radius of 0.75 Å, which is smaller than that of Y^{3+} [103]. The coordination number for In^{3+} is typically 6 or 7 [107–110]. Only a few eight-coordinated In^{3+} complexes are known [111–114]. Owing to its smaller size, In^{3+} does not fit into the coordination cavity of the DOTA-monoamide. Although In^{3+} is shown to be eight-coordinated in the solid state of In(DOTA-monoamide) [114, 115], the carbonyl oxygen may become dissociated in solution [116]. This is particularly true for ^{111}In-labeled DOTA conjugates at the tracer level (10^{-9}–10^{-8} M).

6.2
Difference in Lipophilicity and Solution Properties

The difference in coordination chemistry is also reflected by the difference in the lipophilicity and solution properties. For example, ^{90}Y-SQ169 exists in solution as a mixture of two detectable isomers (cis and trans isomers) as evi-

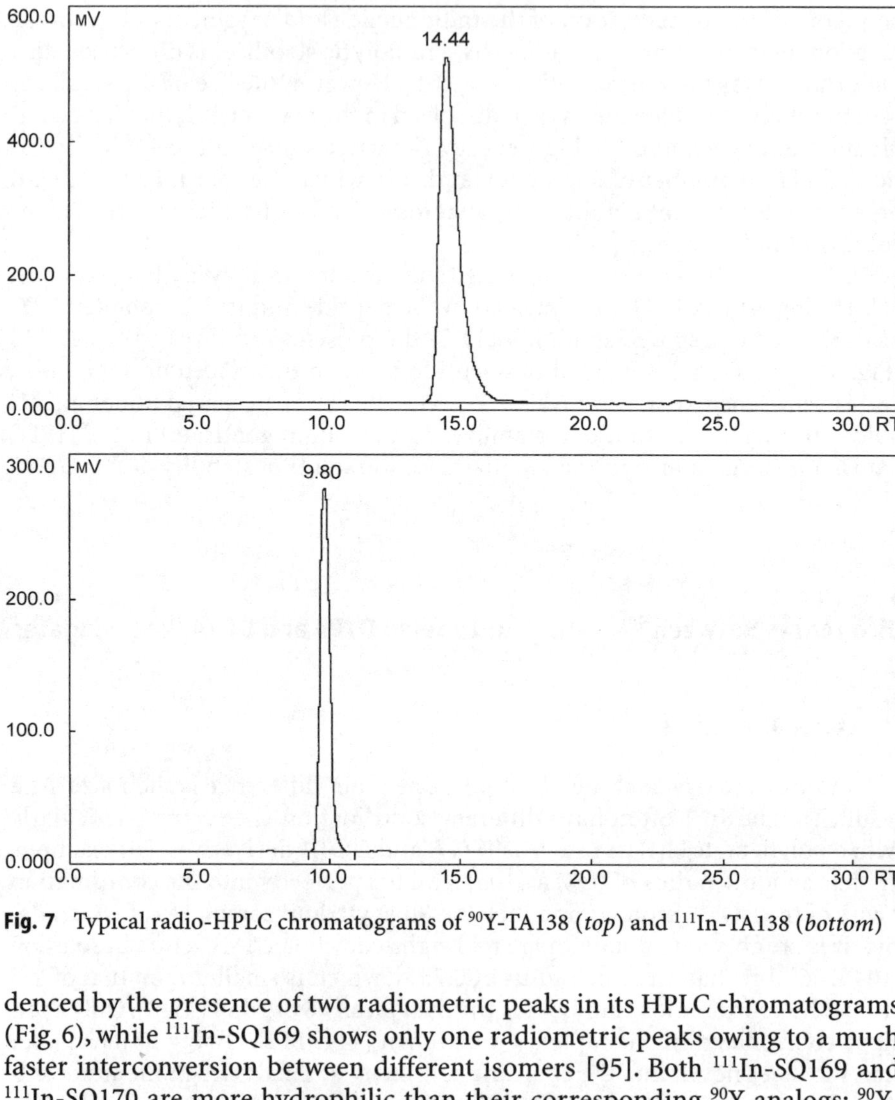

Fig. 7 Typical radio-HPLC chromatograms of ^{90}Y-TA138 (*top*) and ^{111}In-TA138 (*bottom*)

denced by the presence of two radiometric peaks in its HPLC chromatograms (Fig. 6), while ^{111}In-SQ169 shows only one radiometric peaks owing to a much faster interconversion between different isomers [95]. Both ^{111}In-SQ169 and ^{111}In-SQ170 are more hydrophilic than their corresponding ^{90}Y analogs: ^{90}Y-SQ169 and ^{90}Y-SQ170 [94, 95]. Similar results were also observed between ^{111}In-TA138 and ^{90}Y-TA138 (Fig. 7).

7 Bioequivalence Between ^{111}In-TA138 and ^{90}Y-TA138

While ^{90}Y-labeled DOTA bioconjugates are used for tumor radiotherapy, ^{111}In-labeled DOTA bioconjugates are often used as imaging surrogates for biodistribution and dosimetry determination. This is largely due to the fact that ^{90}Y is a pure β-emitter. As discussed previously, ^{90}Y-labeled DTPA and DOTA bioconjugates often show a slight difference in their lipophilicity and solution properties compared with their corresponding ^{111}In analogs owing to different coordination chemistry. The most critical question is whether ^{111}In- and ^{90}Y-labeled bioconjugate are biologically equivalent. This question has to be answered in order to use the ^{111}In-labeled bioconjugate for accurate dosimetry estimation of the corresponding ^{90}Y analog.

Recently, Onthank et al. [117] reported the biodistribution data of ^{111}In-TA138 and ^{90}Y-TA138 in the c-neu Oncomouse model. Despite their differences in lipophilicity and solution structure, the biodistribution data clearly demonstrated that ^{111}In-TA138 and ^{90}Y-TA138 are biologically equivalent with respect to their uptake in tumors and other major organs [117]. Therefore, ^{111}In-TA138 is useful as an imaging surrogate for ^{90}Y-TA138, and should be able to predict the radiation dosimetry of ^{90}Y-TA138, a therapeutic radiopharmaceutical for treatment of rapidly growing tumors.

It should be noted that the HPLC retention time largely depends on the chromatographic conditions, and does not represent the "absolute" value of the hydrophilicity for a specific radiolabeled compound. The hydrophilicity difference between ^{111}In- and ^{90}Y-labeled DOTA conjugates also depends on the relative size of the BM. For large BMs such as antibodies, the slight difference in the radiometal chelate may not cause any significant difference in physical and biological properties of ^{90}Y- and ^{111}In-labeled DOTA conjugates. For small BMs such as peptides, the slight difference in their coordination chemistry may cause significant differences in biological properties of ^{90}Y- and ^{111}In-labeled bioconjugates. Therefore, it is critical to prove that the ^{111}In-labeled DOTA conjugate is biologically equivalent to the corresponding ^{90}Y analog in the same animal model.

8 Radiolabeled Integrin $\alpha_v\beta_3$ Receptor Antagonists for Tumor Radiotherapy

8.1 Radiolabeled RGD-Containing Cyclic Peptides

Peptides are necessary elements in more fundamental biological processes than any other class of molecule, and in many cases the affinities of small peptides for their receptors are significantly greater than those of antibodies or their fragments. They can also tolerate harsher chemical conditions for modification or radiolabeling. Small peptides are easy to synthesize and modify, less likely to be immunogenic, and can have rapid blood clearance. The faster blood clearance often results in better tumor-to-background ratios. In most cases, the primary sites of interactions of the peptides are specific receptors on the surface of the cell membrane (extracellular). All these factors make small peptides excellent candidates for the development of target-specific radiopharmaceuticals.

A dual-isotope (^{125}I and ^{111}In) labeled cyclic peptide conjugate, DTPA–D-[^{125}I]Tyr^4–c(RGDyK) (Fig. 4), was recently reported [118]. In autoradiolography and immunohistochemistry studies, the ^{125}I-labeled analog appeared to bind specifically and with high affinity to integrin $\alpha_v\beta_3$ on neovascular blood vessel sections of different major human cancers, particularly breast and prostate cancers, which overexpress integrin $\alpha_v\beta_3$ receptors. The ^{125}I-labeled analog was found to undergo internalization in human carcinoid Bon cells and rat pancreatic CA20948 tumor cells [118]. The internalization is receptor-specific, and time- and temperature-dependent. The tumor uptake is dependent on the injected dose of unlabeled DTPA conjugate, suggesting that the uptake in the tumor is indeed due to the integrin $\alpha_v\beta_3$ receptor binding. Although the authors proposed that ^{90}Y-DTPA–D-[^{125}I]Tyr^4–c(RGDyK) might be useful for tumor therapy [118], the biodistribution data of ^{111}In-DTPA–D-[^{125}I]Tyr^4–c(RGDyK) showed very low tumor uptake (less than 0.5% ID/g at 24 h post injection).

Recently, Rajopadyhe and coworkers at Bristol-Myers Squibb Medical Imaging reported the synthesis of 6-hydrazinonicotinamide conjugates of a series of RGD-containing cyclic peptides [119–121], which are useful for the development of diagnostic ^{99m}Tc radiopharmaceuticals. The same group also reported the synthesis of a DOTA conjugate (Fig. 4: SU015) and its ^{90}Y and ^{111}In complexes (^{90}Y-SU015 and ^{111}In-SU015) [96, 120]. In the enzyme-linked immunosorbent assay, the DOTA conjugate (SU015) and its indium and yttrium complexes show very high binding affinity, with a IC_{50} value in the nanomolar range, and good selectivity for integrin $\alpha_v\beta_3$. It was found that the attachment of the metal chelate has no significant impact on the receptor binding [120]. In the c-neu Oncomouse tumor model, ^{111}In-SU015 shows a relatively high tumor uptake and long tumor retention time (around 3.0% ID/g at 2 h and 1.5% ID/g at 24 h post injection) [120]. Biodistribution data in the c-neu Oncomouse model show that ^{90}Y-SU015 and ^{111}In-SU015 are biologically equivalent. ^{90}Y-SU015 has been evaluated as a therapeutic radiopharmaceutical in several experimental tumor models, including c-neu Oncomouse, HCT116, and HT460 xenografts [122]. In all cases, the tumor growth is significantly inhibited in mice treated with ^{90}Y-SU015 (2 mCi/kg) on days 7 and 11 after implantation of tumor cells in mice (Fig. 8).

In spite of these promising results, the tumor uptake of ^{90}Y-SU015 is still relatively low (around 1.5% ID/g at 24 h post injection). The biodistribution data in the c-neu Oncomouse model also show that a small part of ^{90}Y-SU015 is excreted via the hepatobiliary route. The radioactivity levels in blood, kidney, and liver are relatively high, which makes ^{90}Y-SU015 less than ideal as a new therapeutic radiopharmaceutical for the treatment of solid tumors.

8.2 Radiolabeled Nonpeptide Antagonists

In order to increase the tumor uptake and improve the pharmacokinetics at the same time, Harris and coworkers [123, 124] recently reported a series of DOTA

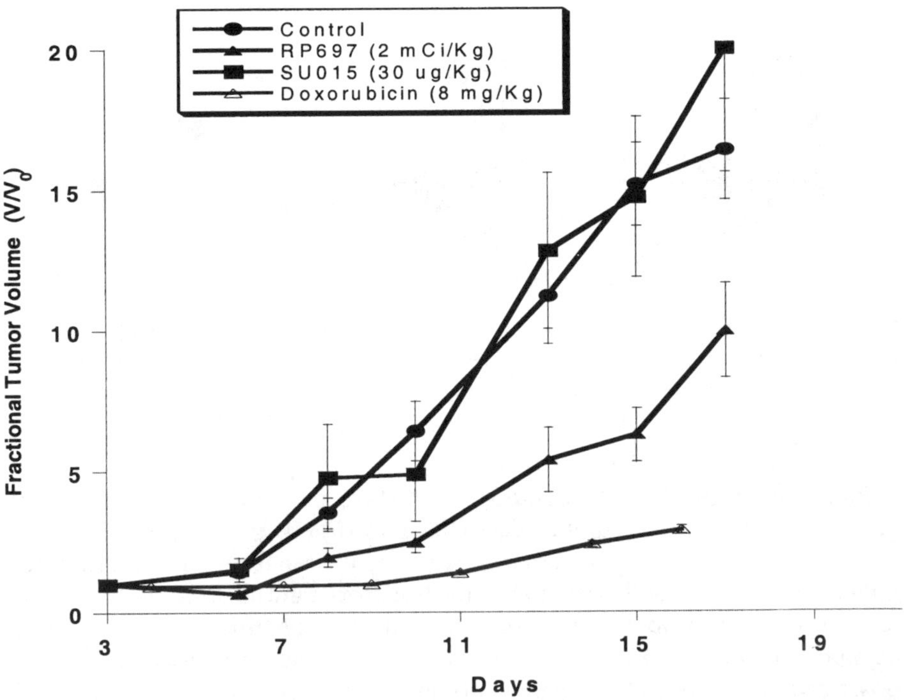

Fig. 8 Tumor growth is significantly inhibited in mice bearing HCT116 xenograft tumors. The mice were treated with ^{90}Y-TA138 (2 mCi/kg) and Doxorubicin (8 mg/kg) on days 7 and 11 after implantation of tumor cells in the mice

conjugates of indazole- and quinolone-based integrin $\alpha_v\beta_3$ receptor antagonists and potential application of their ^{90}Y complexes for tumor radiotherapy. Two examples (TA120 and TA138) of DOTA-nonpeptide conjugates are shown in Fig. 5. TA120 is a DOTA conjugate derived from the indazole scaffold. Although ^{111}In-TA120 demonstrated excellent tumor uptake (12.9% ID/g at 2 h post injection, and 9.2% ID/g at 24 h post injection) in the c-neu Oncomouse model, the kidney uptake was still considered too high (~around 7.5% ID/g at 24 h post injection) for an ideal therapeutic radiopharmaceutical [123]. The uptake in other major organs was lower than that of ^{111}In-SU015 (cyclic peptide dimer). Since TA120 has a limited selectivity for integrin $\alpha_v\beta_3$ over GPIIb/IIIa, ^{90}Y-TA120 is not an ideal agent for radiotherapy.

TA138 (Fig. 5) is a quinolone-based nonpeptide integrin $\alpha_v\beta_3$ receptor antagonist, and has very high binding affinity and specificity for the integrin $\alpha_v\beta_3$ receptor [124]. DOTA is chosen for ^{111}In- and ^{90}Y-chelation owing its ability to form indium and yttrium complexes with high thermodynamic stability and kinetic inertness. The dicysteic acid linker was selected as a pharmacokinetic modifier to increase the hydrophilicity and to improve the renal clearance of the ^{111}In- and ^{90}Y-labeled DOTA-conjugates [124]. Selected images of

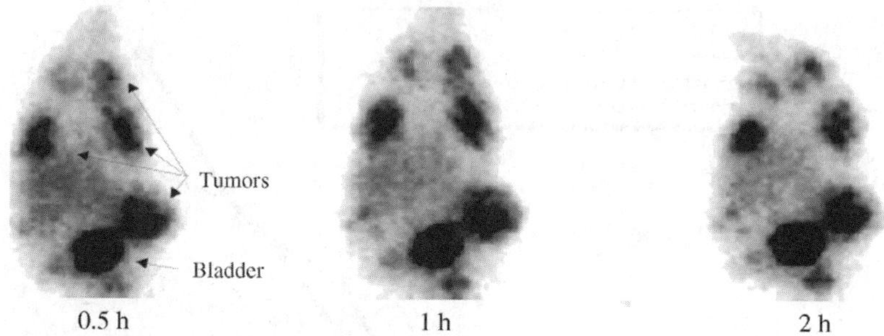

Fig. 9 Representative images of ^{111}In-TA138 at 0.5, 1, and 2 h post injection in the c-neu Oncomouse model. *Arrows* indicate the presence of radioactivity in tumors or the bladder. Theimages have not been filtered

^{111}In-TA138 in the c-neu Oncomouse model are shown in Fig. 9. ^{111}In-TA138 demonstrated very high tumor uptake (around 10% ID/g at 2 h post injection). ^{111}In-TA138 exhibited rapid blood clearance via the renal system [117, 124]. The uptake in the eye and bile were below the limit for quantification by 2 h post injection. The tumor uptake was 3–4 % ID/g at 24 h post injection, and 2–3% at 48 h post injection. HPLC analysis of blood and urine samples at 2 h and 24 h post injection showed that there was no significant metabolism for ^{111}In-TA138.

The therapeutic efficacy of ^{90}Y-TA138 was evaluated in the c-neu Oncomouse model (Fig. 10). Mice with a tumor volume of 318±70 mm^3 were administered with ^{90}Y-TA138 by a single bolus intravenous injection on days 1, 8, and 15. All animals treated with ^{90}Y-TA138 showed positive therapeutic responses. At the lowest dose (15 mCi/m^2), the fractional tumor volume (V/V_o) on day 21 was 2.44 compared with 3.77 for the negative control, representing a tumor growth inhibition of 36%. At the highest dose (90 mCi/m^2), V/V_o was 0.16, corresponding to 96% tumor growth inhibition. There were no significant changes (above 10%) in body weights when comparisons were made with the control. ^{90}Y-TA138 has been selected as a clinical candidate for radiotherapy of rapidly growing tumors.

9 Conclusions

Radiolabeled integrin $\alpha_v\beta_3$ receptor antagonists represent a new class of target-specific radiopharmaceuticals for the treatment of solid tumors. Since integrin $\alpha_v\beta_3$ is highly expressed in rapidly growing tumors and during tumor invasion and metastasis, radiopharmaceuticals targeting integrin $\alpha_v\beta_3$ are most likely to be more specific for rapidly growing and metastatic tumors. There are many factors influencing the tumor uptake and tumor retention of a specific radiotracer. These include the receptor population on endothelial and tumor cells,

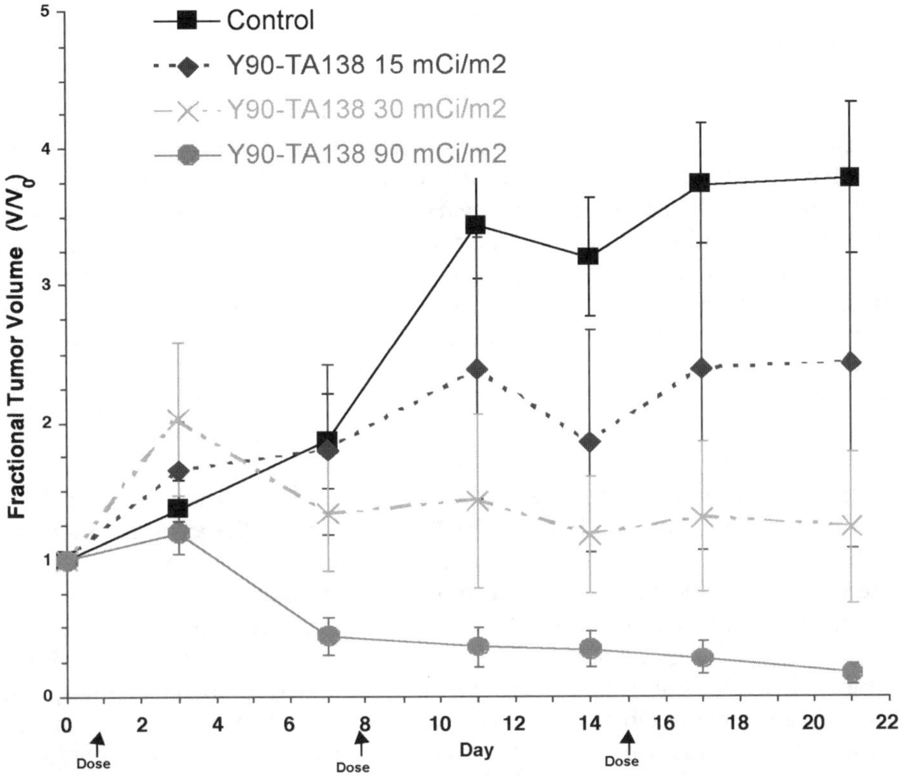

Fig. 10 Dose escalation (15–90 mCi/m^2) study to demonstrate the therapeutic efficacy. The tumor growth is significantly inhibited in mice (c-neu Oncomouse model) treated with ^{90}Y-TA138 on days 1, 8, and 15

receptor binding affinity, lipophilicity, protein binding, receptor binding and dissociation kinetics, and excretion route. Preclinical studies have demonstrated that radiolabeled integrin $\alpha_v\beta_3$ receptor antagonists (peptides and non-peptides) can be used as radiopharmaceuticals for imaging and radiotherapy of solid tumors in experimental animals known to express integrin $\alpha_v\beta_3$ receptors.

However, many questions remain unanswered. In the earlier literature, the overwhelming emphasis was focused on the impact of integrin $\alpha_v\beta_3$ binding affinity on the tumor uptake. Very little attention was paid to the improvement of receptor binding kinetics (association and dissociation rate) and the impact of receptor binding kinetics on the tumor uptake and retention time. There is also very limited information available to link the lipophilicity and protein binding characteristics of radiolabeled integrin $\alpha_v\beta_3$ receptor antagonists directly to their biodistribution and pharmacokinetic properties. While imaging is focused on improvement of the tumor-to-background ratio, radiotherapy relies largely on the tumor uptake and retention time of radiolabeled integrin $\alpha_v\beta_3$ receptor antagonists. An ideal therapeutic radiopharmaceutical targeting integrin $\alpha_v\beta_3$ should be able to selectively accumulate in tumors and have a long tumor retention time in order to achieve the tumoricidal effect without causing significant side effects to normal tissues.

Development of appropriate animal tumor models remains a significant challenge for the preclinical evaluation of radiolabeled integrin $\alpha_v\beta_3$ receptor antagonists. In the earlier literature, various animal models, including tumor-bearing mice and rats, have been used to assess the localization and retention of radiolabeled integrin $\alpha_v\beta_3$ receptor antagonists in tumors. There is often limited information available to demonstrate if integrin $\alpha_v\beta_3$ receptors are overexpressed, what the integrin $\alpha_v\beta_3$ receptor population on endothelial cells and/or tumor cells is, and how the conditions in animals are correlated to those during tumor growth and invasion in humans. Without the proper validation of animal tumor models, it would be very difficult to show if the localization and retention of radiolabeled integrin $\alpha_v\beta_3$ receptor antagonists in tumors are indeed based on specific receptor binding. More importantly, it will be extremely difficult to predict if the radiolabeled integrin $\alpha_v\beta_3$ receptor antagonist will work in humans even though the selected radiopharmaceutical may show significant tumoricidal effects in certain animal models.

Biological evaluation of radiolabeled integrin $\alpha_v\beta_3$ receptor antagonists as therapeutic radiopharmaceuticals is a long and time-consuming process. It requires close collaboration among chemists, biologists, pharmacologists, and oncology clinicians. One of the most critical steps in the whole radiopharmaceutical development process is to screen several different classes of radiolabeled integrin $\alpha_v\beta_3$ receptor antagonists for their tumor uptake, tumor retention time, and pharmacokinetics (excretion route and the rate of excretion preferably via the renal system) in a systematic fashion. Special effort should be made to avoid using the "mix-and-shoot" approach. Without a good understanding of the fundamentals of chemistry, radiochemistry, and biology, it would be extremely difficult to select the most suitable agents for tumor radiotherapy. Once the promising candidates have been identified, they should be evaluated in several different tumor-bearing animal models. Since these agents will be used in patients with rapidly growing and metastatic tumors, the ideal tumor-bearing model would be that which is closest to conditions during tumor growth and metastasis in humans.

Another challenge is the time at which the integrin $\alpha_v\beta_3$-targeted radiopharmaceutical will be used for cancer treatment. Since integrin $\alpha_v\beta_3$ is highly expressed in rapidly growing tumors and during tumor invasion and metastasis, radiopharmaceuticals targeting integrin $\alpha_v\beta_3$ are most likely to be more effective for rapidly growing and metastatic tumors. For slow-growing tumors or tumors with large necrotic cores, these radiopharmaceuticals may not be effective. Therefore, there is an urgent need to develop the most appropriate clinical protocols targeting patients with rapidly growing or metastatic tumors in order to achieve therapeutic efficacy. One can always hope that one day integrin $\alpha_v\beta_3$-targeted therapeutic radiopharmaceuticals will find wide use for the treatment of cancers even though they may not be the "magic bullet". Even if it may not be a single agent, a cocktail of compounds (radiotherapy and chemotherapy) representing the Holy Grail of cancer research may be found in the foreseeable future.

References

1. Schubiger PA, Alberto R, Smith A (1996) Bioconjugate Chem 7:165
2. Volkert WA, Hoffman TJ (1999) Chem Rev 99:2269
3. Van Eijck CHJ, de Jong M, Breeman WAP, Slooter GD, Marquet RL, Krenning EP (1999) Ann Oncol 1999 10(Suppl):S177
4. Heeg MJ, Jurisson S (1999) Acc Chem Res 32:1053
5. Langer M, Beck-Sichinger AG (2000) Curr Med Chem Anti-Cancer Agents 1:71
6. Heppeler H, Froidevaux S, Eberle AN, Maecke HR (2000) Curr Pharm Des 6:971
7. Boerman OC, Oyen WJG, Corstens FHM (2000) Semin Nucl Med 30:195
8. Liu S, Edwards DS (2001) Bioconjugate Chem 12:7
9. Hoffman TJ, Quinn TP, Volkert WA (2001) Nucl Med Biol 28:527
10. Liu S, Edwards DS (2002) Top Current Chem 222:259
11. Weiner RE, Thakur ML (2002) Appl Radiat Isot 57:749
12. Fichna J, Janecka A (2003) Bioconjugate Chem 14:3
13. Illidge TM, Brock S (2000) Curr Pharm Des 6:1399
14. Murray JL (2000) Semin Oncol 27:64
15. Potamianos B, Varvarigou AD, Archimandritis SC (2000) Anticancer Res 20:925
16. Vriesendorp HM, Quadri SM, Borchardt PE (1998) BioDrugs 10:275
17. Kalofonos HP, Karamouzis MV, Epenetos AA (2001) Acta Oncol 40:549
18. Wun T, Kwon DS, Tuscano JM (2001) BioDrugs 15:151
19. White CA, Berfein JR, Grillo-Lopez AJ (2000) Curr Pharm Biotechnol 1:303
20. Witzig TE (2000) Semin Oncol 27(Suppl 12):74
21. Kaminski MS, Zelenetz AD, Press OW, Saleh M, Leonard J, Fehrenhacher L, Lister TA, Stagg RJ, Tidmarsh GF, Kroll S, Wahl RL, Knox SJ, Vose JM (2001) J Clinical Oncol 19:3918
22. Culy CR, Lamb HM (2000) BioDrugs 14:195
23. McDevitt MR, Ma D, Lai LT, Simon J, Borchardt P, Frank RK, Wu K, Pellegrini V, Curcio MJ, Miederer M, Bander NH, Scheinberg DA (2001) Science 294:1537
24. Ballangrud AM, Yang WH, Charlton DE, McDevitt MR, Hamacher KA, Panageas KS, Ma D, Bander NH, Scheinberg DA, Sgouros G (2001) Cancer Res 61:2008
25. McDevitt MR, Ma D, Simon J, Frank RK, Scheinberg DA (2001) Appl Radiat Isot 294:1537
26. Smith-Jones PM, Stolz B, Albert R, Ruser G, Briner U, Maecke HR, Bruns C (1998) Nucl Med Biol 25:181
27. De Jong M, Bernard BF, De Bruin E, Van Gameren A, Bakker WH, Visser TJ, Mäcke HR, Krenning EP (1998) Nucl Med Commun 19:283
28. Heppler A, Froidevaux S, Mäcke HR, Jermann E, Béhé M, Powell P, Hennig M. (1999) Chem Eur J 5:1974
29. De Jong M, Bakker WH, Bernard BF, Valkema R, Kwekkeboom DJ, Reubi JC, Srinivasan A, Schmidt M, Krenning EP (1999) J Nucl Med 40:2081
30. Reubi JC, Waser B, Schaer JC, Laederach U, Erion J, Srinivasam A, Schmidt MA, Bugaj JE (1998) Eur J Nucl Med 25:481
31. Behr TM, Jenner N, Béhé M, Angerstein C, Gratz S, Raue F, Becker W (1999) J Nucl Med 40:1029

32. Behr TM, Jenner N, Béhé M, Angerstein C, Gratz S, Raue F, Becker W (1999) Clin Cancer Res 5:3124 s
33. Kwekkeboom DJ, Bakker WH, Kooj PPM, Erion J, Srinivasan A, de Jong M, Reubi JC, Krenning EP (2000) Eur J Nucl Med 27:1312
34. Breeman WAP, Meardji A, Capello A, Bernard BF, van Eijck CHJ, Krenning EP (2003) Int J Cancer 104:376
35. Waldherer C, Schumacher T, Pless M, Crazzolara A, Maecke HR, Nitzsche EU, Haldemann A, Mueller-Brand J (2001) Nucl Med Commun 22:673
36. Merlo A, Hausmann O, Wasner M, Steiner P, Otte A, Jermann E, Freitag P, Reubi JC, Gratzl O, Mäcke HR (1999) Clin Cancer Res 5:1025
37. Schumacher T, Hofer S, Eichorn K, Wasner M, Zimmerer S, Freitag P, Probst A, Gratzl O, Reubi JC, Maecke HR, Mueller-Brand J, Merlo A (2002) Eur J Nucl Med 29:486
38. Bodei L, Cremonesi M, Zoboli S, Grana C, Bartolomei M, Rocca P, Caracciolo M, Mäcke HR, Chinol M, Paganelli G (2003) Eur J Nucl Med 30:207
39. Jamar F, Barone R, Mathieu I, Walrand S, Labar D, Carlier P, de Camps J, Schran H, Chen T, Smith MC, Bouterfa H, Valkema R, Krening EP, Kvols LK, Pauwels S (2003) Eur J Nucl Med 30:510
40. Kwekkeboom DJ, Bakker WH, Kam BL, Teunissen JJM, Kooij PPM, Herder WW, Feelders RA, Eijck CHJ, Srinivasan A, Erion JL, Krenning EP (2003) Eur J Nucl Med 30:417
41. Anderson CJ, Welch MJ (1999) Chem Rev 99:2219
42. Okarvi SM (1999) Nucl Med Commun 20:1093
43. Liu S, Edwards DS (1999) Chem Rev 99:2235
44. Jurisson S, Lydon JD (1999) Chem Rev 99:2205
45. Behr TM, Gotthardt M, Barth A, Béhé M (2001) Q J Nucl Med 45:189
46. Breeman WAP, de Jong M, Kwekkeboom DJ, Bakker WH, Kooij PPM, Visser TJ, Krenning EP (2001) Eur J Nucl Med 28:1421
47. Virgolini I, Traub T, Novotny C, Leimer M, Füger B, Li SR, Patri P, Pangerl T, Angelberger P, Raderer M, Burggasser G, Andreae F, Kurtaran A, Dudczak R (2002) Curr Pharm Des 8:1781
48. Folkman J (1995) Nat Med 1:27
49. Mousa SA (1998) Drugs Future 23:51
50. Mousa SA (2000) Emerging Ther Targets 4:143
51. Carmeliet P (2000) Nat Med 6:389
52. Meitar D, Crawford SE, Rademaker AW, Cohn SL (1996) J Clin Oncol 14:405
53. Gasparini G, Brooks PC, Biganzoli E, Vermeulen PB, Bonoldi E, Dirix LY, Ranieri G, Miceli R, Cheresh DA (1998) Clin Can Res 4:2625
54. Albelda SM, Mette SA, Elder DE, Stewart R, Damjanovich L, Herlyn M, Buck CA (1990) Cancer Res 50:6757
55. Falcioni R, Cimino L, Gentileschi MP, D'Agnano I, Zupi G, Kennel SJ, Sacchi A (1994) Exp Cell Res 210:113
56. Felding-Haermann B, Mueller BM, Romerdahl CA, Cheresh DA (1992) J Clin Invest 89:2018
57. Sengupla S, Chattopadhyay N, Mitra A, Ray S, Dasgupta S, Chatterjee A (2001) J Exp Clin Cancer Res 20:585
58. Zitzmann S, Ethemann V, Schwab M (2002) Cancer Res 62:5139
59. Weber WA, Haubner R, Vabuliene E, Kuhnast B, Webster HJ, Schwaiger M (2001) Q J Nucl Med 45:179
60. Van de Wiele C, Oltenfreiter R, De Winter O, Signore A, Slegers G, Dieckx RA (2002) Eur J Nucl Med 29:699

61. Sivolapenko GB, Skarlos D, Pectasides D, Stathopoulou E, Milonakis A, Sirmalis G, Stuttle A, Courtenay-Luck NS, Konstantinides K, Epenetos AA (1998) Eur J Nucl Med 25:1383
62. Sutcliffe-Goulden JL, O'Doherty MJ, Marsden PK, Hart IR, Marshall JF, Bansal SS (2002) Eur J Nucl Med 29:754
63. Giannis A, Rübsam F (1997) Angew Chem Int Ed Engl 36:588
64. Haubner R, Finsinger D, Kessler H (1997) Angew Chem Int Ed Engl 36:1374
65. Gottschalk KE, Kessler H (2002) Angew Chem Int Ed Engl 41:1374
66. Huabner R, Wester HJ, Senekowitsch-Schmidtke R, Diefenbach B, Kessler H, Stöcklin G, Schwaiger M (1997) J Labelled Compd Radiopharm 40:383
67. Haubner R, Wester HJ, Reuning U, Senekowisch-Schmidtke R, Diefenbach B, Kessler H, Stöcklin G, Schaiger M (1999) J Nucl Med 40:1061
68. Haubner R, Wester HJ, Weber WA, Mang C, Ziegler SI, Goodman SL, Senekowisch-Schmidtke R, Kessler H, Schwaiger M (2001) Cancer Res 61:1781
69. Haubner R, Wester HJ, Burkhart F, Senekowisch-Schmidtke R, Weber W, Goodman SL, Kessler H, Schwaiger M (2001) J Nucl Med 42:326
70. Su ZF, Liu G, Gupta S, Zhu Z, Rusckowski M, Hnatowich DJ (2002) Bioconjugate Chem 13:561
71. Ogawa M, Jatano K, Oishi S, Kawasumi Y, Fujii N, Kawaguchi M, Doi R, Imamura M, Yamamoto M, Ajito K, Mukai T, Saji H, Ito K (2003) Nucl Med Biol 30:1
72. Moi MK, Meares CF (1988) J Am Chem Soc 110:6266
73. Kline SJ, Betebenner DA, Johnson DK (1991) Bioconjugate Chem 2:26
74. Li M, Meares CF (1993) Bioconjugate Chem 4:275
75. Lewis MR, Raubitschek A, Shively JE (1994) Bioconjugate Chem 5:565
76. Kulis DL, DeNardo SJ, DeNardo GL, O'Donnell RT, Meares CF (1998) J Nucl Med 39:2105
77. Lewis MR, Shively JE (1998) Bioconjugate Chem 9:72
78. Stimmel JB, Stockstill ME, Kull FC Jr (1995) Bioconjugate Chem 6:219
79. Mailet M, Kwok CS, Noujaim AA, Snieckus V (1998) Tetrahedron Lett 39:2659
80. Stimmel JB, Kull FC Jr (1998) Nucl Med Biol 25:117
81. van Hinsbergh VWM, Collen A, Koolwijk P (1999) Ann Oncol 4: S60
82. Brower V (1999) Nature Biol 17:963
83. Brooks PC, Montgomery AMP, Rosenfeld M, Reisenfeld R, Hu T, Klier G, Cheresh DA (1994) Cell 79:1157
84. Giannis A, Rübsam F (1997) Angew Chem Int Ed Engl 36:588
85. Haubner R, Finsinger D, Kessler H (1997) Angew Chem Int Ed Engl 36:1374
86. Drake CJ, Cheresh DA, Little CD (1995) J Cell Sci 108:2655
87. Aumailley M, Gurrath M, Müller G, Calvete J, Timpl, R, Kessler H (1991) FEBS Lett 291:50
88. Miller WH, Keenan RM, Willette RH, Lark MW (2000) Drug Discovery Today 5:397
89. Lange UEW, Backfisch G, Dletzer J, Geneste H, Graef C, Hornberger W, Kling A, Lauterbach A, Subkowski T, Zechel C (2002) Bioorg Med Chem Lett 12:1379
90. Pitts WJ, Wityak J, Smallheer JM, Tobin E, Jetter JW, Buynitsky JB, Hralow PP, Solomon KA, Mousa SA, Wexler RR, Hadhav PK (2002) J Med Chem 43:27
91. Sylyok GA, Gibson C, Goodman L, Hölzemann G, Wiesner M, Kessler H (2000) J Med Chem 43:27
92. Osterkamp F, Ziemer B, Koert U, Wiesner M, Raddatz P, Goodman SL (2000) Chem Eur J 6:666
93. Boturyn D, Dumy P (2001) Tetrohedron Lett 42:2787
94. Liu S, Cheung E, Rajopadhye M, Williams NE, OveroyeK L, Edwards DS (2001) Bioconjugate Chem 12:84
95. Liu S, Edwards DS (2001) Bioconjugate Chem 12:630

96. Liu S, Cheung E, Rajopadhye M, Ziegler MC, Edwards DS (2001) Bioconjugate Chem 12:559
97. Liu S, Edwards DS (2001) Bioconjugate Chem 12:554
98. Garrison WM (1987) Chem Rev 87:381
99. Berger R (1982) Int J Appl Radiat Isot 33:1341
100. Chakrabarti MC, Le N, Paik CH, De Graff WG, Carrasquillo JA (1996) J Nucl Med 37:1384
101. Salako QA, O'Donnell RT, DeNardo SJ (1998) J Nucl Med 39:667
102. Liu S, Ellars CE, Harris TD, Edwards DS (2003) Bioconjugate Chem 14:1030
103. Shannon RD (1976) Acta Crystallogr Sect A 32:751
104. Caravan P, Ellison JJ, McMurry TJ, Lauffer RB (1999) Chem Rev 99:2293
105. Chang CA, Francesconi LC, Malley MF, Kumar K, Cougoutas JZ, Tweedle MF (1993) Inorg Chem 32:3501
106. Kumar K, Chang CA, Francesconi LC, Dischino DD, Malley MF, Cougoutas JZ, Tweedle MF (1994) Inorg Chem 33:3567
107. Yang LW, Liu S, Wong E, Rettig SJ, Orvig C (1995) Inorg Chem 34:2164
108. Liu S, Wong E, Rettig SJ, Orvig C (1993) Inorg Chem 32:4268
109. Liu S, Wong E, Karunaratne V, Rettig SJ, Orvig C (1993) Inorg Chem 32:1756
110. Liu S, Rettig SJ, Orvig C (1992) Inorg Chem 31:5400
111. Riesen A, Kaden TA, Ritter W, Mäcke HR (1989) J Chem Soc Chem Commun 460
112. Mäcke HR, Riesen A, RitterW (1989) J Nucl Med 30:1235
113. Malyrick MA, Ilyhin AB, Petrosyans SP (1994) Main Group Met Chem 17:707
114. Maeke HR, Scherer G, Heppeler A, Henig M (2001) Eur J Nucl Med 28:967
115. Liu S, He ZJ, Fanwick PE (2003) Inorg Chem 42:8831
116. Liu S, Pietryka J, Ellars CE, Edwards DS (2002) Bioconjugate Chem 13:902
117. Onthank DC, Liu S, Silva PJ, Barrett JA, Robinson SP, Edwards DS (2004) Bioconjugate Chem 15:235
118. Van Hagen PM, Breeman WAP, Bernard HF, Schaar M, Mooij CM, Srinivasan A, Schmidt MA, Krenning EP, de Jong M (2000) Int J Cancer (Radiat Oncol Invest) 8:186
119. Rajopadhye M, Harris AR, Nguyen HM, Overoye KL, Bartis J, Liu S, Edwards DS, Barrett JA (2000) J Nucl Med 41:34P
120. Barrett JA, Crocker AC, Onthank DC, Hemingway SJ, Rajopadhye M, Harris AR, Liu S, Bartis J, Edwards DS (2000) J Nucl Med 41:259P
121. Liu S, Edwards DS, Ziegler MC, Harris AR, Hemingway SJ, Barrett JA (2001) Bioconjugate Chem 12:624
122. Janssen M, Oyen WJG, Massuger LFAG, Frielink C, Dijkgraaf I, Edwards DS, Rajopadhye M, Corste, FHM, Boerman OC (2002) Cancer Res 62:6146
123. Harris TD, Kalogeropoulos SA, Edwards DS, Liu S, Onthank DC, Barrett JA (2001) J Labelled Compd Radiopharm 44:S60
124. Harris TD, Kalogeropoulos S, Nguyen T, Liu S, Bartis J, Ellars CE, Edwards DS, Onthank D, Yalamanchili P, Robinson SP, Lazewatsky J, Barrett JA (2003) Cancer Biother Radiopharm 18:627

Author Index Volumes 251–252

Author Index Vols. 26–50 see Vol. 50
Author Index Vols. 51–100 see Vol. 100
Author Index Vols. 101–150 see Vol. 150
Author Index Vols. 151–200 see Vol. 200
Author Index Vols. 201–250 see Vol. 250

The volume numbers are printed in italics

Subject Index

Printing: Krips bv, Meppel
Binding: Litges & Dopf, Heppenheim